# Sustainable Landscape Planning and Natural Resources Management

## IEREK Interdisciplinary Series for Sustainable Development

Sustainable Landscape Planning and Natural Resources Management (SLNR) is a peer-reviewed book series that presents groundbreaking research through the intersection of two main areas that of which are 'Sustainable Landscape Planning' and 'Natural Resources Management'. The series initially covers a wide range of academic disciplines in sustainability, including landscape ecology, green urbanism, urban regeneration, land restoration and renaturation, energy management, and nature conservation. Moreover, it also incorporates the latest developments and research in engineering topics that contribute to more sustainable cities, such as Geosciences, infrastructure, civil engineering, geology, mining, water, and natural resources conservation and management.

By showcasing methods for discovering, managing, and employing natural resources, the series aims to provide approaches that enable the responsible use of these resources. Ultimately, the SLNR series is committed to advancing the United Nations' Sustainable Development Goals for 2030.

The wide range of topics covered in this series are primarily related to sustainable development, involving durable infrastructure projects, civil engineering, urban regeneration, the built, sustainability, climate change mitigation, energy management, geosciences, geology, and the future of cities, all of which depend on the availability of natural resources and effective wealth management. The series also features conceptual and empirical contributions from various interrelated fields of earth science, with an emphasis on the era of new discovery. It focuses on providing practical solutions to ensure sustainable development and provides new case studies that offer concrete examples of how to manage natural resources effectively.

The series encompasses the best research papers from various international conferences organized by IEREK conferences, also accepts proposals from other international conferences, including interdisciplinary contributions. It exhibits innovative approaches and highlights how they can best support both economic and sustainable development through the responsible use of natural resources and global, local, and individual activities for the welfare of all societies on Earth.

The SLNR series is a valuable source for experts in government, international organizations, consultancies, industry, business sectors, and academics. Published in cooperation with IEREK, the Springer SLNR series displays important new research in various sectors, delivering valuable insights and practical solutions to the complex challenges of sustainable development for the readers.

Oumaima Jamal Eddine · Lamia Daghor ·
Mounir Tilaoui · Hicham Chatoui
Editors

# Plant Pathology, Fungal Diversity, and Biotechnological Advances in Agriculture

Springer

*Editors*
Oumaima Jamal Eddine
Department of Chemistry
Faculty of Sciences Ben M'sick
Hassan II University
Casablanca, Morocco

Mounir Tilaoui
Faculty of Sciences
Sidi Mohamed Ben Abdellah University of Fez
Fes, Morocco

Lamia Daghor
Faculty of Health Sciences
Mundiapolis University
Casablanca, Morocco

Hicham Chatoui
Higher Institute of Nursing Professions
and Health Techniques (ISPITS-M)
Ministry of Health
Private University of Marrakech (UPM)
Marrakech, Morocco

ISSN 2948-1910 ISSN 2948-1929 (electronic)
Sustainable Landscape Planning and Natural Resources Management
IEREK Interdisciplinary Series for Sustainable Development
ISBN 978-3-031-97665-0 ISBN 978-3-031-97666-7 (eBook)
https://doi.org/10.1007/978-3-031-97666-7



This Springer imprint is published by the registered company Springer Nature Switzerland AG
The registered company address is: Gewerbestrasse 11, 6330 Cham, Switzerland

# About This Book

This book brings together recent scientific contributions exploring the intersections of plant health, natural bioactive compounds, sustainable innovations in agricultural systems, and resource valorization. Divided into two focused parts: Phytochemicals and Essential Oils—Exploring Bioactivity in Plant Pathology and Fungal Diversity, and Biotechnological Advances in Agriculture and Resource Valorization, this book presents original scientific articles that detail practical approaches and findings from recent research. Seamlessly integrating traditional knowledge with recent advances in phytochemistry, microbiology, and biotechnology, this book is a valuable resource for teachers, researchers, agronomists, and students within academic and research institutions. Addressing key topics such as essential oils with antimicrobial potential, plant disease management, phytoplankton dynamics, and the valorization of agricultural and natural resources, it offers practical insights that contribute meaningfully to scientific knowledge and sustainable agricultural innovation.

# Introduction

This book, *Plant Pathology, Fungal Diversity, and Biotechnological Advances in Agriculture*, presents cutting-edge research on phytochemicals, essential oils, and biotechnological innovations aimed at enhancing agricultural sustainability. It highlights the role of natural compounds in improving plant health, combating pathogens, and optimizing resource utilization.

The first part, 'Phytochemicals and Essential Oils—Exploring Bioactivity in Plant Pathology and Fungal Diversity', explores studies on essential oils from plants like *Ocimum basilicum*, *Eucalyptus*, and *Mentha spicata*, focusing on their chemical profiles and antimicrobial activities. It also investigates the impact of geographical and abiotic factors on phytochemical variability and the role of microalgal phytohormones in managing heavy metal stress.

The second part, 'Biotechnological Advances in Agriculture and Resource Valorization', covers innovations in circular agriculture, such as biovalorization of algae for soil enrichment, the development of the date palm sector in Morocco, and the transformation of food waste into useful materials. It also discusses the seasonal monitoring of phytoplankton and the valorization of gum arabic in sustainable industries.

The book offers valuable insights into the integration of traditional knowledge and modern biotechnologies to address global agricultural challenges, with a focus on regionally specific studies, particularly from Morocco.

# Contents

# About the Editors

**Dr. Oumaima Jamal Eddine** holds a Ph.D. in Materials Science and a master's in Materials Engineering from FSBM, Hassan II University of Casablanca. She is currently Research Associate at FSBM, working on vitreous materials, and serves as Technical Manager at LC2A's Agri-food department, overseeing analyses of pesticides, textiles, pharmaceuticals, and more using LC-MS/MS and GC-MS/MS. With solid teaching experience, she has supervised students at all levels. As Active Member of MAASI, she contributes to conference organization and peer review. She also serves on the editorial boards of the journals *MJQQR* and *JMS: Materials in Electronics.*

**Dr. Lamia Daghor** is Professor at the Faculty of Health Sciences, Mundiapolis University. Previously, she spent five years at the School of Engineering and Innovation (E2IM), Private University of Marrakech. Holding a Ph.D. in marine biology and ecology, her doctoral work focused on phytoplankton and palynological dynamics in Moroccan lagoons. She later completed a postdoctoral fellowship in Ecology at FSBM, Hassan II University, in collaboration with the National Institute of Fisheries Research. Over the past 15 years, she has published 11 articles and three book chapters and presented at over 17 scientific conferences.

**Dr. Mounir Tilaoui** holds a Ph.D. in pharmacology-biochemistry from the Faculty of Science and Technology of Beni Mellal, focusing on the antitumor activity of natural products. After completing his degree, he was Postdoctoral Researcher at Waterford Institute of Technology, Ireland, studying intrinsic apoptosis pathways in cancer therapy. He continued his research at Sultan Moulay Slimane University on natural product-based anticancer drugs and gene expression in triple-negative breast cancer. He is now Full Professor of biochemistry at Sidi Mohamed Ben Abdellah University in Fez, Morocco, with research interests in anticancer drug discovery and cancer pharmacogenomics.

**Dr. Hicham Chatoui** holds a Ph.D. in neurosciences, physiology and eco-health from FST Beni Mellal and FSSM, Cadi Ayyad University. He earned a master's in life sciences and health, specializing in anthropobiology, and began teaching in 2009 at FSSM. From 2013 to 2023, he was Director of the Health and Agro-Industry Engineering Department at the School of Engineering and Innovation, Private University of Marrakech. He is currently Lecturer-Researcher at ISPITS Marrakech, Ministry of Health and Social Protection. Dr. Chatoui has authored several books, articles, and chapters and coordinated numerous international scientific events. His research spans telemedicine, e-health, neurosciences, immunology, nutrition, biomedical engineering, and environmental health impacts.

# Phytochemicals and Essential Oils—Exploring Bioactivity in Plant Pathology and Fungal Diversity

# Diversity in Chemical Composition and Biological Properties of *Ocimum basilicum* Essential Oil: A Comparative Study Across Different Geographical Areas

Wijdane Saguem, Siham Slassi, Oumaima Jamal Eddine, Bahija Mounir, and Said Gmouh

**Abstract**

*Ocimum basilicum* is an aromatic plant with multiple uses in food, medicine, and cosmetics. This chapter compares the chemical composition and antimicrobial activity of essential oils of *Ocimum basilicum* from different geographical regions. The study reveals a significant variation in major compounds such as linalool, eugenol, estragole and methyl cinnamate, which are influenced by geography. In the antimicrobial comparative study, hexane extracts showed the most potent antimicrobial activity, particularly against the *Staphylococcus aureus* strain, with linalool as the main active compound. These results underline the potential of this plant as a natural antimicrobial agent and the importance of optimizing extraction methods.

**Keywords**

*Ocimum basilicum* · Extraction · Essential oil · Linalool · Antimicrobial activity

The chemical composition and the biological activities of Ocimum basilicum change according to geographical regions

W. Saguem · S. Slassi (✉) · O. Jamal Eddine · S. Gmouh
Department of Chemistry, Laboratory of Engineering and Materials (LIMAT), Faculty of Sciences Ben M'sick, University Hassan II of Casablanca, Casablanca, Morocco
e-mail: sihamslassi@gmail.com

O. Jamal Eddine
Multi-laboratory LC2A, Mohammedia, Morocco

B. Mounir
Laboratory of Analytical and Molecular Chemistry (LCAM), Department of Chemistry, Faculty of Sciences Ben M'sick, Hassan II University of Casablanca, Casablanca, Morocco

## 1 Introduction

Basilic (*Ocimum basilicum*), a fragrant herb belonging to the Lamiaceae family, comprises about 160 species, with approximately 65 considered actual members of the genus, while the rest are classified as synonyms (Spence 2024; Ijaz Hussain et al. 2008). Among them is the species commonly known as sweet basil, a name that comes from the Greek word "Basileus", meaning "Royal" or "King". It is called the "King of the herbs" because it is one of the most well-known and widely used species, valued for its aromatic qualities, culinary applications, and medicinal properties (Calderón Bravo et al. 2021). *Ocimum basilicum* is located in many countries worldwide because it can grow in different temperatures and geographical regions (Barickman et al. 2021).

*Ocimum basilicum* exhibits significant variability, both morphologically and chemically. This diversity is reflected in the wide range of leaf shapes and sizes, stem and flower color, and essential oil composition (Ababutain 2019). Based on these morphological traits, *Ocimum basilicum* is generally divided into six main varieties: true basil, characterized by broad, green leaves; small-leaf basil, which has compact and tightly clustered green leaves; lettuce-leaf basil, noted for its large, blistered leaves; purple basil (type A), which features green leaves but purple stems and flowers; purple basil (type B), where all parts of the plant, leaves, stems, and flowers are entirely purple; purple basil (type C), an entirely purple variety with distinctly blistered leaves (Carović-Stanko et al. 2011; Nair et al. 2016; Varga et al. 2017; Akbari et al. 2019).

According to its chemical composition profile, *Ocimum basilicum* can be classified into primary chemotypes, each associated with specific geographic regions and dominant active compounds. European type: located in Europe, the United States, and parts of Arica, this chemotype is primarily characterized by a high percentage of linalool and methyl chavicol; Reunion type: cultivated in Comoros, Seychelles,



O. Jamal Eddine et al. (eds.), *Plant Pathology, Fungal Diversity, and Biotechnological Advances in Agriculture*, Sustainable Landscape Planning and Natural Resources Management, https://doi.org/10.1007/978-3-031-97666-7_1

**Fig. 1** The four chemotypes of *Ocimum basilicum*

various parts of Africa, and Reunion Island, this variety is characterized by the abundant presence of methyl chavicol; Tropical type: Grown primarily in India, Pakistan, Guatemala, Haiti, and certain African regions, this type is rich in methyl cinnamate as its principal active compound; C type: Principally found in North of Africa, Russia, Eastern Europe, and parts of Asia, this type is distinguished by eugenol as its principal active compound (Simon et al. 1999; Marotti et al. 1996; Grayer et al. 1996; Chalchat et al. 1999; Muráriková et al. 2017; Maddi et al. 2019; Abbas Akbari et al. 2018) (Fig. 1).

*Basilicum* is usually used in several fields, such as perfumery, cooking, industrial food, cosmetic, and medicines. Traditionally, *basilicum* has been used as dish seasoning, adding a unique flavor to various culinary preparations (Rehan Nadeem et al. 2022). In addition, *basilicum* is used in traditional medicine in the treatment of headaches, constipation, coughs, diarrhea, warts, and worms. In contrast, its oil is used in the cosmetic industry for the preparation of soaps, shampoos, lotions. It is also used as a perfuming agent due to its special scent (Eunice Bajomo et al. 2022; Reza Aminian et al. 2022; Qamara fatimamudassar et al. 2023).

*Basilicum* possesses diverse significant pharmacological properties that make it a valuable asset in the health field, including antimicrobial (Eriotou et al. 2015; Romano et al. 2022), antioxidant (Sandeep 2017; Nwanjo and Oze 2007; Chinedu and Ofili 2021), antiviral (Özcan and Chalchat 2002; Benabdallah et al. 2020), antispasmodic (Elansary et al. 2020; Marwat et al. 2011), insect-repelling (Teliban et al. 2022; Baba et al. 2012), anticancer (Ilić et al. 2019; Zervoudakis et al. 2015), anti-allergenic (Al Jaouni et al. 2021; Hassan Abdel-razakh et al. 2024) as well as anti-inflammatory properties (Mulugeta et al. 2024; Kumar Joshi et al. 2023). These properties allow basil to play an essential role in treating various disorders and diseases. It is also known for its diuretic properties, which help eliminate toxins and excess fluids from the body (Aminian et al. 2022). Due to its multiple benefits, it is increasingly recognized not only as a culinary herb but also as a natural medicine with health-promoting effects.

The current chapter consists of scientific studies regarding the essential oils of the *Ocimum basilicum* plant from various geographical regions around the word. This study aims to compare the chemical composition and biological activities of *Ocimum basilicum* oils from several areas, and to explore the correlations between environmental conditions, chemical composition, and biological activities.

## 2 Comparative Study of the Chemical Composition of Different Regions

*Ocimum basilicum* is recognized as an aromatic plant used in several field, such as traditional medicine, perfumery, aromatherapy and the culinary arts. The most remarkable feature of this plant is the variability of its essential oil content and composition, which can depend on geographical origin, environmental factors, and extraction methods. Many studies from various regions in the word revealed unique chemical profiles, highlighting the presence of several different chemotypes.

In most of the countries studied, linalool is identified as the dominant compound. It is found in high concentrations in Serbia 53.3%, Morocco 52.98%, and Algeria 52.1%, indicating a substantial prevalence of the linalool chemotype in Mediterranean and Balkan regions (Beatovic et al. 2015; Belkamel et al. 2008; Rezzoug et al. 2019). This compound is one of the main essential fragrance compounds in use today. It's also known for its floral scent and calming, antibacterial, and antifungal properties. In Egypt, linalool is also present at a remarkable level (31.65%), while in India, its presence dropped considerably to 10.83%, suggesting the influence of the environmental conditions (Ahmed et al. 2019; Kumar Kalagatur et al. 2020; Kholiya et al. 2022; Chowdhury et al. 2017).

In contrast, eugenol is the most dominant compound in essential oil in India, which is a phenolic compound with a spicy clove aroma and strong antiseptic properties. This makes Indian basil a unique variety, characterized by a eugenol-rich chemotype typical of tropical zones. Eugenol is also present, but in much lower proportions between 4 and 7%, in countries such as Serbia, Colombia, Congo, and Turkey

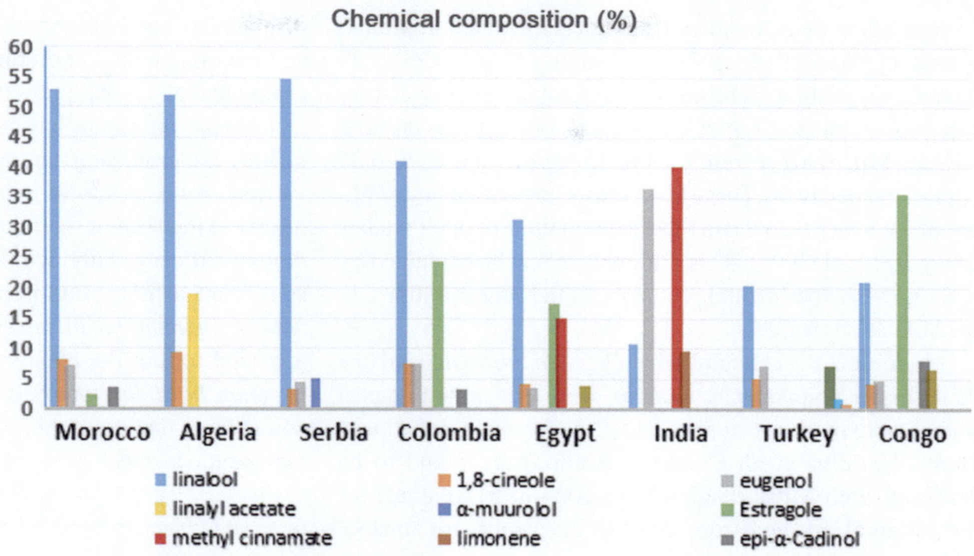


**Fig. 2** Histogram of the chemical composition of essential oil of *Ocimum basilicum* in various regions

(Rezzoug et al. 2019; Dinangayi Tshilanda et al. 2016; Gurkan and Hayaloglu 2023; Calderón Bravo et al. 2021).

Another important compound is estragole (or methyl chavicol), which is present in very high concentrations in Congo (35.72%) and Colombia (24.5%). These high levels indicate an estragole-rich chemotype, characterized by a sweet, aniseed aroma. Egypt also has a moderate estragole content (17.37%), while its presence remains low in Serbia and Morocco (Beatovic et al. 2015; Belkamel et al. 2008; Ahmed et al. 2019; Dinangayi Tshilanda et al. 2016; López-Hernández et al. 2024). This chemical variability underlines that certain chemotypes are more characteristic of specific regions, probably influenced by genetic factors or local practices of cultivation.

1,8-Cineole, also known as eucalyptol, is another significant component in all samples, although its concentration varies according to origin. The highest levels were found in Algeria (9.2%), Morocco (8.15%), and Colombia (7.6%) (Belkamel et al. 2008; Rezzoug et al. 2019; López-Hernández et al. 2024). Known for its refreshing, camphor aroma, 1,8-cineole is also valued for its beneficial effects on the respiratory system.

Although present in smaller quantities, certain minor sesquiterpenes and monoterpenes play a considerable role in the aroma and biological activity of basil essential oil. Compounds such as germacrene-D, camphor, spathulenol, α-humulene, and β-elemene are frequently detected in samples from Colombia, Serbia, and Congo, suggesting greater diversity in these regions (Beatovic et al. 2015; Dinangayi Tshilanda et al. 2016; López-Hernández et al. 2024).

A particularly distinct profile was observed in Turkey, where basil was analyzed from an extract obtained by hot infusion. This sample was characterized by an exceptionally high content of methyl cinnamate (40.26%), a compound with a sweet, balsamic fragrance, also detected in Egypt at a more moderate concentration (15.14%) (Ahmed et al. 2019; Gurkan and Hayaloglu 2023) (Fig. 2).

## 3 Comparative Study of Antimicrobial Activity of *Ocimum basilicum*

Various studies on the antimicrobial activity of *Ocimum basilicum* extracts reveal variable efficacy depending on the type of solvent used in extraction, the targeted strains, and the experimental methods applied. Generally, basil extracts show promising antimicrobial potential, but the results also highlight a complexity linked to the nature of the bioactive compounds extracted and the protocols used.

In the Adigüzel et al. study, the efficacy of hexanolic, ethanolic, methanolic, and extracts was evaluated on 146 bacterial strains representing 55 species, as well as on four types of fungi and yeasts (Adigüzel et al. 2005). Ethanolic and methanolic extracts showed relatively low inhibitory activity, with inhibition rates of 9% and 10%, respectively. The hexanolic extract, on the other hand, showed high antibacterial activity, affecting 13 strains from six bacterial (Acinetobacter, Bacillus, Brucella, Escherichia, Micrococcus and Staphylococcus), as well as three strains of Candida albicans (Ababutain 2019; Backiam et al. 2023).

The minimum inhibitory concentration (MIC) values confirm this trend: hexane exhibited an MIC range of 62.5 to 250 μl/mL, lower than those of the methanolic (250–500 μl/mL) and ethanolic (125–250 μl/mL) extracts, indicating greater efficacy at lower doses. These results are supported by the diameters of inhibition zones, ranging from 7 to 11 mm for hexane, compared to 7 to 17 mm for methanol, and 8 to 16 mm for ethanol. (Adigüzel et al. 2005; Al Abbasy et al. 2015; Ahmet et al. 2005; Silva et al. 2015).

In another study, methanolic extracts exhibited minimum inhibitory concentrations (MICs) ranging from 3.125 to 25 μg/mL, suggesting significant antimicrobial activity at very low concentrations. Ethanolic and aqueous extracts showed slighter MIC ranges, from 6.25 to 25 μg/mL and 12.5 to 25 μg/mL, respectively. These MIC values, expressed in μg/mL, are much lower than those reported in μl/mL in other studies (Adigüzel et al. 2005). This difference could be explained by variations in experimental protocols, initial concentrations, or quantification methods.

In addition, the essential oil of *Ocimum basilicum* and one of its major components, linalool, were also tested against both Staphylococcus aureus and Pseudomonas aeruginosa strains by Silva et al. S. aureus strains were found to be sensitive, with MICs ranging from 1024 to 512 μg/mL for the essential oil, and from 1024 to 32 μg/mL for linalool (Silva et al. 2015). According to the authors, these results indicate moderate activity for the oil, and mild to vigorous activity for linalool. Conversely, results for P. aeruginosa showed that 50% of the strains were resistant to the essential oil. In contrast, the sensitive strains had MICs of 1024 μg/mL, indicating limited activity against this resistant bacterium (Sartoratto et al. 2004).

We note the presence of certain contradictions with earlier studies, demonstrating the power of polar solvents such as methanol and ethanol in extracting antimicrobial compounds. In contrast, the results of Adigüzel et al. show that hexane, a non-polar solvent, enables more effective extraction of active substances against a larger number of strains. This may be explained by the fact that the chemical nature of the secondary metabolites responsible for antimicrobial activity varies from plant to another plant. Therefore the optimal choice of solvent strongly depends on the phytochemical profile of the plant species studied (Naqri et al. 1991; Kroymann 2011; Berini et al. 2018; Francesca González Mera et al. 2019).

In summary, the data of the different antimicrobial studies indicates that the extract of hexane of *Ocimum basilicum* demonstrates the most extensive and potent antimicrobial activity among the tested solvents, but methanol also exhibits significant efficacy. Linalool, as an isolated compound, confirms its central role in the antibacterial potency of *Ocimum basilicum*, particulary against Staphylococcus aureus. These findings underline the importance of further research to identify active compounds and optimize the extraction methods.

## 4 Conclusion

In summary, although studies have been conducted on the *Ocimum basilicum* variety, essential variations were observed in both chemical compositions and biological activities. These divergences highlight the influence of geographical location and environmental conditions on the plant's phytochemical profile and biological properties.

*Ocimum basilicum* has powerful antimicrobial activities, making it a promising natural alternative to synthetic preservatives. Among the various solvent extracts studied, the hexane extract stands for its broad spectrum and potent antimicrobial efficacy, although methanolic extracts also show significant activity. Furthermore, linalool, a major component of *Ocimum basilicum*'s essential oil, has demonstrated a crucial role in the plant's antibacterial potential. Overall, *Ocimum basilicum* presents a valuable source of bioactive compounds that could play a significant role in developing therapeutic agents.

## References

Ababutain, I.M.: Antimicrobial activity and gas chromatography-mass spectrometry (GC-MS) analysis of Saudi Arabian Ocimum basilicum leaves extracts. J. Pure Appl. Microbiol. **13**, 61 (2019)

Abbas Akbari, G., Soltani, E., Binesh, S., Amini, F.: Cold tolerance, productivity and phytochemical diversity in sweet basil (Ocimum basilicum L.) accessions. Industrial Crops Prod. **124**, 677–684 (2018)

Adigüzel, A., Güllüce, M., Sengül, M., Öğütcü, H., Sahin, F., Karaman, I.: Antimicrobial effects of Ocimum basilicum (Labiatae) extract. Turk. J. Biol. **29**, 155–160 (2005)

Ahmed, A.F., Attia, F.A.K., Liu, Z., Li, C., Wei, J., Kang, W.: Antioxidant activity and total phenolic content of essential oils and extracts of sweet basil (Ocimum basilicum L.) plants. Food Sci. Human Wellness **8**, 299–305 (2019)

Ahmet, A., Medine, G., Meryem, S., Hatice, O., Isa, K.: Antimicrobial effects of Ocimum basilicum (Labiate) extracts. Turk. J. Biol. **29**, 155–160 (2005)

Akbari, G.A., Binesh, S., Ramshini, H., Soltani, E., Amini, F., Mirfazeli, M.S.: Selection of basil (Ocimum basilicum L.) full-sib families from diverse landraces. J. Appl. Res. Med. Aromatic Plants **12**, 66–72 (2019)

Al Abbasy, D.W., Pathare, N., Al-Sabahi, J.N., Khan, S.A.: Chemical composition and antibacterial activity of essential oil isolated from Omani basil (Ocimum basilicum Linn.) Asian Pacific J. Trop. Dis. **5**(8), 645–649 (2015)

Al Jaouni, S., Saleh, A.M., Wadaan, M.A.M., Hozzein, W.N., Selim, S., AbdElgawad, H.: Bioactive potential of several actinobacteria isolated from microbiologically barely explored desert habitat, Saudi Arabia. Biology **10**, 235 (2021)

Aminian, A.R., Mohebbati, R., Boskabady, M.H.: The effect of Ocimum basilicum L. and its main ingredients on respiratory disorders: an experimental, preclinical, and clinical review. Front. Pharmacol. **12**, 805391 (2022)

Baba, G., Lawal, A.O., Shariff, H.B.: Mosquito repellent activity and phytochemical characterization of essential oils from Striga Hermonthica, Hyptis Spicigera and Ocimum Basilicum leaf extracts. Br. J. Pharmacol. Toxicol. **3**(2), 43–48 (2012)

Backiam, A.D.S., Duraisamy, S., Karuppaiya, P., Balakrishnan, S., Sathyan, A., Kumarasamy, A., Raju, A.: Analysis of the main bioactive compounds from Ocimum basilicum for their antimicrobial and antioxidant activity. Biotechnol. Appl. Biochem. **70**, 2038–2051 (2023)

Barickman, T.C., Olorunwa, O.J., Sehgal, A., Walne, C.H., Reddy, K.R., Gao, W.: Interactive impacts of temperature and elevated $CO_2$ on basil (Ocimum basilicum L.) root and shoot morphology and growth. Horticulturae **7**, 112 (2021)

Beatovic, D., Krstic-Miloševic, D., Trifunovic, S., Šiljegovic, J., Glamoclija, J., Ristic, M., Jelacic, S.: Chemical composition, antioxidant and antimicrobial activities of the essential oils of twelve Ocimum basilicum L. cultivars grown in Serbia. Records Nat. Prod. **9**(1), 62–75 (2015)

Belkamel, A., Bammi, J., Janneot, V., Belkamel, A., Dehbi, Y., Douira, A.: Évaluation de la biomasse et analyse des huiles essentielles de trois variétés de basilic (Ocimum basilicum L.) cultivées au Maroc. Acta Botanica Gallica **155**(4), 467–476 (2008)

Benabdallah, H., Benchikh, F., Smain, A.: Ethnopharmacological study and evaluation of the antioxidant activity of Ocimum basilicum L. extracts. J. Drug Deliv. Therapeutics **10**(4-s), 132–137 (2020)

Berini, J.L., Brockman, S.A., Hegeman, A.D., Reich, P.B., Muthukrishnan, R., Montgomery, R.A., Forester, J.D.: Combinations of abiotic factors differentially alter production of plant secondary metabolites in five woody plant species in the boreal-temperate transition zone. Front. Plant Sci. **9**, 1257–1273 (2018)

Calderón Bravo, H., Vera Céspedes, N., Zura-Bravo, L., Muñoz, L.A.: Basil seeds as a novel food, source of nutrients and functional ingredients with beneficial properties: a review. Foods **10**, 1467 (2021)

Carović-Stanko, K., Šalinović, A., Grdiša, M., Liber, Z., Kolak, I., Satovic, Z.: Efficiency of morphological trait descriptors in discrimination of Ocimum basilicum L. accessions. Plant Biosyst. **145**, 298–305 (2011)

Chalchat, J.C., Garry, R.P., Sidibe, L., Marama, M.: Aromatic plants of Mali (I): chemical composition of essential oils of Ocimum basilicum L. J. Essent. Oil Res. **11**, 375–380 (1999)

Chinedu, E., Charles, O.: C. Ocimum species: ethnomedicinal uses, phytochemistry and pharmacological importance. Int. J. Curr. Res. Physiol. Pharmacol. **5**(2) (2021)

Chowdhury, T., Mandal, A., Roy, S.C., De Sarker, D.: Diversity of the genus Ocimum (Lamiaceae) through morpho-molecular (RAPD) and chemical (GC–MS) analysis. J. Genetic Eng. Biotechnol. **15**, 275–286 (2017)

Dinangayi Tshilanda, D., Bila Babady, P., Nguwo Vele Onyamboko, D., Muamba Tshilolo Tshiongo, C., Sha-Tshibey Tshibangu, D., Ngbolua, K.T.N., Vuka Tsalu, P., Tshimankinda Mpiana, P.: Chemotype of essential oil of Ocimum basilicum L. from DR Congo and relative in vitro antioxidant potential to the polarity of crude extracts. Asian Pacific J. Trop. Biomed. **6**(12), 1022–1028 (2016)

Elansary, H.O., Szopa, A., Kubica, P., Ekiert, H., El-Ansary, D.O., Al-Mana, F.A., Mahmoud, E.A.: Saudi Rosmarinus officinalis and Ocimum basilicum L. polyphenols and biological activities. Processes **8**, 446 (2020)

Eriotou, E., Anastasiadou, K., Nikolopoulos, D., Koulougliotis, D.: Antimicrobial and free radical scavenging activities of Basil (Ocimum basilicum) essential oil isolated from five plant varieties growing in Greece. Nutr. Food Sci. **5**(3) (2015)

Eunice Bajomo, M., Melanie Aing, S., Lucas Ford, S., Emily Niemeyer, D.: Chemotyping of commercially available basil (Ocimum basilicum L.) varieties: cultivar and morphotype influence phenolic acid composition and antioxidant properties. NFS J. **26**, 1–9 (2022)

Francesca González Mera, I., Estefanía González Falconí, D., Morera Córdova, V.: Secondary metabolites in plants: main classes, phytochemical analysis and pharmacological activities. Revista Bionatura **4**(4), 1000–1009 (2019)

Grayer, R.J., Kite, G.C., Goldstone, F.J., Bryan, S.E., Paton, A., Putiesky, E.: Intraspecific taxonomy and essential oil chemotypes in sweet basil, Ocimum basilicum. Phytochemistry **43**, 1033–1039 (1996)

Gurkan, H., Hayaloglu, A.A.: Changes in volatiles and essential oil composition of three organs (leaf, stem and flower) of purple basil (Ocimum basilicum L.) by GC–MS combined with multivariate statistical approach. Food Chem. Adv. **2**, 100292 (2023)

Hassan Abdel-razakh, H., George Bakari, G., Park, J., Pan, C., Shaaban Hoza, A.: Phenolic contents and antioxidant properties of Bauhinia rufescens, Ocimum basilicum and Salvadora persica, used as medicinal plants in Chad. Molecules **29**(19), 4684 (2024)

Ijaz Hussain, A., Anwar, F., Tufail Hussain Sherazi, S., Przybylski, R.: Chemical composition, antioxidant and antimicrobial activities of basil (Ocimum basilicum) essential oils depends on seasonal variations. Food Chem. **108**, 986–995 (2008)

Ilić, A.S., Antić, M.P., Jelačić, S.C., Šolević Knudsen, T.M.: Chemical composition of the essential oils of three Ocimum basilicum L. cultivars from Serbia. Notulae Botanicae Horti Agrobotanici Cluj-Napoca **47**(2), 347–351 (2019)

Kholiya, S., Punetha, A., Uhan, A.C., Venkatesha, K.T., Kumar, D., Upadhyay, R.K., Padalia, R.C.: Essential oil yield and composition of Ocimum basilicum L. at different phenological stages, plant density and post-harvest drying methods. South Afr. J. Botany **151**, 919–925 (2022)

Kroymann, J.: Natural diversity and adaptation in plant secondary metabolism. Curr. Opin. Plant Biol. **14**, 246–251 (2011)

Kumar Joshi, R., Agarwal, S., Patil, P., Alagarasu, K., Panda, K., Cherian, S., Parashar, D., Roy, S.: Anti-dengue activity of lipophilic fraction of Ocimum basilicum L. Stem. Molecules **28**(3), 144 (2023)

Kumar Kalagatur, N., Gurunathan, S., Reddy Kamasani, J., Gunti, L., Kadirvelu, K., Dhananjaya Mohan, C., Rangappa, S., Prasad, R., Almeida, F., Mudili, V., Siddaia, C.: Inhibitory effect of C. zeylanicum, C. longa, O. basilicum, Z. officinale, and C. martini essential oils on growth and ochratoxin A content of A. ochraceous and P. verrucosum in maize grains. Biotechnol Rep **27**, e00490 (2020)

López-Hernández, M., Criollo-Nuñez, J., Beltran, J.I., Sandoval-Aldana, A.: Chemical composition and antioxidant activity of 'Nufar' basil (Ocimum basilicum L.) essential oil from three municipalities of Tolima, Colombia. Food Chem. Adv. **5**, 100819 (2024)

Maddi, R., Amani, P., Bhavitha, S., Gayathri, T., Lohitha, T.: A review on Ocimum species: Ocimum Americanum L., Ocimum Basilium L., Ocimum Gratissimum L. and Ocimum Tenuiflorum L. Int. J. Res. Ayurveda Pharm. **10**(3), 41–48 (2019)

Marotti, M., Piccaglia, R., Giovanelli, E.: Differences in essential oil composition of basil (Ocimum basilicum L.) Italian cultivars related to morphological characteristics. J. Agric. Food Chem. **44**, 3926–3929 (1996)

Marwat, S.K., Ur-Rehman, F., Shoaib Khan, M., Ghulam, S., Anwar, N., Mustafa, G., Usman, K.: Phytochemical constituents and pharmacological activities of sweet basil-Ocimum Basilicum L. (Lamiaceae). Asian J. Chem. **23**(9), 3773–3782 (2011)

Mulugeta, S.M., Gosztola, B., Radacsi, P.: Diversity in morphology and bioactive compounds among selected Ocimum species. Biochem. Syst. Ecol. **114**, 104826 (2024)

Muráriková, A., Ťažký, A., Neugebauerová, J., Planková, A., Jampílek, J., Mucaji, P., Mikuš, P.: Characterization of essential oil composition in different basil species and pot cultures by a GC-MS method. Molecules **22**, 1221 (2017)

Nair, S.K.P., Ganesan, K., Gulilat Azalewor, H., Letha, N., Banu Gani, S.: Preliminary phytochemical screening and in vitro antioxidant activity of Ethiopian indigenous medicinal plants, Ocimum lamiifolium Hochst. EX Benth and Ocimum Basilicum L. Int. J. Pharm. Sci. Drug Res. **8**, 30–36 (2016)

Naqri, S.A.H., Khan, M.S.Y., Vobora, S.B.: Antibacterial, antifungal and antihelminthic activity on Indian medicinal plants. Fitoterapia **62**, 221–228 (1991)

Nwanjo, H.U., Oze, G.O.: Hypolipidaemic and antioxidant properties of Ocimum gratissimum on diabetic rats. Plant Prod. Res. J. **11**, 1–4 (2007)

Özcan, M., Chalchat, J.C.: Essential oil composition of Ocimum basilicum L. and Ocimum minimum L. in Turkey. Czech J. Food Sci. **20**, 223–228 (2002)

Qamara fatimamudassar, F., Sanaa, A., Naveeda, S., Faizi, S.: Phytochemical characterization, antioxidant activity and antihypertensive evaluation of Ocimum basilicum L. in L-NAME induced hypertensive rats and its correlation analysis. Heliyon **9**, e14644 (2023)

Rehan Nadeem, H., Akhtar, S., Sestili, P., Ismail, T., Neugart, S., Qamar, M., Esatbeyoglu, T.: Toxicity, antioxidant activity, and phytochemicals of Basil (Ocimum basilicum L.) leaves cultivated in Southern Punjab, Pakistan. Foods **11**, 1239 (2022)

Reza Aminian, A., Mohebbati, R., Hossein Boskabady, M.: The effect of Ocimum Basilicum L. and its main ingredients on respiratory disorders: an experimental, preclinical, and clinical review. Front. Pharmacol. **12**, 805391 (2022)

Rezzoug, M., Bakchiche, B., Gherib, A., Roberta, A., FlaminiGuido, Kilinçarslan, Ö., Mammadov, R., Bardaweel, S.K.: Chemical composition and bioactivity of essential oils and Ethanolic extracts of Ocimum basilicum L. and Thymus algeriensis Boiss. & Reut. from the Algerian Saharan Atlas. BMC Complemen. Alt. Med. **19**, 146 (2019)

Romano, R., De Luca, L., Aiello, A., Pagano, R., Di Pierro, P., Pizzolongo, F., Masi, P.: Basil (Ocimum basilicum L.) leaves as a source of bioactive compounds. Foods **11**, 3212 (2022)

Sandeep, P.: The phytochemical constituents, pharmacological and traditional uses of Ocimum Gratissimum L. in tropics. Indo Am. J. Pharm. Sci. **4**(11), 4234–4242 (2017)

Sartoratto, A., Machado, A.L.M., Delarmelina, C., Figueira, G.M., Duarte, M.C., Rehder, V.L.G.: Composition and antimicrobial activity of essential oils from aromatic plants used in Brazil. Braz. J. Microbiol. **35**, 275–280 (2004)

Silva, V.A., Sousa, J.P., Guerra, F.Q.S., Pessôa, H.L.F., Freitas, A.F.R., Alves, L.B.N, Lima, E.O.: Antibacterial activity of Ocimum basilicum essential oil and linalool on bacterial isolates of clinical importance. Int. J. Pharmacogn. Phytochem. Res. **7**(6), 1066–1071 (2015)

Simon, J.E., Morales, M.R., Phippen, W.B., Vieira, R.F., Hao, Z.: Basil: a source of aroma compounds and a popular culinary and ornamental herb. In: Janick, J. (ed.) Perspectives on new crops and new uses, pp. 499–505. ASHS Press, Alexandria, VA

Spence, C.: Sweet basil: an increasingly popular culinary herb. Int. J. Gastronomy Food Sci. **36**, 100927 (2024)

Teliban, G.-C., Burducea, M., Mihalache, G., Zheljazkov, V.D., Dincheva, I., Badjakov, I., Popa, L.D., Bodale, I., Vladut, N.V., Cojocaru, A., Munteanu, N., Stan, T., Caruso, G., Stoleru, V.: Morphological, physiological and quality performances of basil cultivars under different fertilization types. Agronomy **12**, 3219 (2022)

Varga, F., Carović-Stanko, K., Ristić, M., Grdiša, M., Liber, Z., Šatović, Z.: Morphological and biochemical intraspecific characterization of Ocimum basilicum L. Ind. Crops Prod. **109**, 611–618 (2017)

Zervoudakis, G., Salahas, G., Rodi, M.: Nitrogen nutrition effect on aeroponic basil (Ocimum basilicum L.) catalase and lipid peroxidation. Notulae Botanicae Horti Agrobotanici Cluj-Napoca **43**(2), 561–567 (2015)

# Chemical Profiling of Two Eucalyptus Essential Oils: Insecticidal and in Silico Inhibitory Activities on Acetylcholinesterase and Chitin Synthase

Anjoud Harmouzi, Amina Chlouchi, Rabha Afroukh, Najia El Hamzaoui, Fatima Kamari, Mohammed Chrachmy, Hassan Ouallal, Mohamed Chebaibi, Ibrahim Mssillou, and Yassine El Ammari

**Abstract**

The insecticidal and inhibitory activities on acetylcholinesterase (AChE) and chitin synthase are crucial for combating *Sitophilus oryzae L.* (rice weevil), a significant pest responsible for extensive damage to stored grains. These two mechanisms target essential physiological processes in insects, offering an efficient strategy for pest control. This study aimed to validate the pesticidal potential of *Eucalyptus globulus* and *Eucalyptus cinerea* through phytochemical analysis of their essential oils (EOs) and evaluation of their insecticidal efficacy against *S. oryzae*, a common pest in stored cereals. This study has also evaluated the major EO compounds' inhibitory activity on AChE and chitin synthase. Gas chromatography-mass spectrometry, were used to analyze the essential oils indicating 1,8-cineole as the major compound in *E. globulus* (67.87%) and in *E. cinerea* (79.67%). Our EOs demonstarted significant fumigant activity against the rice weevils, *E. globulus* EO's showed the high toxicity with (LD50 = 0.053 µL EO/L of air) than *E. cinerea* EO's (LD50 = 0.17 µL EO/L of air). The results of the in-silico study demonstrated that Terpinen-4-ol and Dihydrocarveol presented the most potent AChE inhibitory, forming hydrogen bonds with active site residues. In other side, p-Cymene and Terpinen-4-ol exhibited the strongest chitin synthase inhibitory. These findings highlight that *E. globulus* and *E. cinerea* essential oils could serve as effective, natural insecticides—targeting key pest defenses like AChE and chitin synthase, offering sustainable and eco-friendly solutions to protect stored cereals.

**Keywords**

Eucalyptus · Essential oils · Chemical composition · Bioinsecticide · Molecular docking.

## 1 Introduction

Unlocking Nature's Defense: Eucalyptus Oils as Dual-Action Insecticidal and Enzyme-Inhibiting Agents

A. Harmouzi (✉) · A. Chlouchi · Y. El Ammari
Organic Chemistry, Catalysis and Environment Laboratory, Sciences Faculty, Ibn Tofail University, Kenitra, Morocco
e-mail: nojoud.harmouzi@gmail.com

A. Harmouzi · A. Chlouchi
Laboratory of Natural Resources and Sustainable Development, Faculty of Sciences, Ibn Tofail University, Kenitra, Morocco

R. Afroukh
Laboratory of Plant, Animal and Agro-Industry Production, Sciences Faculty, Ibn Tofail University, Kenitra, Morocco

N. El Hamzaoui
Higher Institute of Nursing Professions and Health Techniques of Fez-Meknes, Regional Directorate of Health Fes-Meknes, Meknes, Morocco

N. El Hamzaoui · M. Chrachmy · H. Ouallal
Laboratory of Materials Engineering for the Environment and Natural Resources, Faculty of Sciences and Technologies, University of Moulay Ismail, Errachidia, Morocco

F. Kamari · I. Mssillou
Laboratory of Natural Substances, Pharmacology, Environment, Modeling, Health and Quality of Life (SNAMOPEQ), Faculty of Sciences Dhar El Mahraz, Sidi Mohamed Ben Abdellah University, Fez, Morocco

M. Chebaibi
Ministry of Health and Social Protection, Higher Institute of Nursing Professions and Health Techniques, Fez, Morocco

Biomedical and Translational Research Laboratory, FacultyofMedicineandPharmacyofFez, Sidi Mohamed Ben Abdellah University, Fez, Morocco


O. Jamal Eddine et al. (eds.), *Plant Pathology, Fungal Diversity, and Biotechnological Advances in Agriculture*, Sustainable Landscape Planning and Natural Resources Management, https://doi.org/10.1007/978-3-031-97666-7_2

Given the surging need for sustainable and environmentally safe pest control methods worldwide a significant interest in plant-derived bioactive compounds was raised. Essential oils (EOs), which can be defined as natural substances obtained from aromatic plants throughout different processes of extraction, become a potential substitute to synthetic pesticides because of their biodegradability, limited environmental persistence, and low toxicity to non-target organisms. One of them, Eucalyptus species that are particularly appreciated for their rich chemical profiles and diverse biological activities, including insecticidal properties (Huang et al. 2022). The rice weevil, Sitophilus oryzae L., is one of the significant pests causing widespread damage to stockpiled grains (Ren et al. 2022). Fighting this pest is crucial, since it endangers global food security. Two significant mechanisms that can be addressed for efficient pest control are the inhibition of acetylcholinesterase (AChE) and chitin synthase (Ren et al. 2022). The inhibition of acetylcholinesterase, which is an essential enzyme in the insect nervous system, may result in the disruption of neurotransmission, ultimately causing insect mortality (Shao et al. 2020; Zhang et al. 2022). On the other hand, the inhibition of chitin synthase, which plays a crucial role in the production of the insect exoskeleton, can prevent normal molting and development, leading by the way to insect death (Omoniwa et al. 2024).

Numerous studies have exanimated the chemical profiling and biological effects of essential oils obtained from several species of *Eucalyptus*. These researches have accentuated the potential insecticidal, antimicrobial, and medicinal properties of these essential oils. For example, this study has shown that the essential oil of E. globulus and E. cinerea exhibit insecticidal effects against different pests (Barbosa et al. 2016; Ainane et al. 2020). Several studies have looked into the toxicity and acaricidal effects of Eucalyptus essential oils, showing promising potential as a natural bioinsecticide (Atmani-Merabet et al. 2018). Additionally, research on *Eucalyptus globulus* essential oil has highlighted its medicinal properties, particularly its anti-inflammatory and analgesic effects (Sugumar et al. 2014).

*Eucalyptus globulus* and *Eucalyptus cinerea* were chosen as treatment options for the Rice Weevil, *Sitophilus oryzae*, due to their significant bioactivity and chemical composition. Research indicates that these two Eucalyptus species possess compounds, such as 1,8-cineole, that exhibit strong fumigant and repellent activities against storage product pests like the rice weevil (Barbosa et al. 2016). Furthermore, the essential oil's activities as antibacterial and antifungal have been investigated, indicating its potential in combating microbial infections (Rahimi-Nasrabadi et al. 2013). Furthermore, the essential oil has shown promise in repelling mosquitoes (Sheikh et al. 2021). The chemical composition of Eucalyptus species essential oils has been a subject of interest, with studies aiming to characterize its components and assess their biological activities (Salari et al. 2006).

Our study has several objectives: First, we aim to carry out a detailed analysis of the chemical profile of Moroccan *E. globulus* and *E. cinerea* essential oils. Next, we evaluate their potential bioactivity against Sitophilus oryzae, better known as the rice weevil. We also assess how effective these essential oils are as fumigants against this pest. Finally, to better understand the molecular mechanisms behind their activity, we performed an in-silico study using the major components of the essential oils.

## 2 Materials and Methods

### 2.1 Essential Oils Extraction

Leaves of *E. globulus* and *E. cinerea* were sourced in Kenitra region, Morocco. After being dried in a shaded, well-ventilated environment, the EOs were extracted using the hydrodistillation technique with a Clevenger apparatus. Each extraction used 100 g of dried, powdered leaves processed at 100 °C for 3 h. This procedure was repeated three times. The extracted essential oils were dried over anhydrous sodium sulfate and stored in amber vials at 4 °C.

### 2.2 Weevils Population

Adult rice weevils (Sitophilus oryzae L.), members of the Curculionidae family, were collected from a local grain market. They were maintained on rice grains in glass containers under controlled conditions (24 ± 1 °C, 76 ± 5% RH). To promote the development of a uniform population, the insects were kept in complete darkness.

### 2.3 Identification of Chemical Composition

An HP Innowax capillary column (60 m × 0.25 mm × 0.25 μm) was used for the analysis. With system was fitted with a flame ionization detector and linked to an HP GCD analytical setup. The gas chromatography (GC) conditions were set as follows: the oven temperature was initially held at 50 °C for 2 min, then gradually increased by 4 °C per minute until reaching 200 °C, where it was maintained for another 2 min. Neat samples (1 μl) were injected with a split ratio of 1:50. Helium (99.999% purity) served as the carrier gas at a constant flow rate of 1.2 ml/min. Most of the essential oil components were identified by comparing their retention indices—calculated based on a C7–C40 linear alkane series—and their mass spectra. These results were matched against reference data from the Wiley 175.L database (Harmouzi et al. 2016; Lee and Wang 2003).

### 2.4 Bioassays and Data Analysis

To assess the in vivo efficacy of E. globulus and E. cinerea essential oils, we used different doses (0.25, 0.5, 1, and 2 µl/l) were tested against Sitophilus oryzae. Fumigation tests were conducted in plastic boxes, each containing a chiffon gauze pouch with 20 adult Sitophilus oryzae and a filter paper soaked with the essential oils. Three replicates were performed for each concentration.

To evaluate the toxic fumigant activity of the tested essential oils, an analysis of variance was conducted using the square root (percentage mortality) (Arsin) function in ANOVA. Survival probabilities and comparisons of the effects of each tested dose were assessed using the Kaplan–Meier test and the log-rank test, respectively (Lee and Wang 2003). The $LC_{50}$ and $LC_{99}$ values, representing the concentrations required to kill 50% and 99% of the fumigated insects, respectively, were calculated using Probit analysis following Finney's method (Finney 1971).

### 2.5 Molecular Docking Study

Maestro 11.5 (Schrödinger suite) was used to conduct in silico study analyses. The goal of this article is to explore the interaction modes of the EOs main compounds with the active sites of acetylcholinesterase and chitin synthase proteins, specifically, to assess the plant's insecticidal properties.

To study their structures, we downloaded the crystal models of acetylcholinesterase (ID: 6ARY) and chitin synthase (ID: 7STM) from the RCSB Protein Data Bank (El-Aziz et al. 2023; Omoniwa et al. 2024). Using Maestro 11.5's Protein Preparation Wizard, we cleaned up the proteins by adding hydrogen atoms, fixing structural flaws like overlapping or missing atoms, and adjusting hydroxyl groups, water molecules, and amino acids. Finally, we performed a light energy minimization to refine the overall structure (Chebaibi et al. 2024; El Abdali et al. 2023a, b). To prepare ligands we used Maestro 11.5's LigPrep tool (part of the Schrödinger suite). This step converted the 2D structures into 3D models, added hydrogens, corrected bond lengths and angles, and optimized their geometry with the OPLS3 force field. Throughout the process, we maintained the original ionization states and chirality of the molecules (El Abdali et al. 2023a, b). A receptor grid was generated using the Glide docking tool in Maestro 11.5 from the Schrödinger suite. The active site was determined based on the original ligand's coordinates, and ligands were then docked into the protein's X-ray crystal structure to investigate potential interaction conformations. We also defined key parameters, including binding sites, boundaries, rotatable groups, and excluded volume. (El Abdali et al. 2023b). After preparing the ligands, proteins, and receptor grid, molecular docking was carried out using the SP (Standard Precision) mode (El Abdali et al. 2023a; Harmouzi et al. 2024).

## 3 Results and Discussion

### 3.1 Chemical Screening

Table 1 present the essential oil yields and physicochemical properties of *E. globulus* and *E. cinerea.* As shown in Table 1, *E. globulus* produces nearly twice as much essential oil (0.85 ± 0.05%) as *E. cinerea* (0.44 ± 0.05%, with both showing consistent yields. Previous research consistently shows interesting variations in Eos yields between our plants (Barbosa et al. 2016). In a study on *E. globulus,* essential oil yields varied from 1.6 to 2.1% were reported after coppicing (Silva et al. 2011), Studies reported essential oil yields of *E globulus* to be 0.96–2.7% in Morocco (Derwich 2013), 2.68% in Senegal (Bathily et al. 2023), 1.05–2.1% in India (Joshi et al. 2016) and 0.95–1.32% on a fresh weight basis In Ethiopia (Shiferaw et al. 2019). For *Eucalyptus cinerea* EO yields varies between 0.5 and of 5.02% for dried leaves, with variations depending on the season and plant part collected (Derwich 2013). Researches demonstre that the EO yield of eucalyptus leaves can range from 0.1 to 7% in fresh weight (Abdo 2019). Our findings align with previous studies, which showed that *E. globulus* from various regions like Ethiopia, Morocco, Brazil also

**Table 1** Yields and physicochemical characterization of essential oils of *E. globulus* and *E. cinerea*

| Species | Yield (%) | Determination at 20 °C | | |
|---|---|---|---|---|
| | | Refractive index | Optical rotation | Solubility in alcohol (vol at 70%) |
| *Eucalyptus globulus* | 0.85 | 1.452 | + 1°12′ | 0.85 |
| *Eucalyptus cinerea* | 0.44 | 1.450 | + 0°9′ | 0.90 |

produced higher yields than *E. cinerea* (Barbosa et al. 2016; Bisen et al. 1984).

The variation in essential oil yield between *Eucalyptus globulus* and *Eucalyptus cinerea* can be attributed to several factors. Studies have shown that the essential oil composition within the same variety can differ based on geographical region (Barbosa et al. 2016). Also, methods of extraction can also impact the essential oil yield. For example, a study comparing supercritical $CO_2$ extraction and hydrodistillation for *Eucalyptus* species demonstrated differences in yield and chemical composition (Joshi et al. 2016). The choice of extraction method can influence the efficiency of essential oil recovery and the final composition. Moreover, environmental factors such as climate, soil composition, and altitude can influence the growth and chemical profile of *Eucalyptus* species, leading to variations in essential oil yield (Harmouzi et al. 2024).

Results of the physicochemical properties of our two EOs indicate that the specific gravity values closely align with those documented by (Ghaffar et al. 2015). Additionally, the refractive indices, lower than 1.47, confirm the richness of our essential oils in 1,8-cineole (Barbosa et al. 2016). Moreover, there is variability in the optical rotation among the species, ranging from $+ 0°9'$ to $+ 1°12'$.

Table 2 shows GC–MS analysis of the Essential oils, and their retention indices and area percentages. 44 components were identified and representing 98.37% of the *E. globulus* EOs. The major component was 1,8-Cineole also known as eucalyptol, at 67.86%, followed by Limonene (7.12%), Dihydrocarveol (5.05%), cis-Sabinene hydrate (4.04%), e-Terpinen-4-ol (3.15%), and Estragole (2.15%). For the chemical profile of *E. cinerea* essential oil, the major component was also 1,8-Cineole with 79.67%, followed with α-Pinene (4.79%), p-Cymene (2.21%) and Terpinen-4-ol (1.32) (Fig. 1).

Our findings align with certain research outcomes, yet they diverge from others. The essential oils from *E. globulus* leaves have been extensively studied and have been found to contain a variety of compounds, with 1,8-cineole often identified as a major component (Boukhatem et al. 2014; Pereira et al. 2005) Other compounds commonly present in the essential oils of Eucalyptus globulus include α-pinene, limonene, α-terpineol, and globulol (Ghaffar et al. 2015). Studies have shown that the essential oils extracted from Eucalyptus cinerea leaves contain various compounds, with 1,8-cineole also being a prominent component (Ghaffar et al. 2015). Additionally, other compounds such as α-pinene, limonene, α-terpineol, and α-terpinyl acetate have been identified in the essential oils of Eucalyptus cinerea (Silva et al. 2011). The percentage composition of these components can vary depending on factors such as the location, age of the plant, the part of the plant from which the oil is extracted and the extraction method. Essential oils composition contributes to the unique aroma and potential therapeutic properties of *Eucalyptus globulus* essential oil (Boukhatem et al. 2014; Shiferaw et al. 2019).

### 3.2 Insecticidal Activity of Eucaluptus Essential Oils Against Adults of S. oryzae (L.)

The tested concentrations had strong fumigant effects, with their effectiveness varying based on both the dose and how long the weevils were exposed. As Fig. 2 shows, higher concentrations and longer exposure times steadily reduced the weevils' survival rate.

The effect of the essential oils on the rice weevils lasted seven days for *E. globulus* EOs and eight days for *E. cinerea* EOs. Even at the lowest concentration tested, the EOs began killing the insects on the very first day. For the concentration of 2 μl of EO, total insect mortality was achieved by the fourth day. Notably, at lower concentrations, insect deaths gradually increased, leading to the complete elimination of the population by the seventh or eighth day of exposure. The 50% lethal time (LT50) and 99% lethal time (LT99) for adults treated to various doses of EOs varied between 24 h and 4 days and approximately 4 and 8 days, respectively. This vari-ation was dependent on the doses used. Meanwhile, the control group of weevils exhibited an average adult lifespan of 18 days (LT50) and 45 days (LT99). Notably, there was a negative correlation between LT50 and LT99 values and the concentrations tested, indicating a more rapid and potent effect at higher concentrations. For *E. globulus* and *E. cinerea* essential oils, the calculated lethal concentrations LC50 and LC99 exhibit time-dependent variations, gradually decreasing with prolonged fumigation durations. Lethal concentrations LC50 and LC99 varied between 2.78 and 0.053 μl /l air as well as 167.9 and 0.762 μl/l air, respectively.

Research indicates that our two EOs have robust insecticidal properties that can be effective in controlling pests especially the Rice Weevil (Abdelgaleil et al. 2009; Sebei et al. 2015). Studies have demonstrated the insecticidal potential of *E. globulus* EO against various insect species. For example, *E. globulus* EO has been reported to exhibit insecticidal activity against aphids such as *Aphis nerii*, *Aphis gossypii*, and *Aphis fabae* (Russo et al. 2018). Also, *E. globulus* EO has shown an insecticidal propriety against the beetle *Tribolium confusum*, indicating its effectiveness (Russo et al. 2015). Similarly, *E. cinerea* EO has been investigated as biopesticide (Barbosa et al. 2016). While studies

**Table 2** Chemical profile of essential oils

| Number | Compounds | IR | *E. globulus* | *E. cinerea* |
|---|---|---|---|---|
| 1 | Terpinolene | 3.09 | 0.45 | 0.09 |
| 2 | α-Thujene | 3.23 | – | 0.32 |
| 3 | trans-Verbenol | 4.18 | 0.02 | – |
| 4 | **α-Pinene** | 4.44 | 0.65 | **4.79** |
| 5 | Linalool | 4.75 | 0.01 | 0.43 |
| 6 | Terpinolene | 4.80 | – | 0.41 |
| 7 | α-Fenchyl alcohol | 4.83 | – | 0.21 |
| 8 | Spiro[2.4]heptane, 4-methylene- | 4.95 | 0.23 | – |
| 9 | β-Pinene | 5.08 | – | 0.1 |
| 10 | 4-Acetyl-1-methylcyclohexene | 5.12 | 0.02 | – |
| 11 | 1,4-Hexadiene, 5-methyl-3-(1-methylethylidene)- | 5.21 | 0.76 | – |
| 12 | Tricyclo[4.1.0.0(2,7)]heptanes | 5.51 | 0.42 | – |
| 13 | **p-Cymene** | 5.54 | – | **2.21** |
| 14 | 4-Methyl-1,4-heptadiene | 5.56 | 0.07 | – |
| 15 | Dicyclopentadiene, tetrahydro, exo | 5.61 | 0.02 | – |
| 16 | Campholenic aldehyde | 5.62 | – | 0.05 |
| 17 | Isoborneol | 5.64 | 0.04 | 0.2 |
| 18 | **Terpinen-4-ol** | 5.71 | **3.15** | **1.32** |
| 19 | Furfural | 5.83 | 0.09 | – |
| 20 | Myrtenal | 6.12 | 0.20 | 0.6 |
| 21 | **Estragole** | 6.19 | **2.15** | 0.1 |
| 22 | 2,6-Dimethyl-3,5,7-octatriene-2-ol, Z,Z- | 6.22 | 0.67 | – |
| 23 | Nerol | 6.00 | – | 0.20 |
| 24 | Propanal, 2-methyl-3-phenyl- | 6.12 | 0.43 | – |
| 25 | (-)-Carvone | 6.32 | 0.12 | 0.2 |
| 26 | 1,5-Hexadiene, 2,5-dimethyl-3-methylene- | 6.76 | 0.02 | – |
| 27 | **Dihydrocarveol** | 7.33 | **5.05** | - |
| 28 | α-Terpinen-7-al | 7.45 | 0.81 | - |
| 29 | Bornyl acetate | 7.50 | 0.32 | - |
| 30 | 1Methyltricyclo[2.2.1.0(2,6)]heptane | 7.64 | 0.34 | - |
| 31 | p-Menth-8-ene, 3-methylene- | 7.74 | 0.01 | - |
| 32 | α-Longipinene | 8.08 | 0.04 | - |
| 33 | α-Santalol | 8.16 | 0.03 | 0.06 |
| 34 | **cis-Sabinene hydrate** | 8.44 | **4.04** | - |
| 35 | α-Himachalene | 8.59 | 0.23 | - |
| 36 | 8-Oxabicyclo[5.1.0]oct-5-en-2-ol,1,4,4-trimethyl- | 8.65 | 0.06 | - |
| 37 | α-Sinensal | 8.95 | 0.18 | 1.13 |
| 38 | **1,8-Cineole** | 9.56 | **67.87** | **79.67** |
| 39 | α-Terpineol | 10.54 | 0.10 | 0.07 |
| 40 | Limonene | 10.64 | 7.12 | 0.67 |
| 41 | O-Trifluoroacetyl-isopulegol | 10.71 | 0.12 | – |
| 42 | Ocimene | 10.73 | 0.38 | – |
| 43 | Epiglobulol | 10.92 | – | 0.43 |
| 44 | Isothujol | 11.09 | 0.07 | – |
| 45 | Thunbergol | 11.12 | 0.02 | – |

(continued)

**Table 2** (continued)

| Number | Compounds | IR | *E. globulus* | *E. cinerea* |
|---|---|---|---|---|
| 46 | Cycloheptene, 5-ethylidene-1-methyl- | 12.23 | 0.21 | – |
| 47 | Camphene | 12.71 | 0.04 | – |
| 48 | α-Farnesene | 13.25 | 0.38 | – |
| 49 | β-Terpinene | 13.41 | 1.34 | 0.21 |
| 50 | β-Elemenone | 13.53 | 0.06 | 0.50 |
| 51 | Geranyllinalool | 13.47 | 0.04 | – |
| 52 | Tris(tert-butyldimethylsilyloxy)arsane | 13.65 | 0.02 | – |

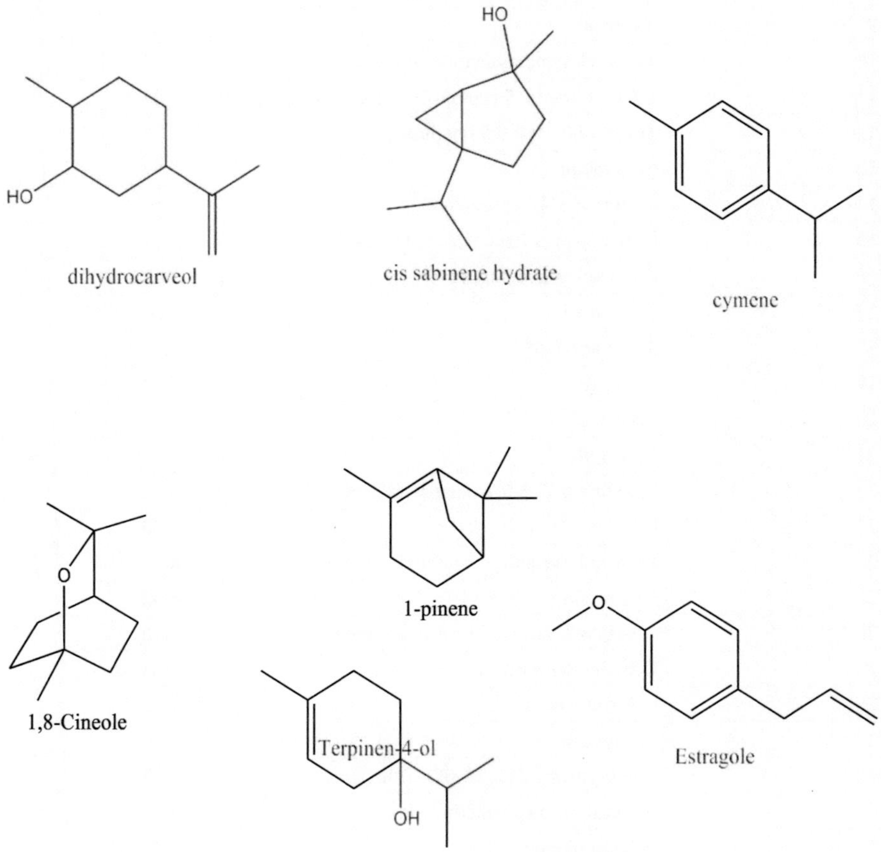


**Fig. 1** Main compounds from the essential oils of *E. globulus* and *E.* cinerea

on its activity against *Sitophilus oryzae L.* are limited. Overall, this study demonstrates that both *E. globulus* and *E. cinerea* EOs present their promising insecticidal activities against pests (Barbosa et al. 2016).

## 3.3 Molecular Docking

Insecticides work by attacking critical biological systems in insects, through two key mechanisms: inhibition of acetylcholinesterase (AChE) and disruption of chitin synthesis. AChE inhibitors, such as organophosphates and carbamates, inhibit acetylcholine degradation, leading to continuous nerve signaling, which results in overstimulation, paralysis, and rapid death of the insect. Meanwhile, chitin synthesis inhibitors hinder the development of chitin, which is an essential component of the insect's exoskeleton and gut lining, provoking developmental defects and eventual mortality. These mechanisms contribute to effective pest control, with AChE inhibitors acting promptly and chitin synthesis inhibitors providing long-term population control, while decreasing non-target toxicity and resistance risks.

Our computational analysis revealed that all the compounds we tested were able to block both acetylcholinesterase (AChE) and chitin synthase. Among these, Terpinen-4-ol, Dihydrocarveol, and p-Cymene were identified as the most effective AChE inhibitors, with glide g-scores of − 6.476, − 6.340, and − 5.880 kcal/mol, respectively (Table 3).

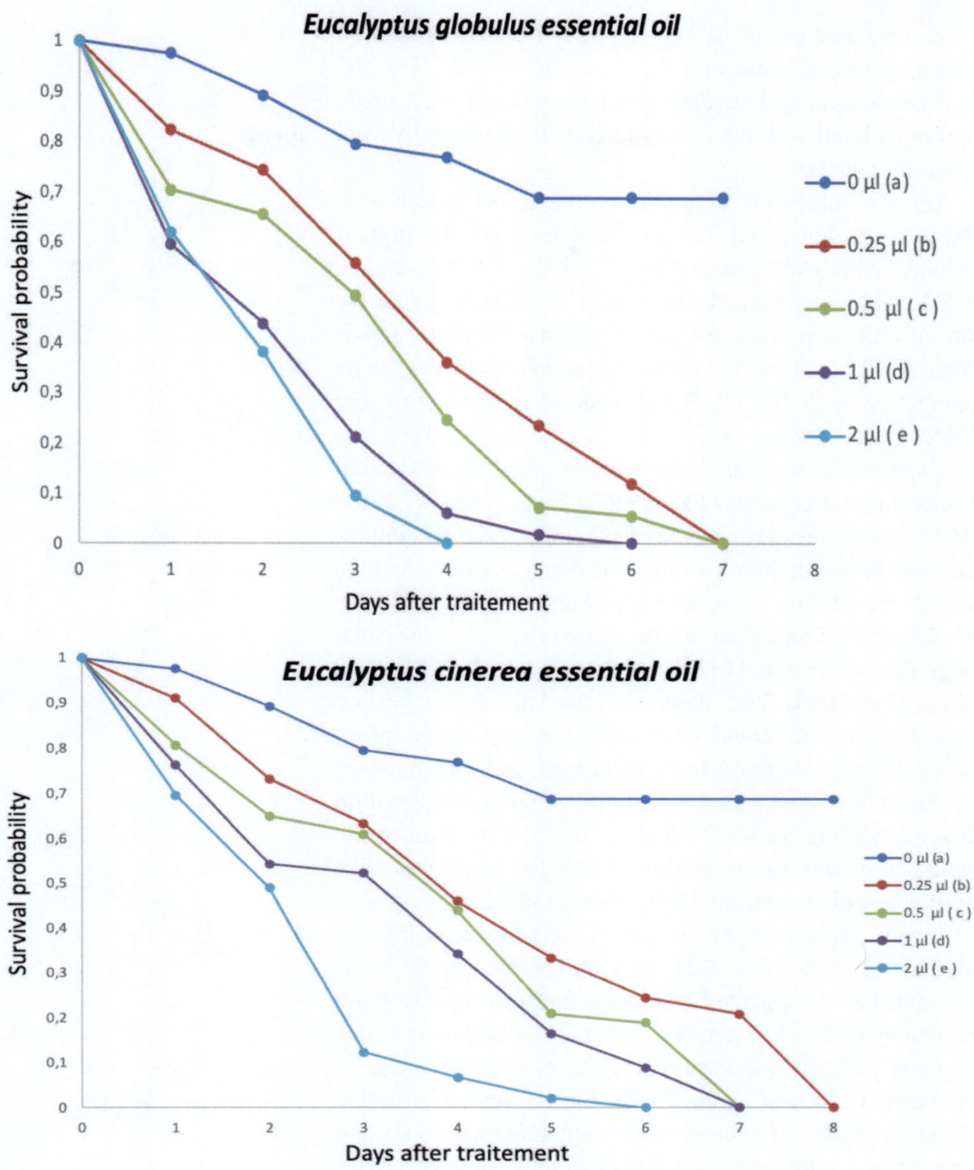


**Fig. 2** Survival rate of S. oryzae (L.) adults fumigated by *E. globulus* and *E. cinerea* essential oils

**Table 3** Ligand docking studies with various receptor types

| | Glide gscore (kcal/mol) | |
|---|---|---|
| | Acetylcholinesterase (PDB: 6ARY) | Chitin synthase (PDB: 7STM) |
| 1,8-Cineole | − 4.665 | − 4.18 |
| alpha-PINENE | − 5.731 | − 3.611 |
| cis-Sabinene hydrate | − 5.804 | − 5.555 |
| Dihydrocarveol | − 6.34 | − 4.886 |
| Estragole | − 4.923 | − 3.328 |
| Limonene | − 5.002 | − 2.96 |
| p-Cymene | − 5.88 | − 5.892 |
| Terpinen-4-ol | − 6.476 | − 5.481 |

In the active site of AChE, Terpinen-4-ol demonstrated a single hydrogen bond with the TYR 282 residue (Fig. 1a, b). Simultaneously, Dihydrocarveol also established a single hydrogen bond with the CYS 447 residue of this active site (Figs. 2b and 3a).

For the inhibition of chitin synthase, p-Cymene, cis-Sabinene hydrate, and Terpinen-4-ol revealed the highest activity, with glide g-scores of − 5.892, − 5.555, and − 5.481 kcal/mol, respectively (Table 3). Within the active site of chitin synthase, p-Cymene did not form any bonds (Fig. 2c). In contrast, cis-Sabinene hydrate formed a single interaction with the GLU 321 residue in the active site (Figs. 2d and 4d).

Compounds such as Terpinen-4-ol and Dihydrocarveol interacting with residues like TYR 282 and CYS 447 in the AChE active site. Previous studies, emphasized the importance of targeting unique cysteine residues in insect AChEs, which are absent in vertebrates (Polsinelli et al. 2010). Methanethiosulfonate molecules demonstrated irreversible inhibition of insect AChEs, sparing mammalian enzymes (Pang et al. 2009). This aligns with our findings, suggesting these compounds could offer selective insecticidal properties. AChE inhibitors (e.g., donepezil and galantamine) demonstrated strong binding affinities and stability within enzyme binding pockets (Ali et al. 2017). The attached file similarly reports strong binding scores for compounds like Terpinen-4-ol, supporting their potential efficacy.

In our study compounds such as cis-Sabinene hydrate interacting with GLU 321 in chitin synthase. Previous research has documented resistance mechanisms involving mutations in the *chs*1 gene (e.g., I1043M mutation in *Culex pipiens*) and overexpression leading to increased cuticle thickness (Lucchesi et al. 2022). Earlier studies modeled chitin synthase structures using computational tools and investigated inhibitors like nikkomycin Z and polyoxins for their binding affinities (Zhang and Yan Zhu 2013; Ranganathan et al. 2020). The attached file similarly uses docking approaches to explore interactions but focuses on natural compounds, expanding the scope of potential inhibitors.

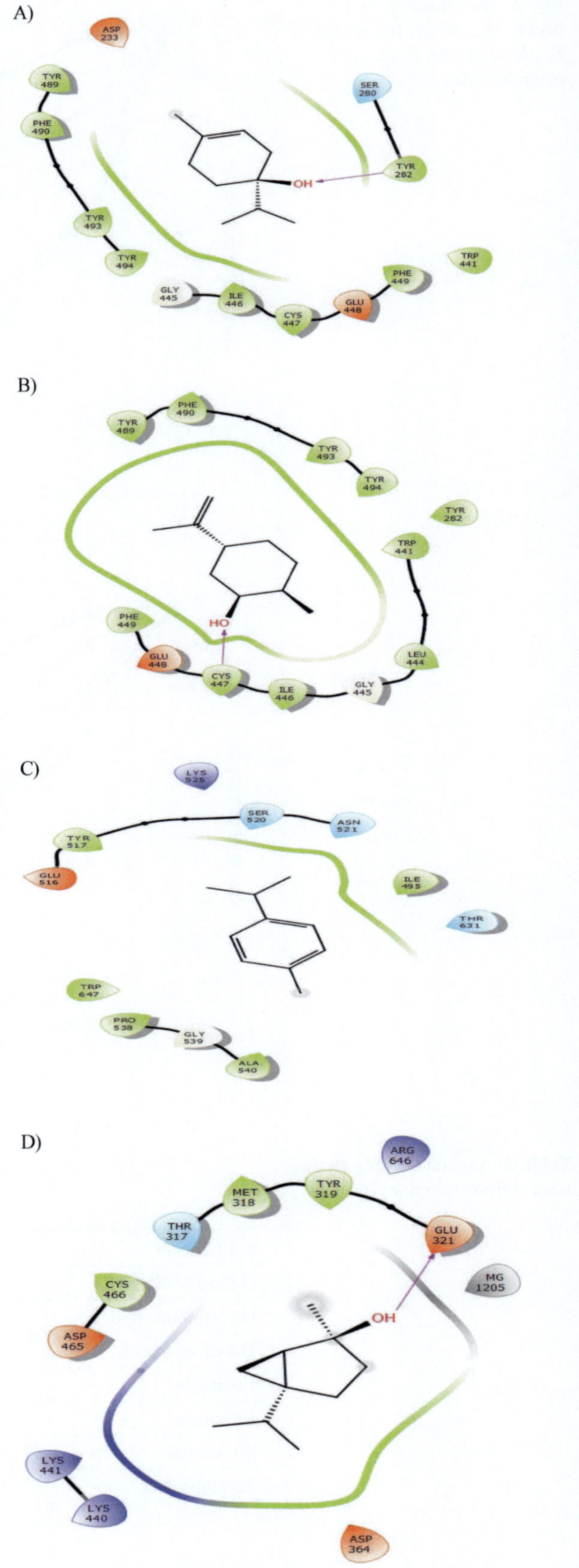


**Fig. 3** The 2D viewer of ligands' interactions with the active site. **a** and **b** Terpinen-4-ol, and dihydrocarveol interactions with the active site of acetylcholinesterase. **c** and **d** p-Cymene, cisSabinene hydrate interactions with the active site of chitin synthase

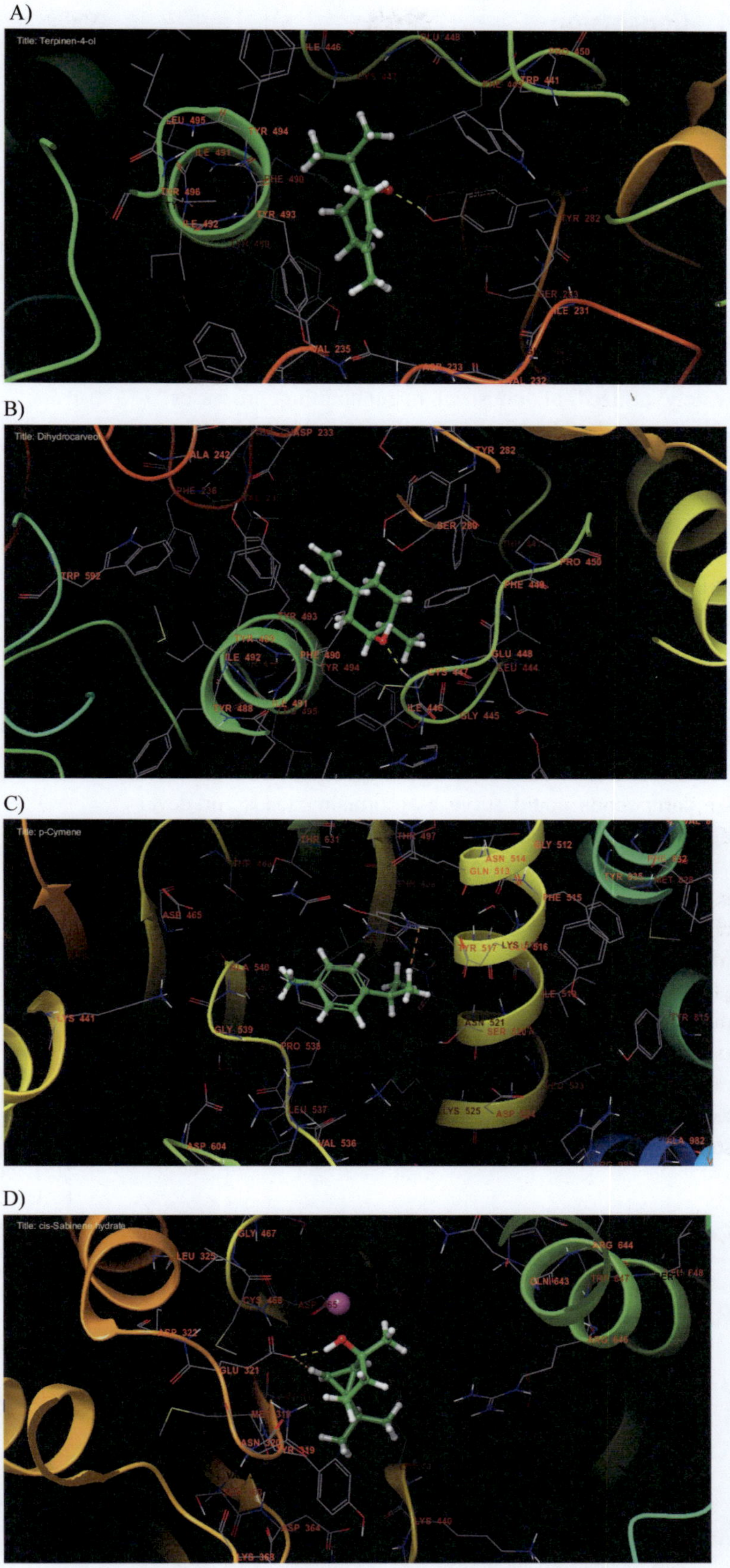


**Fig. 4** The 3D viewer of ligands' interactions with the active site. **a** and **b** Terpinen-4-ol, and dihydrocarveol interactions with the active site of acetylcholinesterase. **c** and **d** p-Cymene, cisSabinene hydrate interactions with the active site of chitin synthase

## 4 Conclusion

In our study, we conducted a phytochemical analysis of two species of *Eucalyptus*, *E. globulus* and *E. cinerea*, determining the chemical composition of their essential oils and evaluating their insecticidal properties. Our findings indicate that the yield and chemical profile of essential oils vary between eucalyptus species, with both exhibiting diverse chemical profiles dominated primarily by 1,8-cineole. Moroccan eucalyptus emerges as a significant source of this compound alongside potential synergistic interactions with minor compounds. A vital component widely utilized in industrial manufacturing processes. In silico study highlights the significant inhibitory effects of the tested compounds against acetylcholinesterase (AChE) and chitin synthase. Among these, Terpinen-4-ol, Dihydrocarveol, and p-Cymene were identified as the most potent AChE inhibitors, with favorable glide g-scores indicating strong binding affinity. Notably, Terpinen-4-ol and Dihydrocarveol formed key hydrogen bonds with specific residues within the AChE active site, enhancing their inhibitory potential. For chitin synthase, p-Cymene, cis-Sabinene hydrate, and Terpinen-4-ol exhibited the highest inhibitory activities, although p-Cymene did not establish any bonds in the active site. These findings suggest that these essential oil compounds could serve as promising leads for developing novel inhibitors targeting AChE and chitin synthase, with potential applications in pest control and therapeutic strategies. Ongoing research aims to isolate, identify, and characterize the bioactive compounds within this plant. Moreover, employing eucalyptus essential oils for fumigation offers a viable approach to controlling S. oryzae populations, especially in scenarios involving large quantities of stored seeds where fumigation can be applied without relocation. This underscores the potential of Eucaluptus oils as an alternative to synthetic pesticides for safeguarding stored cereals and protecting crops.

**Conflicts of Interest** The authors declare no conflicts of interest.

## References


Abdelgaleil, S.A.M., Mohamed, M.I.E., Badawy, M.E.I., El-Arami, S.A.A.: Fumigant and contact toxicities of monoterpenes to sitophilus oryzae (L.) and tribolium castaneum (Herbst) and their inhibitory effects on acetylcholinesterase activity. J. Chem. Ecol. **35**(5), 518–525 (2009). https://doi.org/10.1007/S10886-009-9635-3/METRICS

Abdo, B.M.: Physico-chemical profile and antioxidant activities of Eucalyptus globulus Labill and Eucalyptus citriodora essential oils in Ethiopia. Med. Aromatic Plants **8**(2), 1–4 (2019). https://doi.org/10.35248/2167-0412.19.8.332

Ainane, A., Mohamed Abdoul-Latif, F., Mohamed Abdoul-Latif, T., & Ainane, T. (2020). Evaluation of biological activities of two essential oils as a safe environmental bioinsecticides: Case of Eucalyptus globulus and Rosmarinus officinalis. *Przegląd Naukowy Inżynieria i Kształtowanie Środowiska, 29*(4), 544–556. https://doi.org/10.22630/PNIKS.2020.29.4.47

Ali, M.R., Sadoqi, M., Møller, S.G., Boutajangout, A., Mezei, M.: Assessing the binding of cholinesterase inhibitors by docking and molecular dynamics studies. J. Mol. Graph. Model. **76**, 36–42 (2017). https://doi.org/10.1016/J.JMGM.2017.06.027

Atmani-Merabet, G., Belkhiri, A., Dems, M. A., Lalaouna, A., Khalfaoui, Z., & Mosbah, B. (2018). Chemical composition, toxicity, and acaricidal activity of Eucalyptus globulus essential oil from Algeria. *Current Issues in Pharmacy and Medical Sciences, 31*(2), 89–93. https://doi.org/10.1515/CIPMS-2018-0017

Barbosa, L.C.A., Filomeno, C.A., Teixeira, R.R.: Chemical variability and biological activities of Eucalyptus spp. essential oils. Molecules **21**(12), 1671 (2016). https://doi.org/10.3390/MOLECULES21121671

Bathily, M., Ngom, B., Mbengue, M., Gassama, D.: Evaluation of the inhibitory action of essential oil from Eucalyptus globulus leaves on the corrosion of mild carbon steel in 1M HCl medium. Ovidius Univ. Ann. Chem. **34**(1), 1–7 (2023). https://doi.org/10.2478/AUOC-2023-0001

Bisen, S.S., Prabhu, V.V., Theagarajan, K.S.: Studies on the relationship of cineole content of the oil and stomata characters of eucalyptus hybrid (E. tereticornis) leaf. Indian Forester **110**(6), 577–581 (1984). https://doi.org/10.36808/IF/1984/V110I6/10508

Boukhatem, M.N., Ferhat, M.A., Kameli, A., Saidi, F., Walid, K., Mohamed, S.B.: Quality assessment of the essential oil from *Eucalyptus globulus* Labill of Blida (Algeria) origin. Int. Lett. Chem. Phys. Astronomy **36**, 303–315 (2014). https://doi.org/10.18052/WWW.SCIPRESS.COM/ILCPA.36.303

Chebaibi, M., Mssillou, I., Allali, A., Bourhia, M., Bousta, D., Gonçalves, R.F.B., Hoummani, H., Aboul-Soud, M.A.M., Augustyniak, M., Giesy, J.P., Achour, S.: Antiviral activities of compounds derived from medicinal plants against SARS-CoV-2 based on molecular docking of proteases. J. Biol. Biomed. Res. **1**(1), 10–30 (2024). https://doi.org/10.69998/J2BR1

Derwich, E.: Antifungal activity and gas chromatography coupled with mass spectrometry (GC-MS) leaf oil analysis of essential oils extracted from Eucalyptus globulus (Myrtaceae) of north centre region of Morocco. Afr. J. Pharm. Pharmacol **7**(19), 1157–1162 (2013). https://doi.org/10.5897/AJPP11.231

El Abdali, Y., Beniaich, G., Mahraz, A.M., El Moussaoui, A., Bin Jardan, Y.A., Akhazzane, M., Chebaibi, M., Nafidi, H.A., Eloutassi, N., Bourhia, M., Bouia, A.: Antibacterial, antioxidant, and in silico NADPH oxidase inhibition studies of essential oils of Lavandula dentata against foodborne pathogens. Evidence-Based Complementary and Alternative Medicine: ECAM, 2023 (2023a). https://doi.org/10.1155/2023/9766002

El Abdali, Y., Mahraz, A.M., Beniaich, G., Mssillou, I., Chebaibi, M., Bin Jardan, Y.A., Lahkimi, A., Nafidi, H.A., Aboul-Soud, M.A.M., Bourhia, M., Bouia, A.: Essential oils of Origanum compactum Benth: chemical characterization, in vitro, in silico, antioxidant, and antibacterial activities. Open Chem. **21**(1) (2023b). https://doi.org/10.1515/CHEM-2022-0282/ASSET/GRAPHIC/J_CHEM-2022-0282_FIG_006.JPG

El-Aziz, F.E.Z.A.A., Hifney, A.F., Mohany, M., Al-Rejaie, S.S., Banach, A., Sayed, A.M.: Insecticidal activity of brown seaweed (Sargassum latifolium) extract as potential chitin synthase inhibitors: Toxicokinetic and molecular docking approaches. S. Afr. J. Bot. **160**, 645–656 (2023). https://doi.org/10.1016/J.SAJB.2023.07.058

Finney, D.J.: Probit Analysis, 3rd ed. J. Pharma. Sci. **60**(9), 1432–1432 (1971). https://doi.org/10.1002/JPS.2600600940

Ghaffar, A., Yameen, M., Kiran, S., Kamal, S., Jalal, F., Munir, B., Saleem, S., Rafiq, N., Ahmad, A., Saba, I., Jabbar, A.: Chemical composition and in-vitro evaluation of the antimicrobial and antioxidant activities of essential oils extracted from seven eucalyptus species. Molecules **20**(11), 20487 (2015). https://doi.org/10.3390/MOLECULES201119706

Harmouzi, A., Boughdad, A., El Ammari, Y., Chaouch, A.: Chemical composition and toxicity of Moroccan Tetraclinis articulata and Juniperus phoenicea essential oils against Aphis citricola Goot, 1912 (Homoptera, Aphididae). Res. Chem. Intermediates **42**(9), 7185–7197 (2016). https://doi.org/10.1007/S11164-016-2528-5/METRICS

Harmouzi, A., EL Ammari, Y., Mssillou, I., Chlouchi, A., Lim, A., Abdelaziz Shahat, A., Chebaibi, M.: Insecticidal potential of essential oils from Ammi visnaga L. and Trachyspermum ammi L. against Sitophilus oryzae (L.) and in silico study of their major constituents. Horticulturae **10**(7), 722 (2024). https://doi.org/10.3390/HORTICULTURAE10070722

Huang, S., Huang, H., Xie, J., Wang, F., Fan, S., Yang, M., Zheng, C., Han, L., & Zhang, D. (2022). The latest research progress on the prevention of storage pests by natural products: Species, mechanisms, and sources of inspiration. *Arabian Journal of Chemistry, 15*(11), 104189. https://doi.org/10.1016/J.ARABJC.2022.104189

Joshi, A., Sharma, A., Bachheti, R.K., Pandey, D.P.: A comparative study of the chemical composition of the essential oil from Eucalyptus globulus growing in Dehradun (India) and around the world. Oriental J. Chem. **32**(1), 331–340 (2016). https://doi.org/10.13005/OJC/320137

Lee, E.T., Wang, J.W.: Statistical methods for survival data analysis (2003). https://doi.org/10.1002/0471458546

Lucchesi, V., Grimaldi, L., Mastrantonio, V., Porretta, D., Di Bella, L., Ruspandini, T., Di Salvo, M.L., Vontas, J., Bellini, R., Negri, A., Epis, S., Caccia, S., Bandi, C., Urbanelli, S.: Cuticle modifications and over-expression of the chitin-synthase gene in diflubenzuron-resistant phenotype. Insects **13**(12), 1109 (2022). https://doi.org/10.3390/INSECTS13121109/S1

Omoniwa, B.P., Johnson, T.O., Acho, M.A., Nwonuma, C.O., SojiOmoniwa, O., Oluwafemi, A.G., Tomo, B.E., Odejobi, K., Alejolowo, O., Oludipe, E.O.: Larvicidal and adulticidal activities of essential oil of Helianthus annuus (L.) seed and molecular docking of its GCMS constituents against Aedes aegypti acetylcholinesterase. Inform. Med. Unlocked **45**, 101446 (2024). https://doi.org/10.1016/J.IMU.2024.101446

Pang, Y.P., Singh, S.K., Gao, Y., Lassiter, T.L., Mishra, R.K., Zhu, K.Y., Brimijoin, S.: Selective and irreversible inhibitors of aphid Acetylcholinesterases: steps toward human-safe insecticides. PLoS ONE **4**(2), e4349 (2009). https://doi.org/10.1371/JOURNAL.PONE.0004349

Pereira, S.I., Freire, C.S.R., Neto, C.P., Silvestre, A.J.D., Silva, A.M.S.: Chemical composition of the essential oil distilled from the fruits of Eucalyptus globulus grown in Portugal. Flavour Fragr. J. **20**(4), 407–409 (2005). https://doi.org/10.1002/FFJ.1444

Polsinelli, G.A., Singh, S.K., Mishra, R.K., Suranyi, R., Ragsdale, D.W., Pang, Y.P., Brimijoin, S.: Insect-specific irreversible inhibitors of acetylcholinesterase in pests including the bed bug, the eastern yellowjacket, German and American cockroaches, and the confused flour beetle. Chem. Biol. Interact. **187**(1–3), 142–147 (2010). https://doi.org/10.1016/J.CBI.2010.01.036

Rahimi-Nasrabadi, M., Pourmortazavi, S. M., Nazarian, S., Ahmadi, F., & Batooli, H. (2013). Chemical composition, antioxidant, and antibacterial activities of the essential oil and methanol extracts of eucalyptus oleosa leaves. *International Journal of Food Properties, 16*(5), 1080–1091. https://doi.org/10.1080/10942912.2011.558227

Ranganathan, S., Ampasala, D.R., Palaka, B.K., Ilavarasi, A.V., Patidar, I., Poovadan, L.P., Sapam, T.D.: In silico binding profile analysis and in vitro investigation on chitin synthase substrate and inhibitors from maize stem borer, Chilo partellus. Current Comp. Aided Drug des. **17**(7), 881–895 (2020). https://doi.org/10.2174/1573409916666201013150920

Ren, J., Yuan, H., Liu, X., Yu, Z., Meng, F., Xiong, L., Yang, N., Li, Y., Li, Z., & Fan, Z. (2022). Novel fluorinated aniline anthranilic diamides improved insecticidal activity targeting the ryanodine receptor. *Journal of Agricultural and Food Chemistry, 70*(34), 10453–10465. https://doi.org/10.1021/ACS.JAFC.2C03134/SUPPL_FILE/JF2C03134_SI_001.PDF

Russo, S., Cabrera, N., Chludil, H., Yaber-Grass, M., Leicach, S.: Insecticidal activity of young and mature leaves essential oil from Eucalyptus globulus Labill. against Tribolium confusum Jacquelin du Val (Coleoptera: Tenebrionidae). Chilean J. Agric. Res. **75**(3), 375–379 (2015). https://doi.org/10.4067/S0718-58392015000400015

Russo, S., Yaber Grass, M.A., Fontana, H.C., Leonelli, E.: Insecticidal activity of essential oil from Eucalyptus globulus against Aphis nerii (Boyer) and Gynaikothrips ficorum (Marchal). AgriScientia **35**(1), 63–67 (2018). https://doi.org/10.31047/1668.298X.V1.N35.20458

Salari, M. H., Amine, G., Shirazi, M. H., Hafezi, R., & Mohammadypour, M. (2006). Antibacterial effects of Eucalyptus globulus leaf extract on pathogenic bacteria isolated from specimens of patients with respiratory tract disorders. *Clinical Microbiology and Infection, 12*(2), 194–196. https://doi.org/10.1111/J.1469-0691.2005.01284.X

Sebei, K., Sakouhi, F., Herchi, W., Khouja, M.L., Boukhchina, S.: Chemical composition and antibacterial activities of seven Eucalyptus species essential oils leaves. Biol. Res. **48**(1), 1–5 (2015). https://doi.org/10.1186/0717-6287-48-7/TABLES/3

Shao, Z. M., Li, Y. J., Ding, J. H., Liu, Z. X., Zhang, X. R., Wang, J., Sheng, S., & Wu, F. A. (2020). Identification, characterization, and functional analysis of chitin synthase genes in glyphodes pyloalis walker (Lepidoptera: Pyralidae). *International Journal of Molecular Sciences, 21*, 4656. https://doi.org/10.3390/IJMS21134656

Sheikh, Z., Amani, A., Basseri, H. R., Kazemi, S. H. M., Sedaghat, M. M., Azam, K., Azizi, M., & Amirmohammadi, F. (2021). Repellent Efficacy of Eucalyptus globulus and Syzygium aromaticum Essential Oils against Malaria Vector, Anopheles stephensi (Diptera: Culicidae). *Iranian Journal of Public Health, 50*(8), 1668–1677. https://doi.org/10.18502/IJPH.V50I8.6813

Shiferaw, Y., Kassahun, A., Tedla, A., Feleke, G., Abebe, A.A.: Investigation of essential oil composition variation with age of eucalyptus globulus growing in Ethiopia. Nat. Prod. Chem. Res. **7**(2), 1–4 (2019). https://doi.org/10.35248/2329-6836.19.7.360

Silva, S.M., Abe, S.Y., Murakami, F.S., Frensch, G., Marques, F.A., Nakashima, T.: Essential oils from different plant parts of eucalyptus cinerea F. Muell. ex Benth. (Myrtaceae) as a source of 1,8-cineole and their bioactivities. Pharmaceuticals (Basel, Switzerland) **4**(12), 1535–1550 (2011). https://doi.org/10.3390/PH4121535

Sugumar, S., Clarke, S. K., Nirmala, M. J., Tyagi, B. K., Mukherjee, A., & Chandrasekaran, N. (2014). Nanoemulsion of eucalyptus oil and its larvicidal activity against Culex quinquefasciatus. *Bulletin of Entomological Research, 104*(3), 393–402. https://doi.org/10.1017/S0007485313000710

Zhang, X., Yan Zhu, K.: Biochemical characterization of chitin synthase activity and inhibition in the African malaria mosquito, Anopheles gambiae. Insect Sci. **20**(2), 158–166 (2013). https://doi.org/10.1111/J.1744-7917.2012.01568.X

Zhang, L., Chen, R., Li, X., Xu, X., Xu, Z., Cheng, J., Wang, Y., Li, Y., Shao, X. S., & Li, Z. (2022). Synthesis, insecticidal activities, and 3D-QASR of N-Pyridylpyrazole Amide derivatives containing a phthalimide as potential ryanodine receptor activators. *Journal of Agricultural and Food Chemistry, 70*(39), 12651–12662. https://doi.org/10.1021/ACS.JAFC.2C03971/SUPPL_FILE/JF2C03971_SI_001.PDF

# Microalgal Phytohormones Functions in Microalgae and Higher Plants: Advancements in the Management of Heavy Metals and Future Microalgae Cultures Challenges

Salma Latique and Imane El-Baggar

**Abstract**

The microalgae are presented as one of the most promising sources of raw materials. They have special properties which make them suitable for bio-technical applications. In the past decade, increasing microalgae research has focused on phytohormones studies. Thus, the utilization of microalgae as possible biofertilizers, biostimulants, and biopesticides is therefore explored in this paper. It also details the phytohormone's uses and the possibilities of increasing microalgae biomass through them. The impact of phytohormones on primary and secondary metabolites. Moreover, the results of the present review suggest that refocusing the target of microalgal research from biofertilizer manufacturing to even more significant biorefinery co-products can serve to help support their sustainability as a valuable resource. In addition, cooperative innovation should be promoted to reduce the high investment in microalgae biomass utilization. This review is expected to contribute to the development of a roadmap to the potential role of microalgae in future biorefinery industries.

**Keywords**

Microalgae · Higher plants · Phytohormones · Agriculture · Heavy metals (HVs) · Economic feasibility

S. Latique (✉)
Laboratory of Drug Sciences, Faculty of Medicine, Pharmacy and Dentistry, Sidi Mohamed Ben Abdellah University, Fez, Morocco
e-mail: salma.latique@usmba.ac.ma

S. Latique · I. El-Baggar
Laboratory of Biotechnology, Conservation and Valorization of Bioresources, Faculty of Sciences Dhar El Mahraz, Sidi Mohamed Ben Abdellah University, Fez, Morocco

## 1 Introduction

The recognition of climate change as a prominent global menace with extensive repercussions is widespread. These include the annual increase in global carbon dioxide emissions, climate change, and the loss of natural resources (Show et al. 2017). This is why establishing sustainable industrial processes is a need of the hour and is one of the significant challenges worldwide for humankind. Microalgae are already considered a key and effective biological method for global carbon capture and storage (Moreira and Pires 2016; Alami et al. 2021; Onyeaka et al. 2021). Microalgae and other photosynthetic microorganisms are amazing little creatures: They grow to maturity rapidly, effectively converting $CO_2$ into biomass, and are nutritional powerhouses.

They are characterized by high proteins, vitamins, healthy fats, pigments, and essential minerals (Salama et al. 2014). Because they produce such a wide variety of bioactive compounds, they are perfect for biorefinery processes, where every part of their biomass can be put to good use (Stirk and van Staden 2020). Even more interesting is that microalgae produce plant hormones like auxins, gibberellins, and jasmonic acid that aid in their growth and stress adaptation (Romanenko et al. 2015).

Phytohormones are central to the physiological functions of cells, influencing functions such as cell division and the effectiveness of photosynthetic mechanisms (Wang et al. 2021). Hence, the fusion of phytohormones and abiotic stressors could regulate the development and metabolism of microalgae, which may promote the synthesis of metabolites and lipids (Zhao et al. 2019). Nevertheless, especially in outdoor production systems, this might increase the effectiveness of biofuel consumption and assist in lowering production costs (Borowitzka 2016; Stirk and van Staden 2020). This review explores these questions, focusing on the potential of phytohormones on the microalgal side: what they are,



O. Jamal Eddine et al. (eds.), *Plant Pathology, Fungal Diversity, and Biotechnological Advances in Agriculture*, Sustainable Landscape Planning and Natural Resources Management, https://doi.org/10.1007/978-3-031-97666-7_3

how they work, and their ability to assist crops in coping with heavy metal stress and revolutionize agriculture. We further discuss using phytohormone treatments to maximize microalgae cultivation processes, improving biomass quality and yield for biotechnology. Collectively, these advances could enable finally lab research and real-world microalgae-based products.

## 2 Phytohormone Profiles in Microalgae

As it happens, microalgae also share much with plants, producing their hormones. These small power plants come oozing with auxins, cytokinins (CK), gibberellins (GA), and growth regulators that you would find in higher life forms of plants. These hormones are the masters for controlling all, from cell division to the development of microalgae (Hashtroudi et al. 2013; Nguyen et al. 2020a, b; Alvarez et al. 2021).

### 2.1 Auxin

It has been reported that microalgae excrete mainly two types of these: indole-3-acetic acid (IAA) and indole-3-butanoic acid (IBA), which are known to be major players in their growth (Romanenko et al. 2015). Hence, one of the extensively researched auxins, Indole-3-acetic acid (IAA), is synthesized by cyanobacteria as a tryptophan-dependent by-product (Chittapun et al. 2018). *Chlorella minutissima* was identified to have the precursors tryptophan (TRP) and anthranilate, indicating that tryptophan-dependent pathways may occur in microalgae (Stirk et al. 2014; Stirk and van Staden 2020). Nevertheless, the quantity of auxins varied among different species. For instance, *C. minutissima* has been shown to contain tryptophan (TRP) and indole-3-acetamide (IAM) at higher amounts than IAA (Stirk et al. 2014). Moreover, *Chlamydomonas reinhardtii* exhibited the highest recorded quantity of IAA, while *Coelastrum microporum* demonstrated the most significant amount of IAM (Romanenko et al. 2015; Stirk et al. 2013a). In fact, *Nostoc calcicola* and *Anabaena vaginicola* showed an IBA predominance (Hashtroudi et al. 2013). Microalgae rely on tryptophan as the key starting material for producing auxins through multiple biosynthetic pathways, some of the mechanisms via which auxin is synthesized (Kiseleva et al. 2012; Stirk and van Staden 2020). Interestingly, various algal species across different evolutionary groups all employ similar enzymatic tools like tryptamine and indole-3-acetonitrile for these conversions (Romanenko et al. 2015; Kiseleva et al. 2012) (Figs. 1 and 2). This shared biochemical machinery suggests that auxin production evolved early and has been conserved throughout algal diversification.

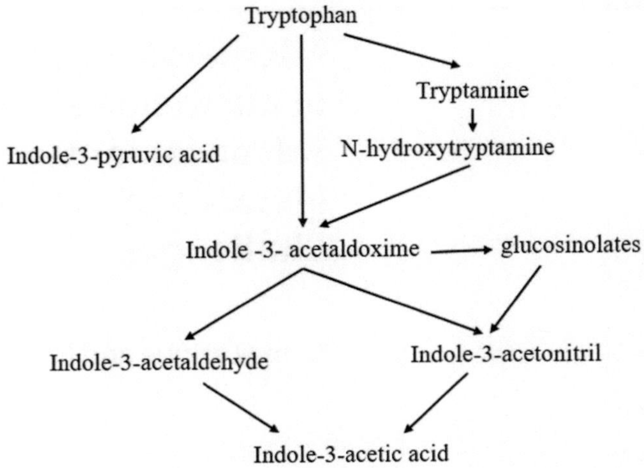


**Fig. 1** Tryptophan—dependent pathway of IAA biosynthesis (Kiseleva et al. 2012)

### 2.2 Cytokinins

These fascinating compounds start as simple purine molecules (think adenine/adenosine) but get a functional upgrade when either an isopentenyl or aromatic group attaches at their sixth position (Romanenko et al. 2015). They come in two flavors: the isoprenoid team (featuring tZ, cZ, DHZ, and iP) and the aromatic squad (hydroxylated topolins and BA) (Sakakibara 2006; Stirk and van Staden 2020). However, it gets interesting here that these molecules shine when they transform into active forms like ribosides, glucosides, or methylthio-derivatives (Mok and Mok 2001; Kondrat'eva and Shelepova 2010). Similar to fingerprints, different microalgae species show unique patterns of these cytokinin varieties, hinting at their specialized roles in algal biology.

Studies of the CKs endogenous content in some representatives of the genus *Chlorella* revealed that higher concentrations of iP derivatives than cZ derivatives were found, and very few levels of tZ were also reported (Jirásková et al. 2009; Stirk and van Staden 2020). Additional investigations in *C. minutissima* affirmed and augmented the earlier findings regarding the cytokinin profile. This microalga shows a higher content of cZRMP derivatives than cZ derivatives content (Stirk et al. 2014). In *Euglena gracilis*, cZ derivatives predominated cytokinins over iP isomers, however, tZ derivatives only in trace amounts and aromatic BA (Noble

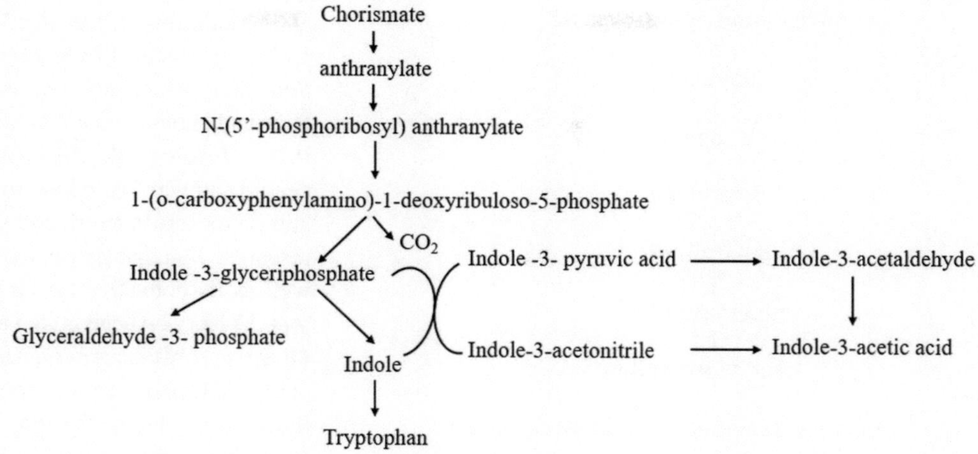


**Fig. 2** Tryptophan–independent pathway of IAA biosynthesis (Kiseleva et al. 2012)

et al. 2014). The major contributors to the cytokinin pool in many Cyanobacteria and microalgae were cis-Z types and 2-methylthio derivatives, primarily because of cytokinins coupled to tRNA. This suggests that tRNA hydrolysis is a crucial source of cytokinins in microalgae (Žižková et al. 2017; Stirk and van Staden 2020).

## 2.3 Gibberellins

Tetracyclic diterpene carboxylic acids are known as gibberellins (GAs) (Stirk and van Staden 2020). *Chlamydomonas reinhardtii* and 31 other microalgal species had 19 GA variants discovered thanks to the LC–MS technology (Stirk et al. 2013a; Romanenko et al. 2015). Gibberellins are also present in *Chlorella minutissima* with 19 gibberellins forms (Stirk et al. 2014, 2019). The biologically active form present was GA (1, 3, 4, 5, 6, and 7), although GA6 is the dominant active and stable form in all studied microalgae (Fig. 3) (Stirk et al. 2013a, 2014, 2019).

## 2.4 Abscisic Acid

Abscisic acid (ABA) is a compound of sesquiterpene nature comprising 15 carbon atoms (Wang et al. 2021). Lower ABA levels were reported in *Chlorella minutissima* (Stirk et al. 2014). ABA was also detected in the cells of 39 microalgae species spanning 11 divisions. The methylerythritol-phosphate (MEP) pathway synthesizes it from the precursor isopentenyl. The concentration of ABA is controlled by a careful equilibrium between production and catabolism (Fig. 4). This mechanism is also related to the one in higher plants (Kiseleva et al. 2012; Stirk and van Staden 2020).

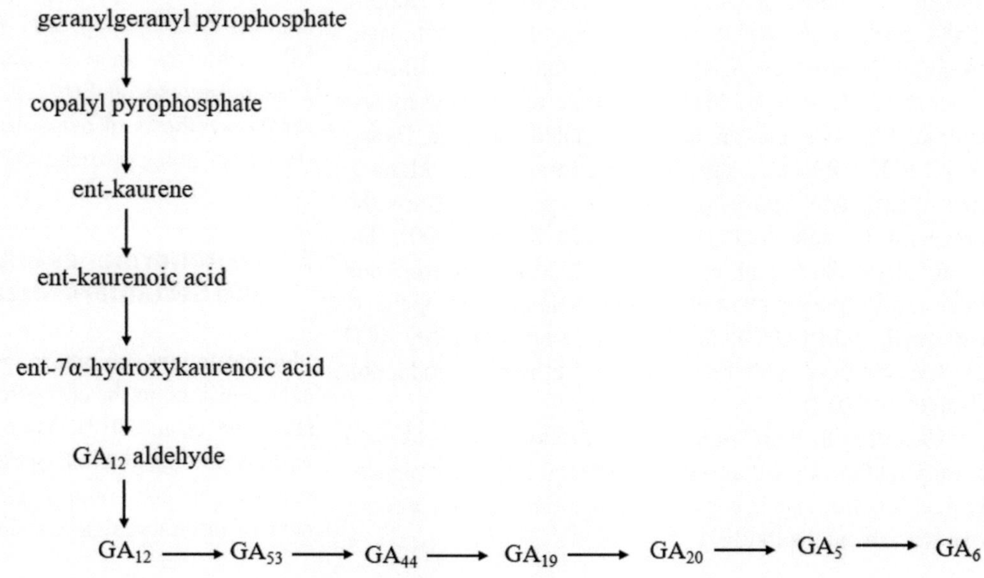


**Fig. 3** Gibberellin biosynthesis (Kiseleva et al. 2012; Stirk et al. 2013b)

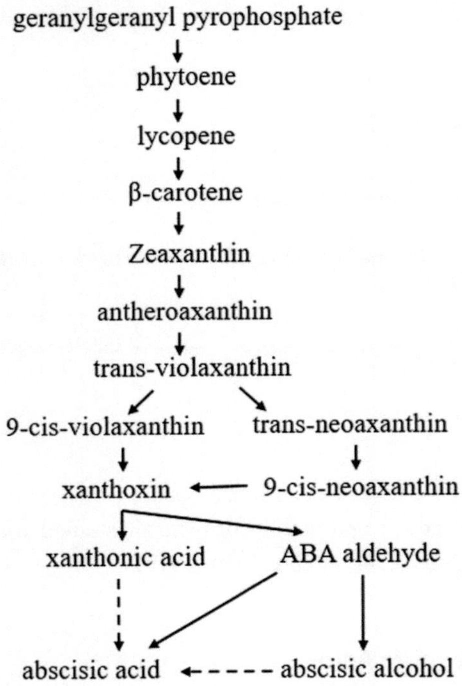


**Fig. 4** Abscisic acid (ABA) biosynthesis (Kiseleva et al. 2012)

## 2.5 Others Phytohormones

According to Stirk et al. (2018) and Stirk and van Staden (2020), brassinosteroids are polyhydroxylated steroid hormones with a 5 α-cholestane structure. Brassinosteroids have structurally varied side chains and A and B rings (Oklestkova et al. 2015; Stirk et al. 2018). These variations include the degree of oxidation of the B-ring, the presence of hydroxyl (OH) groups at the A-ring, and the presence of substituents at the side chain's C-22, C-23, and C-24 (Bajguz and Tretyn 2003; Stirk et al. 2018). Although some of these metabolites exhibit biological activity, there are already 70 known metabolites of brassinosteroids (Bajguz 2009; Stirk et al. 2018). In particular, the direct biosynthesis of castasterone (CS) and brassinolide (BL) (Bartwal et al. 2013; Stirk et al. 2018). A simple unsaturated hydrocarbon with two carbon atoms is called ethylene (Wang et al. 2022). Recently, ethylene was found in the filamentous charophyte *Spirogyra pratensis*, produced from the precursor L-methionine (Bleecker and Kende 2000; Han et al. 2018; Stirk and van Staden 2020). Moreover, one intermediate precursor of ethylene is aminocyclopropane-1-carboxylic acid (ACC). In *Haematococcus pluvialis*, ACC oxidase catalyzes the final phase of ethylene production (Han et al. 2018).

According to Wang et al. (2020) jasmonic acid (JA) is a plant phytohormone that regulates several important physiological functions such as growth, development, and response to environmental stimuli.

In addition, it has also been shown that microalgae, including species of *Dunaliella* species, and the prokaryote *Spirulina*, which are key components of aquatic ecosystems and photosynthetic organisms, can produce and respond to JA (Jusoh et al. 2015; Wang et al. 2022). The octadecanoid pathway is used to produce jasmonates in chloroplasts and peroxisomes (Stirk and van Staden 2020). According to research, JA controls microalgal growth and metabolism as well as its defensive systems against infections and predators. JA has also been linked to the control of the formation of bioactive substances in microalgae, including pigments, fatty acids, and secondary metabolites (Cui et al. 2021). Overall, the finding that microalgae produce and signal JA shows the potential for exploiting these organisms as a source of bioactive substances and for creating novel approaches to enhancing their growth and productivity (Czerpak et al. 2006).

Salicylic acid (SA) is a phytohormone present in higher plants and responsible for overseeing plant growth, development, and reactions to both biotic and abiotic stresses (Arif et al. 2020). While studies on the presence of SA in microalgae are scarce, some studies have suggested that SA may be present in these organisms. For example, studies published in 2010, 2015, and 2019 reported the detection of salicylic acid in microalgae such as *Scenedesmus quadricauda*, *Chlamydomonas reinhardtii*, and *C. minutissim* (Kováčik et al. 2010; Zuo et al. 2015; Stirk et al. 2019). However, since SA is an aromatic molecule, microalgae can only produce it via the shikimate pathway, which produces only phenylalanine and tyrosine as predecessors (Cheynier et al. 2013; Caretto et al. 2015; Stirk and van Staden 2020).

However, both bacteria and microorganisms, as well as microalgae, can synthesize isoprenoids, employing isopentenyl diphosphate (IPP) and its isomer dimethylallyl diphosphate (DMAPP) as a basic building unit (Paniagua-Michel et al. 2012). Recent studies demonstrate that for the de novo biosynthesis of cytosolic sterols, the marine diatoms *Nitzschia* sp. and *Phaeodactylum* sp. utilize the MVA pathway, whereas *Chlorella* sp., *Chlamydomonas* sp., *Scenedesmus* sp., and *Dunaliella* sp. follow the MEP pathway for the synthesis of cytosolic isoprenoids, including sterols (Fig. 5) (Paniagua-Michel et al. 2012).

# 3 Phytohormones Impact Primary and Secondary Metabolites

Phytohormones are natural growth regulators of plants and can contribute to the cost-effective production of microalgae (Kozlova et al. 2017). Microalgae are valuable sources of various products including biofuels, nutraceuticals, and food supplements for animals. However, the price competitiveness of the microalgae products remains a major challenge

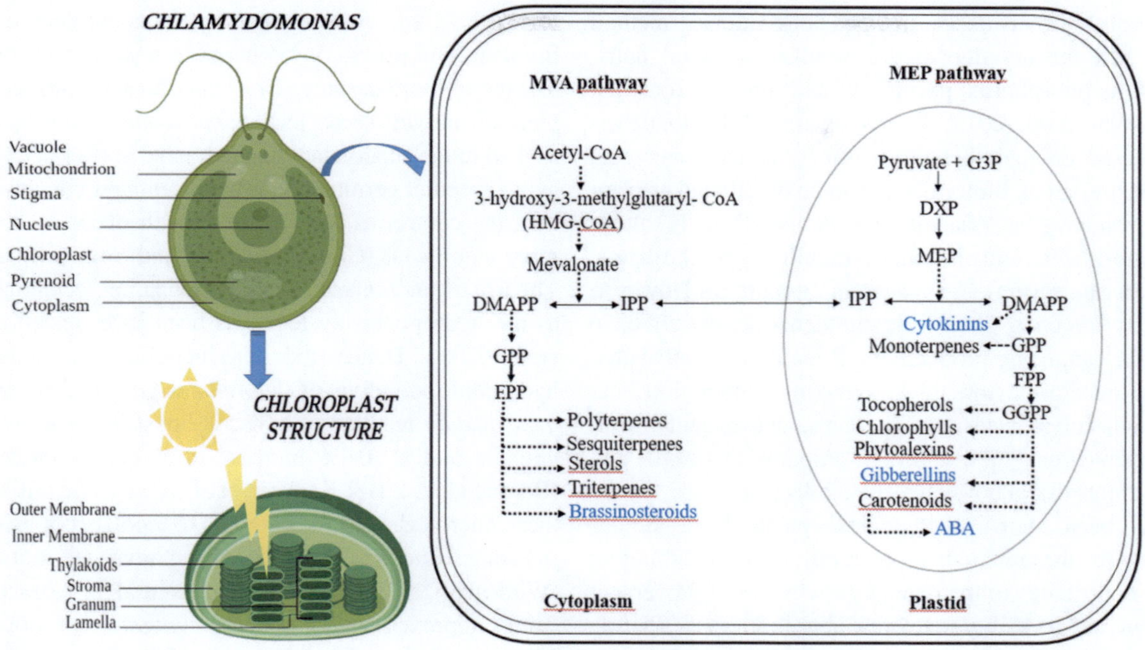


**Fig. 5** The mevalonate (MVA) and methylerythritol phosphate (MEP) pathways in eukaryotic cells, as well as most algae (modified Paniagua-Michel et al. 2012)

in their commercialization (Wang et al. 2020). Additionally, as suggested by certain authors, phytohormones can stimulate the synthesis of lipids or pigments in microalgae by modulating internal biochemical pathways (Lu and Xu 2015; Sun et al. 2018), which act as agents that promote growth during the cultivation process. One example of a phytohormone that has been demonstrated to yield positive outcomes on microalgal growth is auxin (Bajguz and Piotrowska-Niczyporuk 2013). Auxins are plant hormones that control cell division and elongation. Auxins have been shown in studies to increase biomass yield, improve nutrient uptake efficiency, and boost bioactive compound synthesis in microalgae cultures (Bajguz and Piotrowska-Niczyporuk 2013; Salama et al. 2014). Furthermore, Supplementation of 2 μM ABA to *Scenedesmus quadricauda* culture resulted 2.1-fold increase in dry biomass production at 24 h, whereas fatty acid (FA) synthesis was also affected upon supplementation of 5 μM ABA and a build-up of saturated FA (12%) was seen with a 11% decrease in unsaturated FA content, under nitrogen (Sulochana and Arumugam 2016; Nguyen et al. 2020b). Moreover, a combination of phytohormones is also commonly practiced to increase algal propagation for potential value-added products. It was also found that Kinetin and 2,4-D favor the growth of *Dunaliella salina* and *Haematococcus pluvialis* (de Jesus Raposo and de Morais 2013), and *Skeletonema costatum's* can also be accelerated by using GA + NAA (Han et al. 2018). In addition, the growth of *Chlorella sorokiniana* was also increased by 138% and 136% from that of the control, respectively, under two sets of co-treatments of NAA + GA 3 and NAA + zeatin. Furthermore, some combinations of NAA and 2-phenylacetic as well as NAA and zeatin + GA3 increased the yield of biomass and chlorophyll a concerning control *Chlorella vulgaris* in particular combined treatments including brassinolide (BL) + auxin (IAA) or trans-zeatin (Bajguz and Piotrowska-Niczyporuk 2013, 2014), the highest level of biochemical parameters, such as protein monosaccharides and chlorophylls show that application of phytohormones activated the proton pumps and an antioxidative mechanism (Yu et al. 2017, 2018). In summary, Phytohormones possess the potential to substantially cut down the costs of microalgae production by boosting growth and productivity (Kozlova et al. 2017). Additional research is required to pinpoint the most efficacious phytohormones and optimize their use in microalgal cultivation for maximum cost savings.

## 4 Phytohormones Roles as Regulators in Plants

Numerous investigations have revealed that various microalgal extracts have a significant potential to enhance the development of agricultural production. Microalgae and cyanobacteria constitute a significant reservoir of biologically active compounds. Considering this rich composition of diverse metabolites, microalgae, and cyanobacteria products can be classified as biofertilizers (biomass acting on soil), biostimulants (biomass acting on plants), and biopesticides (biomass acting on pathogenic organisms) (Gonçalves 2021).

Generally, biofertilizers in the conventional method improve soil fertility due to the provision of vital nutrients such as phosphorus, potassium, and nitrogen for plant growth (Win et al. 2018; Ronga et al. 2019; Gonçalves 2021). Based on their beneficial effects and the microorganism assimilation, biofertilizers can be distributed as plant growth-promoting bacteria, nitrogen fixators, P and K solubilizing biofertilizers, and phosphorus mobilizing biofertilizers, which presents several advantages in agriculture. However, by directly affecting plants, biostimulants are an effective method for enhancing productivity. Benefits were attributed to the presence of phytoregulators (auxins, gibberellins, and cytokinins), fulvic acids, humic acids, antioxidants, vitamins, proteins, enzymes, and high amounts of macro- and micronutrients (Tarakhovskaya et al. 2007).

It has been shown that several plants have similar responses to the microalgae extracts. This biostimulant capacity contributes to the overall growth and improvement of the plant and its ability to tolerate abiotic stresses (Sharma et al. 2014; Blanke 2016; Michalak et al. 2016; Kapoore et al. 2021). In addition, it was studied if biostimulants from microalgae could be produced by adding free nutrients such as sewage and waste digestate obtained from anaerobic digestion (Renuka et al. 2018; Stiles et al. 2018; Kapoore et al. 2021).

On the other hand, biopesticides are known for their activity against phytopathogens and pathogenic organisms. Some microalgal compounds present antimicrobial, antiviral, antiprotozoal, and antifungal activities that promote crop development (Hernández-Carlos and Gamboa-Angulo 2011).

According to recent research, phytohormones play multiple regulatory functions in microalgae and higher plants (Lu and Xu 2015; Kapoore et al. 2021). In this wise, the use of microalgal phytohormones as biostimulants is outstanding. Henceforth, there are numerous ways to employ microalgae extract going forward, including foliar spray, soil treatment, seed priming, hydroponics, and carriers, which use living microalgae cells (Renuka et al. 2018).

In addition to the research of Hernández-Carlos and Gamboa-Angulo (2011), numerous other investigations delved into the impacts of microalgal biofertilizers on plants, yielding a comprehensive grasp of the favorable roles of microalgal biofertilizers in ecofriendly agriculture (Peng and Bruns 2019; Zou et al. 2020).

Microalgae and cyanobacteria are known to produce auxin and cytokinin-like compounds, however, because they fix nitrogen in the soil, cyanobacteria like *Anabaena* sp., *Nostoc* sp., and *Oscillatoria angustissima* get an additional potential and have a biofertilizer impact (Win et al. 2018; Kapoore et al. 2021).

Recent work proved the positive impact of microalgae on seed priming. Before germination, plant seeds are immersed for 18 to 24 h in water containing a small amount of a biostimulant. This process is known as seed soaking (or priming). The use of *Acutodesmus dimorphus* and *Scenedesmus dimorphus* on tomato seeds (*Solanum lycopersicum* var. Roma) soaked in the supernatant and extracted at over 0.75 g mL$^{-1}$ led to speedier germination rate percentages two days before than the control as well as the growth of lateral roots was more significant (Garcia-Gonzalez and Sommerfeld 2016). The use of microalgae in other application techniques, such as the hydroponics system, has been investigated (Sharma et al. 2014). Hence, the introduction of digestate to the hydroponic co-culture of *Chlorella vulgaris* and *Scenedesmus quadricauda* resulted in a 32.5% boost in the growth of C. vulgaris and a 30.4% increase in *S. quadricauda* growth (Barone et al. 2018; Kapoore et al. 2021). The utilization of these microalgae has the potential to modify root growth and structure. (Lucini et al. 2015). Using advanced root imaging (WinRhizo), scientists found that microalgae extract significantly improved sugar beet (*Beta vulgaris* L.) root growth. When applied at 2 mL/L or 4 mL/L doses, the extracts increased total root length, surface area, and root tip numbers, all with strong statistical significance ($p < 0.01$), demonstrating their potential to enhance root system development (Barone et al. 2018).

The focus of the current study is on phytohormones, which are produced by the secondary metabolism of microalgae. These investigations revealed that biostimulants impact plants (Dineshkumar et al. 2020; González-Pérez et al. 2022). Auxin is believed to exert a substantial regulatory influence on diverse facets of plant growth, encompassing cell division, tropisms, apical dominance, senescence, abscission, and flowering, according to the literature (Werner et al. 2007). Furthermore, while dealing with auxins, cytokinins are especially important for cell division, capillary morphogenesis, apical predominance, and nutrient uptake (Teale et al. 2006). Gibberellic acid activation triggers the hydrolytic enzymes amylase and protease, which play a key role in the mobilization of reserve substances and the migration of the endosperm. Thus, the germination of seeds is significantly impacted by this process. Moreover, it promotes break-up of dormancy, lengthening and growth of stems, cell division, and, consequently, expansion of leaves (Taiz and Zeiger 2010; González-Pérez et al. 2022).

Nowadays, as everyone knows, metal contamination in the environment is a worldwide problem, damaging agricultural systems (Bücker-Neto et al. 2017). In natural ecosystems, heavy metals are divided into two groups: non-essential elements of plant growth like chromium, arsenic, cadmium, lead, mercury, etc. Other heavy metals required for plant growth include zinc, copper, nickel, etc. However, a high metal concentration can severely cause a toxic effect on plants. In this sense, most higher plants have gradually formed some tolerance mechanisms in the heavy metals stress environment

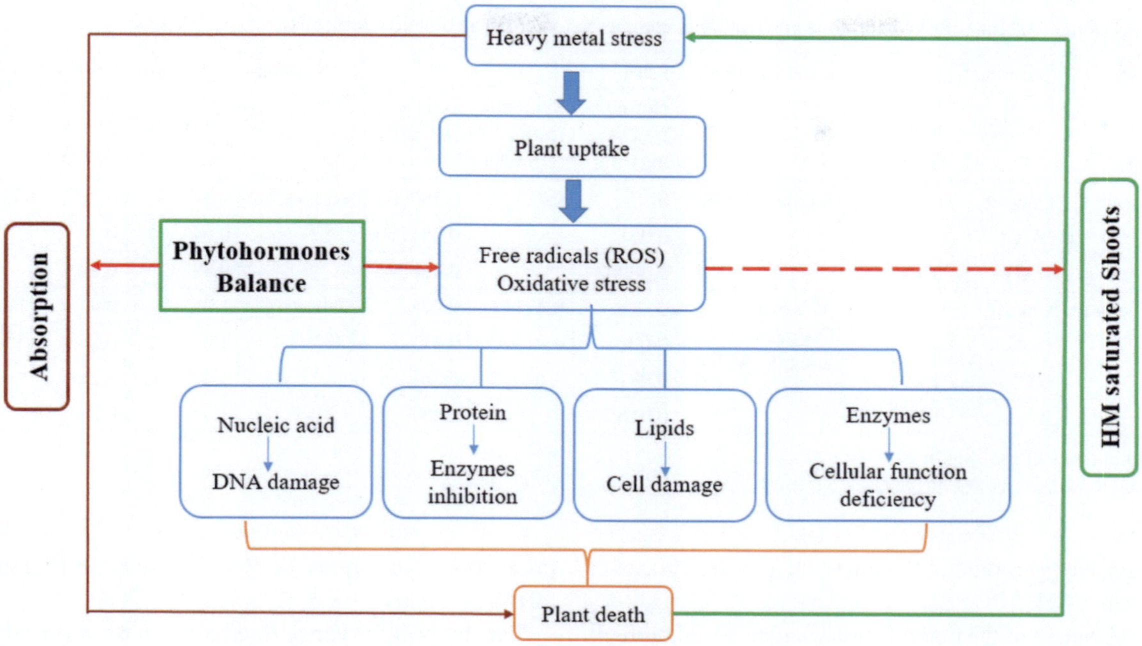


**Fig. 6** Mechanism of action of heavy metal toxicity in the plant (modified Alengebawy et al. 2021)

like the phytohormones signaling molecules (Yu et al. 2015; Bücker-Neto et al. 2017).

Generally, heavy metals can interfere with phytohormone synthesis, transport, and signaling in plants, causing changes in growth and development (Bücker-Neto et al. 2017). Therefore, heavy metals such as lead and cadmium, have been shown to suppress the production of auxin, a key plant hormone involved in cell division and elongation, resulting in stunted growth (Choudhary et al. 2012; Rady 2012; Ramakrishna and Seeta Ram Rao 2013; Ostrowski et al. 2016).

Heavy metals don't just damage plants directly, they also disrupt key hormones like abscisic acid (ABA), cytokinins, and gibberellins, which regulate stress responses and growth (Fariduddin et al. 2015). In some cases, metal exposure triggers excessive hormone production, further harming the plant. While we know these interactions are significant, the full impact of heavy metals on hormone signaling and how this alters plant development—remains unclear and requires deeper study (Bücker-Neto et al. 2017; Alengebawy et al. 2021).

Furthermore, heavy metals are particularly harmful to plants because they induce the formation of reactive oxygen species (ROS). These ROS molecules damage essential cellular structures—they oxidize lipids, break down proteins, degrade photosynthetic pigments, and harm DNA. This oxidative damage results in lipid peroxidation, depolarization of the cell membrane, non-functioning enzymatic systems, and ultimately the death of the cell (Hegedüs et al. 2001; Piotrowska-Niczyporuk et al. 2012) (Fig. 6).

## 5 The Use of Phytohormones for Large-Scale Microalgae Production: An Economic Feasibility Study

Microalgae have gained increasing interest as a viable, eco-friendly, and a source of renewable green energy (Ishika et al. 2017). Furthermore, microalgae are a known prospective source of environmentally friendly biomass for biofuel production due to their capacity to sequester $CO_2$ into bioenergy through photosynthesis (Borowitzka and Moheimani 2013; Onyeaka et al. 2021). Recent studies are currently focused on scaling up the process and figuring out whether making fuel with feedstock produced from microalgae is industrially possible. Each new technology developed in a bio-based economy might well, however, only be useful if it can successfully address the following criteria: first, is it technically feasible, which means that all transitional systems it uses from the raw material to the completed product are attainable; second, is it industrially viable and profitable; and third, is it sustainable and eco-friendly, which means that the advance tools has an environmentally safe and friendly impact (Thomassen et al. 2018; Sarma et al. 2021). To make algae culturing feasible, the biomass productivity of the algae has to exceed 30 g $m^{-2}$ per day (Bhattacharya and Goswami 2020). Thus, the production of algal biomass can be more efficient if less expensive substrates such as carbon-rich wastewater or industrial effluent are employed. At present time, the greatest challenge to be faced with microalgal cultures is the reduction of the growth medium cost (Braun and Colla 2022). Nevertheless, the growing and harvesting practices are also the

**Table 1** Economic viability of phytohormone application in commercial microalgae biomass production

| Microalgae species | Treatment | Amount of hormones (mg/L) | Chemical grade | Price | | | Cultivation time | Maximum cell density (g/L) | Cost (US $/g) biomass | References |
|---|---|---|---|---|---|---|---|---|---|---|
| | | | | Hormone (US $/mg) | Synthetic media (US $/L) | Total (US $/L) | | | | |
| *Scenedesmus obliquus* | Control | 0 | Sigma—Aldrich | 0 | 0.25* | 0.250000 | Exponential phase | 0.32 | 0.78 | Salama et al. (2014) |
| | IAA | 1.75 | | 0.001 | | 0.251750 | | 0.64 | 0.39 | |
| | DAH | 2.15 | | 0.00002 | | 0.250043 | | 0.83 | 0.30 | |
| *Chlamydomonas reinhardtii* | Control | 0 | American Chemical Society (ACS) | 0 | 0.471** | 0.471 | Stationary phase | 1.05 | 0.45 | Park et al. (2013) |
| | IAA | 3 | | 0.002 | | 0.477 | | 1.69 | 0.28 | |
| | GA3 | 1 | | 0.031 | | 0.508 | | 1.76 | 0.28 | |
| | Kinetin | 1 | | 0.016 | | 0.487 | | 1.77 | 0.28 | |

*Cultivated in Bold's Basal Medium
**Cultivated in Tris–Acetate-Phosphate

most expensive procedure in the microalgae-based processes (Sarma et al. 2021).

The assessment of the potential microalgae-based biofuels for biofuels or other valuable products is to compare the novel products with the standard biofuels production (Kim 2011) since oil production cost is increasing (Reijnders and Huijbregts 2008). Therefore, life cycle assessments (LCAs) are needed to evaluate the environmental impact of algal biomass production.

Mathematical modeling is employed to predict the microalgae growing process by supplying it strategy. There is evidence that phytohormones, in some studies, influence growth of microalgae (Kapoore et al. 2021; González-Pérez et al. 2022).

The subsequent formula was employed to compute the expense associated with microalgal biomass obtained in a medium enriched with phytohormones (Salama et al. 2014):

$$Cost(\text{US }\$) = \frac{g}{biomass} = \frac{((A \times B) + C)}{D}$$

where A, B, C, and D are the concentration of each phytohormone added to a unit of culture medium (mg $L^{-1}$), the price of the substrate (US \$ $mg^{-1}$), the standard price of the culture medium (US \$ $L^{-1}$), and the grown biomass (g $L^{-1}$), respectively. The results are reported in Table 1.

The advantaged community increase of phytohormones has caused the economic practicability of biomass production in the phytohormone-absorptive medium to be far greater than, as compared with those in the control (medium not containing phytohormones) (Park et al. 2013). As reported in economics, the culturing medium contributes significantly to the high cost of microalgal biomass. Hence, replacing the conventional medium of various effluents (rich in sources of carbon, nitrogen, and phosphorus) and phytohormones (especially IAA) for energy-effective microalgal biomass formation could be a potential strategy for large-scale microalgal production that require further investigation (Salama et al. 2014).

For the past 50 years, microalgae have been raised on a vast scale (Mata et al. 2010). Currently, 1150 ha in Australia is used for β carotene production from Dunaliella salina. The same product is also widely produced in Israel and the USA (Borowitzka 2005). In 2017, microalgae-based products had a global market value of 32.60 billion dollars, however, by 2026, that market is expected to grow to 53.43 billion dollars (Rahman 2020).

Even though microalgae research has recently been performed all over the world, China's companies and universities have been discovered to be second only to the USA in the number of patents relating to microalga biorefineries that have been filed. These investigations are essential because substances like pigments would quickly deteriorate as a result of light, temperature, and other microorganisms. Finally, it is essential to establish a viable method for quick and affordable fabrication as well as thorough research of economic and ecological conditions to build high-value products from microalgae. Additionally, to evaluate the method's sustainability, the life cycle of elevated chemicals from microalgae can be analyzed. Consequently, further efforts should be undertaken to minimize the energy expenses involved with guiding throughout a sustainably huge processing stage for extracting high-value chemicals from microalgae and to lessen the dependence on petroleum-derived products (Bhattacharya and Goswami 2020).

## 6 Conclusions and Recommendations

In conclusion, the use of dry biomass, aqueous cellular extracts, and live microalgae culture may not only promote plant growth but also boost the viability of algal production systems. To enhance and protect crops worldwide as climate change worsens, new solutions will be needed. In response to the challenge of meeting growing food demands within resource constraints, microalgae prove to be a viable alternative for boosting and safeguarding agricultural productivity. This not only offers economic and environmental benefits but also provides advantages for both farmers and algae growers (Garcia-Gonzalez and Sommerfeld 2016). To manufacture several bio-based products from a single phytoplankton strain, researchers are now concentrating on optimizing and simplifying the resource and energy balances. However, it appears that ineffective multiproduct downstream processing methods and low biomass productivity have been the main obstacles to the development of a profitable microalgae biorefinery. This problem can be solved by doing in-depth research to produce genetically engineered, quickly growing microalgae strains and by developing simple, economically viable downstream processing methods including phytohormones in the culture medium. These functional studies can also lead to the discovery of a new pathway for heavy metals (HM) binding substances and hormone-related components, bringing up an interesting model, to study the interaction between phytohormones and algal physiology against HM stress. Therefore, thorough research on a few of these issues would lead to the future establishment of a commercially viable microalgae biorefinery.

**Competing Interests** The author declares that they have no conflict of interest.

## References


Alami, A.H., Alasad, S., Ali, M., Alshamsi, M.: Investigating algae for $CO_2$ capture and accumulation and simultaneous production of biomass for biodiesel production. Sci. Total. Environ. **759**, 143529 (2021). https://doi.org/10.1016/j.scitotenv.2020.143529

Alengebawy, A., Abdelkhalek, S.T., Qureshi, S.R., Wang, M.-Q.: Heavy metals and pesticides toxicity in agricultural soil and plants: ecological risks and human health implications. Toxics **9**, 42 (2021). https://doi.org/10.3390/toxics9030042

Alvarez, A.L., Weyers, S.L., Goemann, H.M., et al.: Microalgae, soil, and plants: a critical review of microalgae as renewable resources for agriculture. Algal Res. **54**, 102200 (2021). https://doi.org/10.1016/j.algal.2021.102200

Arif, Y., Sami, F., Siddiqui, H., et al.: Salicylic acid in relation to other phytohormones in plant: a study towards physiology and signal transduction under challenging environment. Environ. Exp. Bot. **175**, 104040 (2020). https://doi.org/10.1016/j.envexpbot.2020.104040

Bajguz, A.: Isolation and characterization of brassinosteroids from algal cultures of Chlorella vulgaris Beijerinck (Trebouxiophyceae). J. Plant Physiol. **166**, 1946–1949 (2009). https://doi.org/10.1016/j.jplph.2009.05.003

Bajguz, A., Piotrowska-Niczyporuk, A.: Synergistic effect of auxins and brassinosteroids on the growth and regulation of metabolite content in the green alga Chlorella vulgaris (Trebouxiophyceae). Plant Physiol. Biochem. **71**, 290–297 (2013). https://doi.org/10.1016/j.plaphy.2013.08.003

Bajguz, A., Tretyn, A.: The chemical characteristic and distribution of brassinosteroids in plants. Phytochemistry **62**, 1027–1046 (2003). https://doi.org/10.1016/S0031-9422(02)00656-8

Bajguz, A., Piotrowska-Niczyporuk, A.: Interactive effect of brassinosteroids and cytokinins on growth, chlorophyll, monosaccharide and protein content in the green alga *Chlorella vulgaris* (Trebouxiophyceae). Plant Physiol. Biochem. **80**, 176–83 (2014). https://doi.org/10.1016/j.plaphy.2014.04.009

Barone, V., Baglieri, A., Stevanato, P., et al.: Root morphological and molecular responses induced by microalgae extracts in sugar beet (Beta vulgaris L.). J. Appl. Phycol. **30**, 1061–1071 (2018). https://doi.org/10.1007/s10811-017-1283-3

Bartwal, A., Mall, R., Lohani, P., et al.: Role of secondary metabolites and brassinosteroids in plant defense against environmental stresses. J. Plant Growth Regul. **32**, 216–232 (2013). https://doi.org/10.1007/s00344-012-9272-x

Bhattacharya, M., Goswami, S.: Microalgae—a green multi-product biorefinery for future industrial prospects. Biocatal. Agric. Biotechnol. **25**, 101580 (2020). https://doi.org/10.1016/j.bcab.2020.101580

Blanke, M.: Biostimulantien–von Algenextrakten bis Shrimpsschalen–eine Branche (er-)findet sich. Erwerbs-Obstbau **58**, 81–87 (2016). https://doi.org/10.1007/s10341-016-0276-8

Bleecker, A.B., Kende, H.: Ethylene: a gaseous signal molecule in plants. Annu. Rev. Cell Dev. Biol. **16**, 1–18 (2000). https://doi.org/10.1146/annurev.cellbio.16.1.1

Borowitzka, M.A.: Culturing microalgae in outdoor ponds. In: Algal Culturing Techniques, pp. 205–218. Elsevier (2005)

Borowitzka, M.A.: Algal physiology and large-scale outdoor cultures of microalgae. In: Borowitzka, M.A., Beardall, J., Raven, J.A. (eds.) The Physiology of Microalgae, pp. 601–652. Springer International Publishing, Cham (2016)

Borowitzka, M.A., Moheimani, N.R.: Sustainable biofuels from algae. Mitig. Adapt. Strateg. Glob. Change **18**, 13–25 (2013). https://doi.org/10.1007/s11027-010-9271-9

Braun, J.C.A., Colla, L.M.: Use of microalgae for the development of biofertilizers and biostimulants. BioEnergy Res (2022). https://doi.org/10.1007/s12155-022-10456-8

Bücker-Neto, L., Paiva, A.L.S., Machado, R.D., et al.: Interactions between plant hormones and heavy metals responses. Genet. Mol. Biol. **40**, 373–386 (2017). https://doi.org/10.1590/1678-4685-gmb-2016-0087

Caretto, S., Linsalata, V., Colella, G., et al.: Carbon fluxes between primary metabolism and phenolic pathway in plant tissues under stress. Int. J. Mol. Sci. **16**, 26378–26394 (2015). https://doi.org/10.3390/ijms161125967

Cheynier, V., Comte, G., Davies, K.M., et al.: Plant phenolics: recent advances on their biosynthesis, genetics, and ecophysiology. Plant Physiol. Biochem. **72**, 1–20 (2013). https://doi.org/10.1016/j.plaphy.2013.05.009

Chittapun, S., Limbipichai, S., Amnuaysin, N., et al.: Effects of using cyanobacteria and fertilizer on growth and yield of rice, Pathum Thani I: a pot experiment. J. Appl. Phycol. **30**, 79–85 (2018). https://doi.org/10.1007/s10811-017-1138-y

Choudhary, S.P., Kanwar, M., Bhardwaj, R., Yu, J.-Q., Tran, L.-S.P.: Chromium stress mitigation by polyamine-brassinosteroid application involves phytohormonal and physiological strategies in *Raphanus sativus* L. PLoS One **7**(3), e33210 (2012). https://doi.org/10.1371/journal.pone.0033210

Cui, N., Xiao, J.M., Feng, Y.J., Zhao, Y.T., Yu, X.Y., Xu, J.W., Li, T., Zhao. P.: Antioxidants enhance lipid productivity in *Heveochlorella* sp. Yu. Algal Res. **55** (2021). https://doi.org/10.1016/j.algal.2021.102235

Czerpak, R., Piotrowska, A., Szulecka, K.: Jasmonic acid affects changes in the growth and some components content in alga *Chlorella vulgaris*. Acta Physiol. Plant **28** 195–203 (2006). https://doi.org/10.1007/BF02706531

Dineshkumar, R., Subramanian, J., Arumugam, A., et al.: Exploring the microalgae biofertilizer effect on onion cultivation by field experiment. Waste Biomass Valorization **11**, 77–87 (2020). https://doi.org/10.1007/s12649-018-0466-8

Fariduddin, Q., Ahmed, M., Mir, B.A., et al.: 24-Epibrassinolide mitigates the adverse effects of manganese induced toxicity through improved antioxidant system and photosynthetic attributes in Brassica juncea. Environ. Sci. Pollut. Res. **22**, 11349–11359 (2015). https://doi.org/10.1007/s11356-015-4339-4

Garcia-Gonzalez, J., Sommerfeld, M.: Biofertilizer and biostimulant properties of the microalga Acutodesmus dimorphus. J. Appl. Phycol. **28**, 1051–1061 (2016). https://doi.org/10.1007/s10811-015-0625-2

Gonçalves, A.L.: The use of microalgae and cyanobacteria in the improvement of agricultural practices: a review on their biofertilising, biostimulating and biopesticide roles. Appl. Sci. **11**, 871 (2021). https://doi.org/10.3390/app11020871

González-Pérez, B.K., Rivas-Castillo, A.M., Valdez-Calderón, A., Gayosso-Morales, M.A.: Microalgae as biostimulants: a new approach in agriculture. World J. Microbiol. Biotechnol. **38**, 4 (2022). https://doi.org/10.1007/s11274-021-03192-2

Han, X., Zeng, H., Bartocci, P., et al.: Phytohormones and effects on growth and metabolites of microalgae: a review. Fermentation **4**, 25 (2018). https://doi.org/10.3390/fermentation4020025

Hashtroudi, M.S., Ghassempour, A., Riahi, H., et al.: Endogenous auxins in plant growth-promoting Cyanobacteria—Anabaena vaginicola and Nostoc calcicola. J. Appl. Phycol. **25**, 379–386 (2013). https://doi.org/10.1007/s10811-012-9872-7

Hegedüs, A., Erdei, S., Horváth, G.: Comparative studies of $H_2O_2$ detoxifying enzymes in green and greening barley seedlings under cadmium stress. Plant Sci. **160**, 1085–1093 (2001). https://doi.org/10.1016/S0168-9452(01)00330-2

Hernández-Carlos, B., Gamboa-Angulo, M.M.: Metabolites from freshwater aquatic microalgae and fungi as potential natural pesticides. Phytochem. Rev. **10**, 261–286 (2011). https://doi.org/10.1007/s11101-010-9192-y

Ishika, T., Moheimani, N.R., Bahri, P.A.: Sustainable saline microalgae co-cultivation for biofuel production: a critical review. Renew. Sustain. Energy Rev. **78**, 356–368 (2017). https://doi.org/10.1016/j.rser.2017.04.110

Jirásková, D., Poulíčková, A., Novák, O., et al.: High-throughput screening technology for monitoring phytohormone production in microalgae. J. Phycol. **45**, 108–118 (2009). https://doi.org/10.1111/j.1529-8817.2008.00615.x

Jusoh, M., Loh, S.H., Chuah, T.S., et al.: Elucidating the role of jasmonic acid in oil accumulation, fatty acid composition and gene expression in Chlorella vulgaris (Trebouxiophyceae) during early stationary growth phase. Algal Res. **9**, 14–20 (2015). https://doi.org/10.1016/j.algal.2015.02.020

Kapoore, R.V., Wood, E.E., Llewellyn, C.A.: Algae biostimulants: a critical look at microalgal biostimulants for sustainable agricultural practices. Biotechnol. Adv. **49**, 107754 (2021). https://doi.org/10.1016/j.biotechadv.2021.107754

Kim, S., Dale, B.E.: Indirect land use change for biofuels: testing predictions and improving analytical methodologies. Biomass Bioenergy **35**, 3235–3240 (2011). https://doi.org/10.1016/j.biombioe.2011.04.039

Kiseleva, A.A., Tarachovskaya, E.R., Shishova, M.F.: Biosynthesis of phytohormones in algae. Russ. J. Plant Physiol. **59**, 595–610 (2012). https://doi.org/10.1134/S1021443712050081

Kondrat'eva, V.V., Shelepova, O.V.: The relationship between cytokinins and the amount of nitrogen in the wintering organs of herbaceous plants. Biol. Bull. **37**, 579–584 (2010). https://doi.org/10.1134/S1062359010060051

Kováčik, J., Klejdus, B., Hedbavny, J., Backor, M.: Effect of copper and salicylic acid on phenolic metabolites and free amino acids in *Scenedesmus quadricauda* (Chlorophyceae). Plant Sci. **178**, 307–311 (2010). https://doi.org/10.1016/j.plantsci.2010.01.009

Kozlova, T.A., Hardy, B.P., Krishna, P., Levin, D.B.: Effect of phytohormones on growth and accumulation of pigments and fatty acids in the microalgae Scenedesmus quadricauda. Algal Res. **27**, 325–334 (2017). https://doi.org/10.1016/j.algal.2017.09.020

Lu, Y., Xu, J.: Phytohormones in microalgae: a new opportunity for microalgal biotechnology? Trends Plant Sci. **20**, 273–282 (2015). https://doi.org/10.1016/j.tplants.2015.01.006

Lucini, L., Rouphael, Y., Cardarelli, M., et al.: The effect of a plant-derived biostimulant on metabolic profiling and crop performance of lettuce grown under saline conditions. Sci. Hortic. **182**, 124–133 (2015). https://doi.org/10.1016/j.scienta.2014.11.022

de Jesus Raposo, M.F., de Morais, R.M.S.C.: Influence of the growth regulators kinetin and 2,4-d on the growth of two chlorophyte microalgae, Haematococcus pluvialis and Dunaliella salina. J. Basic Appl. Sci. **9**, 302–308 (2013). https://doi.org/10.6000/1927-5129.2013.09.40

Mata, T.M., Martins, A.A., NidiaS, C.: Microalgae for biodiesel production and other applications: a review. Renew. Sustain. Energy Rev. **14**, 217–232 (2010). https://doi.org/10.1016/j.rser.2009.07.020

Michalak, I., Górka, B., Wieczorek, P.P., et al.: Supercritical fluid extraction of algae enhances levels of biologically active compounds promoting plant growth. Eur. J. Phycol. **51**, 243–252 (2016). https://doi.org/10.1080/09670262.2015.1134813

Mok, D.W., Mok, M.C.: Cytokinin metabolism and action. Annu. Rev. Plant. Physiol. Plant. Mol. Biol. **52**, 89–118 (2001). https://doi.org/10.1146/annurev.arplant.52.1.89

Moreira, D., Pires, J.C.M.: Atmospheric $CO_2$ capture by algae: negative carbon dioxide emission path. Bioresour. Technol. **215**, 371–379 (2016). https://doi.org/10.1016/j.biortech.2016.03.060

Nguyen, H.N., Kisiala, A.B., Emery, R.J.N.: The roles of phytohormones in metal stress regulation in microalgae. J. Appl. Phycol. **32**, 3817–3829 (2020a). https://doi.org/10.1007/s10811-020-02271-5

Nguyen, T.Q., Sesin, V., Kisiala, A., Emery, R.J.N.: The role of phytohormones in enhancing metal remediation capacity of algae. Bull. Environ. Contam. Toxicol. **105**, 671–678 (2020b). https://doi.org/10.1007/s00128-020-02880-3

Noble, A., Kisiala, A., Galer, A., et al.: Euglena gracilis (Euglenophyceae) produces abscisic acid and cytokinins and responds to their exogenous application singly and in combination with other growth regulators. Eur. J. Phycol. **49**, 244–254 (2014). https://doi.org/10.1080/09670262.2014.911353

Oklestkova, J., Rárová, L., Kvasnica, M., Strnad, M.: Brassinosteroids: synthesis and biological activities. Phytochem. Rev. **14**, 1053–1072 (2015). https://doi.org/10.1007/s11101-015-9446-9

Onyeaka, H., Miri, T., Obileke, K., et al.: Minimizing carbon footprint via microalgae as a biological capture. Carbon Capture Sci. Technol. **1**, 100007 (2021). https://doi.org/10.1016/j.ccst.2021.100007

Ostrowski, N., Roy, A., Kumta, P.N.: Magnesium phosphate cement systems for hard tissue applications: a review. ACS Biomater. Sci. Eng. **2**(7), 1067–1083 (2016). https://doi.org/10.1021/acsbiomaterials.6b00056

Paniagua-Michel, J., Olmos-Soto, J., Ruiz, M.A.: Pathways of carotenoid biosynthesis in bacteria and microalgae. In: Barredo, J.-L. (ed.) Microbial Carotenoids from Bacteria and Microalgae, pp. 1–12. Humana Press, Totowa, NJ (2012)

Park, W.-K., Yoo, G., Moon, M., et al.: Phytohormone supplementation significantly increases growth of Chlamydomonas reinhardtii cultivated for biodiesel production. Appl. Biochem. Biotechnol. **171**, 1128–1142 (2013). https://doi.org/10.1007/s12010-013-0386-9

Peng, X., Bruns, M.A.: Development of a nitrogen-fixing cyanobacterial consortium for surface stabilization of agricultural soils. J. Appl.

Phycol. **31**, 1047–1056 (2019). https://doi.org/10.1007/s10811-018-1597-9

Piotrowska-Niczyporuk, A., Bajguz, A., Zambrzycka, E., Godlewska-Żyłkiewicz, B.: Phytohormones as regulators of heavy metal biosorption and toxicity in green alga Chlorella vulgaris (Chlorophyceae). Plant Physiol. Biochem. **52**, 52–65 (2012). https://doi.org/10.1016/j.plaphy.2011.11.009

Rady, M.M.: A novel organo-mineral fertilizer can mitigate salinity stress effects for tomato production on reclaimed saline soil. South Afr. J. Bot. **81**, 8–14 (2012). https://doi.org/10.1016/j.sajb.2012.03.013

Rahman, K.M.: Food and high-value products from microalgae: market opportunities and challenges. In: Alam, Md.A., Xu, J.-L., Wang, Z. (eds.) Microalgae Biotechnology for Food, Health and High Value Products, pp. 3–27. Springer Singapore, Singapore (2020)

Ramakrishna, B., Seeta Ram Rao, S.: 24-Epibrassinolide maintains elevated redox state of AsA and GSH in radish (*Raphanus sativus* L.) seedlings under zinc stress. Acta Physiol. Plant **35**, 1291–1302 (2013). https://doi.org/10.1007/s11738-012-1168-7

Reijnders, L., Huijbregts, M.A.J.: Biogenic greenhouse gas emissions linked to the life cycles of biodiesel derived from European rapeseed and Brazilian soybeans. J. Clean. Prod. **16**(18), 1943–1948 (2008). https://doi.org/10.1016/j.jclepro.2008.01.012

Renuka, N., Guldhe, A., Prasanna, R., et al.: Microalgae as multi-functional options in modern agriculture: current trends, prospects, and challenges. Biotechnol. Adv. **36**, 1255–1273 (2018). https://doi.org/10.1016/j.biotechadv.2018.04.004

Romanenko, E.A., Kosakovskaya, I.V., Romanenko, P.A.: Phytohormones of microalgae: biological role and involvement in the regulation of physiological processes. Pt I. Auxins, abscisic acid, ethylene. Int. J. Algae **17**, 275–289 (2015). https://doi.org/10.1615/InterJAlgae.v17.i3.80

Ronga, D., Biazzi, E., Parati, K., et al.: Microalgal biostimulants and biofertilisers in crop productions. Agronomy **9**, 192 (2019). https://doi.org/10.3390/agronomy9040192

Sakakibara, H.: CYTOKININS: activity, biosynthesis, and translocation. Annu. Rev. Plant Biol. **57**, 431–449 (2006). https://doi.org/10.1146/annurev.arplant.57.032905.105231

Salama, E.-S., Kabra, A.N., Ji, M.-K., et al.: Enhancement of microalgae growth and fatty acid content under the influence of phytohormones. Bioresour. Technol. **172**, 97–103 (2014). https://doi.org/10.1016/j.biortech.2014.09.002

Sarma, S., Sharma, S., Rudakiya, D., et al.: Valorization of microalgae biomass into bioproducts promoting circular bioeconomy: a holistic approach of bioremediation and biorefinery. 3 Biotech **11**, 378 (2021). https://doi.org/10.1007/s13205-021-02911-8

Sharma, K., Li, Y., Schenk, P.M.: UV-C-mediated lipid induction and settling, a step change towards economical microalgal biodiesel production. Green Chem. **16**, 3539–3548 (2014). https://doi.org/10.1039/C4GC00552J

Show, P., Tang, M., Nagarajan, D., et al.: A holistic approach to managing microalgae for biofuel applications. Int. J. Mol. Sci. **18**, 215 (2017). https://doi.org/10.3390/ijms18010215

Stiles, W.A., Styles, D., Chapman, S.P., Esteves, S., Bywater, A., Melville, L., Silkina, A., Lupatsch, I., Grünewald, C.F., Lovitt, R., Chaloner, T.: Using microalgae in the circular economy to valorise anaerobic digestate: challenges and opportunities. Bioresour. Technol. **267**, 732–742 (2018). https://doi.org/10.1016/j.biortech.2018.07.100

Stirk, W.A., van Staden, J.: Potential of phytohormones as a strategy to improve microalgae productivity for biotechnological applications. Biotechnol. Adv. **44**, 107612 (2020). https://doi.org/10.1016/j.biotechadv.2020.107612

Stirk, W.A., Ördög, V., Novák, O., et al.: Auxin and cytokinin relationships in 24 microalgal strains 1. J. Phycol. **49**, 459–467 (2013a). https://doi.org/10.1111/jpy.12061

Stirk, W.A., Bálint, P., Tarkowská, D., Novák, O., Strnad, M., Ördög, V., Van Staden, J.: Hormone profiles in microalgae: gibberellins and brassinosteroids. Plant Physiol. Biochem. **70**, 348–353 (2013b). https://doi.org/10.1016/j.plaphy.2013.05.037

Stirk, W.A., Bálint, P., Tarkowská, D., et al.: Effect of light on growth and endogenous hormones in Chlorella minutissima (Trebouxiophyceae). Plant Physiol. Biochem. **79**, 66–76 (2014). https://doi.org/10.1016/j.plaphy.2014.03.005

Stirk, W.A., Bálint, P., Tarkowská, D., et al.: Endogenous brassinosteroids in microalgae exposed to salt and low-temperature stress. Eur. J. Phycol. **53**, 273–279 (2018). https://doi.org/10.1080/09670262.2018.1441447

Stirk, W.A., Tarkowská, D., Gruz, J., et al.: Effect of gibberellins on growth and biochemical constituents in Chlorella minutissima (Trebouxiophyceae). South Afr. J. Bot. **126**, 92–98 (2019). https://doi.org/10.1016/j.sajb.2019.05.001

Sulochana, S.B., Arumugam, M.: Influence of abscisic acid on growth, biomass and lipid yield of Scenedesmus quadricauda under nitrogen starved condition. Bioresour. Technol. **213**, 198–203 (2016). https://doi.org/10.1016/j.biortech.2016.02.078

Sun, X.-M., Ren, L.-J., Zhao, Q.-Y., et al.: Microalgae for the production of lipid and carotenoids: a review with a focus on stress regulation and adaptation. Biotechnol. Biofuels **11**, 272 (2018). https://doi.org/10.1186/s13068-018-1275-9

Taiz, L., Zeiger, E.: Plant physiology, 5th edn. Sinauer Associates, Inc., Publishers, Sunderland, MA (2010). https://scholar.google.com/scholar_lookup?title=Plant%20physiology&author=L%20Taiz&author=E%20Zeiger&publication_year=2010&

Tarakhovskaya, E.R., Maslov, Yu.I., Shishova, M.F.: Phytohormones in algae. Russ. J. Plant Physiol. **54**, 163–170 (2007). https://doi.org/10.1134/S1021443707020021

Teale, W.D., Paponov, I.A., Palme, K.: Auxin in action: signalling, transport and the control of plant growth and development. Nat. Rev. Mol. Cell Biol. **7**, 847–859 (2006). https://doi.org/10.1038/nrm2020

Thomassen, G., Van Dael, M., Van Passel, S.: The potential of microalgae biorefineries in Belgium and India: an environmental techno-economic assessment. Bioresour. Technol. **267**, 271–280 (2018). https://doi.org/10.1016/j.biortech.2018.07.037

Wang, J., Song, L., Gong, X., et al.: Functions of jasmonic acid in plant regulation and response to abiotic stress. Int. J. Mol. Sci. **21**, 1446 (2020). https://doi.org/10.3390/ijms21041446

Wang, K., Khoo, K.S., Chew, K.W., et al.: Microalgae: the future supply house of biohydrogen and biogas. Front Energy Res **9**, 660399 (2021). https://doi.org/10.3389/fenrg.2021.660399

Wang, S.-K., Yang, K.-X., Zhu, Y.-R., et al.: One-step co-cultivation and flocculation of microalgae with filamentous fungi to valorize starch wastewater into high-value biomass. Bioresour. Technol. **361**, 127625 (2022). https://doi.org/10.1016/j.biortech.2022.127625

Werner, T., IKöllmer, I., Bartrina, I., Holst, K., Schmülling, T.: New insights into the biology of cytokinin degradation. Plant Biol. **8**, 1–12 (2007). https://doi.org/10.1055/s-2006-923928

Win, T.T., Barone, G.D., Secundo, F., Fu, P.: Algal biofertilizers and plant growth stimulants for sustainable agriculture. Ind. Biotechnol. **14**, 203–211 (2018). https://doi.org/10.1089/ind.2018.0010

Yu, X., Chen, L., Zhang, W.: Chemicals to enhance microalgal growth and accumulation of high-value bioproducts. Front. Microbiol. **6** (2015). https://doi.org/10.3389/fmicb.2015.00056

Yu, Z., Pei, H., Jiang, L., et al.: Phytohormone addition coupled with nitrogen depletion almost tripled the lipid productivities in two algae. Bioresour. Technol. **247**, 904–914 (2018). https://doi.org/10.1016/j.biortech.2017.09.192

Yu, Z., Song, M., Pei, H., et al.: The effects of combined agricultural phytohormones on the growth, carbon partitioning and cell morphology of two screened algae. Bioresour. Technol. **239**, 87–96 (2017). https://doi.org/10.1016/j.biortech.2017.04.120

Zhao, Y., Wang, H.-P., Han, B., Yu, X.: Coupling of abiotic stresses and phytohormones for the production of lipids and high-value by-products by microalgae: a review. Bioresour. Technol. **274**, 549–556 (2019). https://doi.org/10.1016/j.biortech.2018.12.030
Žižková, E., Kubeš, M., Dobrev, P.I., et al.: Control of cytokinin and auxin homeostasis in cyanobacteria and algae. Ann. Bot. **119**, 151–166 (2017). https://doi.org/10.1093/aob/mcw194
Zuo, S.P., Fang, Z.S., Yang, S.Y., Wan, K., Han, Y.J.: Effect of allelopathic potential from selected aquatic macrophytes on algal interaction in the polluted water. Biochem. Syst. Ecol. **61**, 133–138 (2015). https://doi.org/10.1016/j.bse.2015.06.011
Zou, W., Li, X., Li, C., et al.: Influence of size and phase on the biodegradation, excretion, and phytotoxicity persistence of single-layer molybdenum disulfide. Environ. Sci. Technol. **54**, 12295–12306 (2020). https://doi.org/10.1021/acs.est.0c02642

# Comparative Study of Chemical Composition, Antibacterial Activity, and Molecular Docking of Three Anethole-Rich Essential Oils

Amina Chlouchi, Anjoud Harmouzi, Rabha Afroukh, Hajar El Omari, Asmaa Oubihi, Mohamed Chebaibi, Brahim Mssillou, and Nour El Hoda Mustaphi

**Abstract**

Essential oils (EOs) from *Pimpinella anisum* (anise), *Foeniculum vulgare* (fennel), and *Illicium verum* (star anise) are widely used in Moroccan traditional medicine and cuisine. Given the global challenge of antibiotic resistance, this study investigates their antibacterial potential through chemical profiling, antimicrobial assays, and molecular docking. GC–MS analysis revealed *trans*-anethole (57–67%) as the dominant compound, alongside estragole (6–9%), fenchone, d-limonene, and α-pinene. Antibacterial testing against six pathogens demonstrated strong inhibition, particularly against *Staphylococcus aureus*, *Staphylococcus. epidermidis*, *Acinetobacter baumannii*, and *Enterobacter cloacae. F. vulgare* EO showed notable efficacy against *S. epidermidis*. Molecular docking explored interactions between key bioactive compounds and *Staphylococcus aureus* nucleoside diphosphate kinase, beta-ketoacyl-[acyl carrier protein] synthase from *Escherichia coli* and *E. coli* gyrase B. *Trans*-anethole formed a hydrogen bond with THR 300 and π–π stacking with HIE 298, exhibiting the highest binding affinity (glide score: − 6.18 kcal/mol). Fenchone (− 6.13 kcal/mol) and γ-himachalene (− 5.9 kcal/mol) also displayed strong interactions. These findings highlight the therapeutic potential of anise, fennel, and star anise EOs as natural antimicrobial agents, supported by their bioactive composition and mechanistic interactions. Further research could optimize their application against resistant pathogens.

Nature's Anethole Power: Antibacterial Potency and Molecular Insights from Essential Oils

A. Chlouchi (✉) · A. Harmouzi (✉)
Organic Chemistry, Catalysis and Environmental Laboratory, Sciences Faculty, Ibn Tofail University, Kenitra, Morocco
e-mail: amina.chlouchi@uit.ac.ma

A. Harmouzi
e-mail: nojoud.harmouzi@gmail.com

A. Chlouchi
Higher National School of Chemistry, Ibn Tofail University, Kenitra, Morocco

R. Afroukh
Laboratory of Plant, Animal and Agro-Industry Production, Sciences Faculty, Ibn Tofail University, Kenitra, Morocco

H. El Omari
Natural Resources Management and Development Team, Laboratory of Health and Environmental, Sciences Faculty, Moulay Ismail University, Meknes, Morocco

Higher Institute of Nursing Professions and Technical Health Fez, Meknes, Morocco

A. Harmouzi · A. Oubihi
Natural Resources and Sustainable Development Laboratory, Department of Biology, Sciences Faculty, Ibn Tofail University, Kenitra, Morocco

M. Chebaibi
Ministry of Health and Social Protection, Higher Institute of Nursing Professions and Health Techniques, Fez, Morocco

Biomedical and Translational Research Laboratory, Faculty of Medicine and Pharmacy of Fez, Sidi Mohamed Ben Abdellah University, Fez, Morocco

B. Mssillou
Laboratory of Natural Substances, Pharmacology, Environment, Modeling, Health and Quality of Life (SNAMOPEQ), Sidi Mohamed Ben Abdellah University, Fez, Morocco

N. E. H. Mustaphi
Laboratory of Heterocyclic Organic Chemistry, Faculty of Sciences, Mohammed V University I, Rabat, Morocco



O. Jamal Eddine et al. (eds.), *Plant Pathology, Fungal Diversity, and Biotechnological Advances in Agriculture*,
Sustainable Landscape Planning and Natural Resources Management, https://doi.org/10.1007/978-3-031-97666-7_4

**Keywords**

Essential oils (EOs) · Anethole · Chemical composition · Antibacterial activity · Molecular docking

## 1 Introduction

The comparative study of essential oils (EOs) derived from volatile aromatic plants has attracted considerable attention in recent years owing to their diverse biological activities, especially their antibacterial and antifungal properties, along with their complex chemical compositions. This study focuses on EOs from three prominent members of the *Apiaceae* family: *Pimpinella anisum* (anise), *Foeniculum vulgare* (fennel), and *Illicium verum* (star anise). These plants are not only known for their culinary uses but also for their health-promoting qualities, predominantly attributed to their rich content of bioactive compounds like anethole, which is the principal component in their essential oils (Rehman et al. 2016).

The antibacterial efficacy of these EOs has been substantiated across multiple studies. For instance, star anise essential oil is renowned as highly potent antibacterial agent, effective against a wide range of pathogens. This remarkable property is attributed to its abundant trans-anethole content, which can exceed 90% (Loi et al. 2023; Trâm et al. 2021). Research indicates that this compound plays a significant role in inhibiting not only bacterial growth but also fungal strains (Huang et al. 2010; Hamed and Al-Wasify 2023). Fennel EOs have demonstrated notable antibacterial activity against some Gram-positive bacteria. In fact, studies have clearly shown that its effectiveness surpasses that of standard antibiotics like ampicillin, particularly in combating certain strains (Shahat et al. 2011; Mota et al. 2015). Anise Essential oil exhibits similar antibacterial properties, making these EOs formidable agents against foodborne pathogens and other infectious microorganisms (Mohammed and Ebraheem 2020; Shahrajabian et al. 2021).

Furthermore, these essential oils vary widely in chemical composition, with each plant contributing a unique volatile profile. For example, fennel EO contains significant amounts of estragole and trans-anethole (Shahat et al. 2011; Mota et al. 2015), while star anise also features compounds like α-pinene and limonene alongside its predominant anethole (Trâm et al. 2021; Shahrajabian et al. 2021). Such variations affect not only the oils' biological activities but also their applications in pharmacotherapy and food preservation.

Molecular docking studies have helped to understand the mechanisms involved in the antibacterial activity of these EOs. Interactions of key compounds like anethole with bacterial cell components have been explored to elucidate their modes of action, revealing potential targets for developing novel antimicrobial agents (Vecchio et al. 2016; Shamsudeen and Palffy 2024). These studies suggest that the efficacy of EOs can be enhanced through various extraction methods and formulations, pointing towards a dynamic area of research in natural product chemistry.

This study analyzes and compares the chemical composition of three Eos—*Foeniculum vulgare, Illicium verum, and Pimpinella anisum*—and evaluates their antibacterial properties. Molecular docking studies were conducted to explore interactions between their major bioactive compounds and key bacterial target proteins—*Staphylococcus aureus nucleoside diphosphate kinase, Escherichia coli beta-ketoacyl synthase*, and *E. coli gyrase B*—providing mechanistic insights into their antimicrobial potential. This research highlights the promise of these EOs as natural antimicrobials and supports future applications in the fields of food preservation, pharmacology, and alternative medicine by correlating chemical profiles with bioactivity.

## 2 Materials and Methods

### 2.1 Plant Material

Three seed samples were obtained from a local herb shop: Fennel and green anise were grown in Morocco, while star anise was imported from China. To preserve their volatile compounds, the seeds were crushed immediately before extraction.

### 2.2 Essential Oils Extraction

According to the methodology described in our previous work, the essential oils were extracted by hydrodistillation with a Clevenger apparatus (El Kamari et al. 2023). For this process, 300 g of plant material were immersed in distilled water within a 3-L flask and subjected to boiling (100 °C) for 4 h. The volatile essential oil was collected via condensation, dehydrated using sodium sulfate ($Na_2SO_4$), and stored in amber glass bottles at + 4 °C to prevent degradation until further analysis.

### 2.3 Gas Chromatography-Mass Spectrometry (GC–MS)

Chromatographic separations are conducted using a Perkin ElmerTM GC-680 gas chromatograph coupled to a Q-8 MS mass spectrometer featuring an ion trap. Fragmentation is achieved through electronic impact at 70 eV. A 30 m × 0.25 mm i.d., 0.25 μm-film Rxi®-5sil capillary column (1,4-bis(dimethylsiloxy)phenylene dimethyl polysiloxane) was used, with the oven temperature programmed from 60 °C to 310 °C at 4 °C/min. The flame ionization detector (FID) (PerkinElmer, Waltham, MA, USA) was held at 310 °C. Helium served as the carrier gas with a steady flow rate of 1 mL/min and samples introduced in split-injection mode, Identification of individual constituents was achieved by comparing their mass spectra against the NIST database and Adams' reference tables (Adams 2007), together with confirmation via calculated Kovats retention indices.

## 2.4 Microorganisms Studied

To evaluate the antimicrobial efficacy of our three seeds extracts, we tested them against two Gram-positive bacterial strains—*Staphylococcus aureus* and *Staphylococcus epidermidis*- and four Gram-negative bacterial strains—*Escherichia coli, Acinetobacter baumannii, Klebsiella pneumoniae*, and *Enterobacter cloacae.* All isolates, known for their high levels of antibiotic resistance, were obtained from the Microbiology Department of the Medical Analysis Laboratory. These bacteria are cultured and sustained on nutrient agar via frequent subculturing (Oubihi et al. 2020).

## 2.5 Microbiological Procedure

### 2.5.1 Disc Diffusion

The essential oils' antibacterial properties are evaluated using the disc diffusion method (Candan et al. 2003). First, a suspension of the microorganism with an optical density equivalent to 1 McFarland was evenly spread on Mueller Hinton Agar (MHA) Petri dishes. Next, a 6 mm diameter filter paper disc saturated with 15 μL of essential oil was placed onto the inoculated surfaces. Petri dishes are incubated at 4 °C for 2 h, then transferred to 37 °C for 24 h for bacteria, or 30 °C for 48 h for yeasts. Inhibition zones are measured in millimeters to determine antimicrobial effectiveness. Penicillin (5 μg/disc) and Amoxicillin (25 μg/disc) serve as positive controls for bacteria. To ensure reproducibility, all tests are performed in triplicate (Ait benlabchir et al. 2024).

### 2.5.2 The Minimum Inhibitory Concentrations (MIC)

The minimum inhibitory concentrations (MIC) are determined using the agar dilution method (Elwardani et al. 2024). The essential oil is emulsified in a 0.2% agar solution and diluted to concentrations of 100, 40, 20, 10, 5, 3.3 and 2 μL/mL. In each test tube, containing 13.5 mL of sterilized solid medium (autoclaved at 121 °C), 1.5 mL of each dilution is aseptically added to achieve final concentrations of 10, 4, 2, 1, 0.5, 0.33, and 0.2 μL/mL. The mixture is rapidly transferred to sterile Petri dishes. Control samples, consisting exclusively of the culture medium and the 0.2% agar solution, are meticulously prepared in a similar procedure. Inoculation is carried out by streaking with a calibrated platinum loop to collect a consistent volume of a 24-h culture broth. Incubation is required for 24 h at 37 °C. Control samples, consisting only of the culture medium and the 0.2% agar solution, are similarly prepared. Inoculation is performed by streaking with a calibrated platinum loop to collect a consistent volume of a 24-h culture broth. Incubation follows for 24 h at 37 °C. All experiments are carried out in triplicate to minimize variability.

## 2.6 Molecular Docking Study

### 2.6.1 Protein Preparation

The crystal structures of the target proteins were obtained from the RCSB database. These proteins include *Staphylococcus aureus* nucleoside diphosphate kinase (PDB ID: 3Q8U), beta-ketoacyl-[acyl carrier protein] synthase from *Escherichia coli* (PDB ID: 1FJ4) and *E. coli* gyrase B (PDB ID: 3G7E) (Lafraxo et al. 2022). The Protein Preparation Wizard in Maestro 11.5 was used. It involves preprocessing, refinement, and reduction (Mssillou et al. 2021). Hydrogen atoms were added, and any structural issues were resolved by reorienting hydroxyl groups, water molecules, and amino acids. The proteins were then minimally adjusted for optimization (Amrati et al. 2023; Chebaibi et al. 2024a).

### 2.6.2 Ligand Preparation

The major constituents identified in *Foeniculum vulgare*, *Pimpinella anisum*, and *Illicium verum* were resourced from PUBCHEM database. Ligands were prepared using the LigPrep module (v11.5, Schrödinger Suite, Maestro). The protocol included the following steps (Mssillou et al. 2024; Tourabi et al. 2023):

- Structural conversion: 2D ligand representations were transformed into 3D atomic coordinates.
- Hydrogen addition: Protonation states were assigned at pH 7.4 ± 0.5 to reflect physiological conditions.
- Steric optimization: Bond lengths and angles were corrected using the OPLS3 force field to resolve structural clashes.
- Energy minimization: Conformational energy was reduced via gradient descent.
- Stereochemistry preservation: Specified chiral centers were maintained without epimerization.

### 2.6.3 Receptor Grid Generation

Receptor grids were created using Glide docking module (Schrödinger, LLC) to define active sites based on the coordinates of the co-crystallized ligands. The ligands were positioned within the X-ray crystal structures of the target proteins to explore potential binding conformations. During grid generation, the locations and boundaries of the active sites were specified, rotatable groups were defined, and excluded volumes were set to refine the docking precision. (Kumar et al. 2022; Meryem et al. 2024).

### 2.6.4 Performing Molecular Docking

Docking was performed in Standard Precision (SP) mode following the preparation of ligands and proteins and construction of the grids. This mode was selected to evaluate the binding interactions between molecules. Binding energies were calculated in kcal $mol^{-1}$, alongside internal energy, root mean square deviation (RMSD), desolvation energy, hydrogen bonding, lipophilic interactions, and π-π stacking interactions. This method quickly analyzes protein–ligand interactions and provides detailed insights into binding mechanisms (Chebaibi et al. 2024b; El Abdali et al. 2023).

## 2.7 Statistical Analysis

Data were analyzed using GraphPad Prism 8 (GraphPad Software, San Diego, CA, USA). Mean values and standard deviations (± SD) were calculated. Statistical comparisons were performed using one-way analysis of variance (ANOVA), followed by Tukey's post hoc test for multiple comparisons. A *p*-value of $< 0.05$ was considered statistically significant.

# 3 Results and Discussion

## 3.1 Chemical Composition of Essential Oils

Table 1 summarizes the extraction yields of the three essential oils obtained by hydro-distilling dried seeds.

The extraction rate of *Foeniculum vulgare* essential oil (FVEO) was 2.06% (v/w), which aligns with previously reported values in the literature. (Anwar et al. 2009; Sabzi-Nojadeh et al. 2021) For instance, a study on fennel cultivated in various regions of Turkey documented yields ranging from 1.12 to 2.35% (Aslan et al. 2019). Additionally, another study investigating different extraction methods for *Foeniculum vulgare* reported yields between 0.50 and 4.60% (Simić et al. 2018).

**Table 1** Essential oils yield of the three seeds, anise, fennel and star anise

| Seeds | EOs yield (%) | SE |
|---|---|---|
| Anise | 1.56 | ± 0.07 |
| Fennel | 2.06 | ± 0.08 |
| Star anise | 3.24 | ± 0.10 |

The obtained amount of *Pimpinella anisum* essential oil (PAEO) was 1.56% (v/w), which is consistent with findings from other Mediterranean studies (Saibi et al. 2012). However, this yield is significantly lower than those reported in other countries (Sharifi et al. 2008; Zheljazkov et al. 2013; Pezo et al. 2022). Research confirms that PAEO yields can vary considerably based on geographical location, extraction methods, and cultivation practices, with some regions achieving higher yields than others. According to Javidnia et al. (2007), PAEO yields ranged from 2.90 to 3.70%, with the highest yields obtained when anise fruits were dried using a microwave oven.

Regarding *Illicium verum* essential oil (IVEO), the yield was 3.42% (v/w). This yield is within the typical range reported in various studies, with IVEO yields fluctuating between 2.20% and 4.50% across different harvesting stages. Factors such as light exposure, sowing date, and irrigation practices have been noted to influence the yield and anethole content of these plants (Li et al. 2020; Outemsaa et al. 2021).

Our study definitively identified twenty-five compounds representing over 96% of the total EOs in *Foeniculum vulgare* (see Table 2). The major components were trans-anethole (56.9%), fenchone (12.8%), estragole (9%), and d-limonene (8.8%). Additionally, fennel contains a variety of other monoterpene hydrocarbons in smaller amounts, including α- and β-pinene, camphene, camphor, terpinolene, δ-terpinene, α-phellandrene, β-myrecene, terpinen-4-ol, anisylacetone and more. This study's results align precisely with those from previous research (Raal et al. 2011; Alam et al. 2019).

Fennel essential oils are thoroughly researched because of its diverse applications in medicine, food, and agriculture. These investigations have shed light on the primary compounds found in these oils and their potential biological activities. One research study definitively found that FVEO's main components are estragole, α-pinene, and α-phellandrene: 55.3%, 11.3%, and 9.1% respectively (Dadalioglu and Evrendlilek 2004). Another research (Shahat et al. 2011) focused on organically grown fennel cultivars, pinpointing estragole as a predominant component, underscoring its significance in fennel's essential oil composition. It is clear from the findings of other studies (Coşge et al. 2008; Shojaii and Fard 2012) that trans-anethole is present in high amounts in sweet and bitter fennel. It makes up 95.25% and 75.13% of sweet and bitter fennel, respectively. This confirms the complex chemical makeup of this plant.

GC–MS analysis of PAEO revealed forty constituents, accounting for over 93% of the essential oil's composition

**Table 2** Chemical composition of the essential oils isolated from *Pimpinella anisum, Foeniculum vulgare* and *illicium verum* analyzed by GC/MS

| N | Compounds | Retention time (RT) (min) | Area | | |
|---|---|---|---|---|---|
| | | | Anise | Fennel | Star anise |
| 1 | Hexanal | 3.154 | – | 0.028 | 0.025 |
| 2 | Heptanal | 4.388 | 0.024 | – | – |
| 3 | β-Thujene | 4.745 | 0.012 | – | 0.120 |
| 4 | **α-Pinene** | 4.851 | 0.225 | **2.053** | **2.174** |
| 5 | Camphene | 5.085 | 0.170 | 0.580 | 0.128 |
| 6 | β-Pinene | 5.464 | 0.423 | 0.286 | 0.569 |
| 7 | β-Myrcene | 5.560 | 0.065 | 1.887 | 1.165 |
| 8 | Octanal | 5.747 | 0.018 | – | – |
| 9 | α-Phellandrene | 5.847 | 0.039 | 0.614 | – |
| 10 | **β-Carene** | 5.901 | – | – | **2.078** |
| 11 | α–Terpinene | 5.994 | – | – | 0.827 |
| 12 | p-Cymene | 6.080 | 0.202 | 0.056 | – |
| 13 | **D-Limonene** | 6.141 | 0.145 | **8.776** | **10.343** |
| 14 | ɣ-Terpinene | 6.277 | – | 1.170 | 0.866 |
| 15 | p-Menth-2-enol | 6.286 | – | – | 0.990 |
| 16 | β-Ocimene | 6.347 | – | 0.038 | – |
| 17 | Trans-Linalool oxide | 6.705 | – | – | 0.057 |
| 18 | **Fenchone** | 6.955 | 0.753 | **12.772** | – |
| 19 | Linalool | 7.076 | 0.809 | – | 1.951 |
| 20 | Fenchol | 7.495 | – | 0.056 | 0.041 |
| 21 | 3-Isopropenyl-4-methyl-4-vinyl-cyclohexene | 7.645 | 0.337 | – | – |
| 22 | Camphor | 7.740 | 0.110 | 1.111 | – |
| 23 | endo-Borneol | 8.076 | 0.041 | – | 0.060 |
| 24 | Terpinen-4-ol | 8.160 | 0.250 | 0.383 | 1.283 |
| 25 | **Estragole** | 8.435 | **7.040** | **9.045** | **6.375** |
| 26 | Fenchyl acetate | 8.593 | – | 0.297 | – |
| 27 | Nerol | 8.647 | – | – | 0.065 |
| 28 | Cumaldehyde | 8.957 | 0.177 | – | – |
| 29 | **T. Anethole** | 9.557 | **65.861** | **56.884** | **66.880** |
| 30 | Mesitylethanone | 10.544 | 0.094 | – | – |
| 31 | Benzeneacetic acid | 10.642 | 0.123 | – | – |
| 32 | Methyleugenol | 10.713 | – | 0.072 | – |
| 33 | β-Elemene | 10.737 | 0.380 | – | – |
| 34 | Caryophyllene | 11.085 | 0.073 | 0.038 | 0.818 |
| 35 | T-alpha-Bergamotene | 11.109 | – | 0.015 | 0.348 |
| 36 | α Farnesene | 11.128 | 0.205 | – | 0.209 |
| 37 | D-Germacrene | 11.171 | 0.042 | – | – |
| 38 | (E)-β-Famesene | 11.257 | 0.383 | 0.043 | 0.627 |
| 39 | α-Himachalene | 11.418 | 1.744 | – | – |
| 40 | α-Humulene | 11.422 | – | – | 0.119 |
| 41 | Aromandendrene | 11.470 | 0.042 | – | 0.084 |
| 42 | β-Copaene | 11.551 | – | – | 0.022 |
| 43 | Anisylacetone | 11.700 | – | 0.259 | 0.070 |
| 44 | **ɣ-Himachalene** | 11.725 | **3.946** | – | – |

(continued)

**Table 2** (continued)

| N | Compounds | Retention time (RT) (min) | Area | | |
|---|---|---|---|---|---|
| | | | Anise | Fennel | Star anise |
| 45 | **α-Longipinene** | 11.761 | **3.810** | – | – |
| 46 | β-Bisabolene | 11.854 | 0.110 | – | 0.309 |
| 47 | δ-Amorphene | 12.003 | – | 0.103 | – |
| 48 | b-Sesquiphellandrene | 12.031 | 0.660 | – | – |
| 49 | Isospathulenol | 12.638 | 0.264 | – | – |
| 50 | α–Cadinol | 13.376 | – | 0.024 | 0.013 |
| 51 | Foeniculin | 13.489 | 0.610 | – | 0.139 |
| 52 | **Pseudoisoeugenol 2-methylbutanoate** | 14.887 | **1.878** | – | – |
| 53 | Thellungianin | 15.378 | 1.956 | – | – |
| 54 | 9-Octadecenoic acid (Z) | 15.912 | – | 0.055 | – |
| 55 | 11-Hexadecenoic acid (Z) | 15.944 | 0.043 | – | – |
| 56 | Methyl palustrate | 18.594 | 0.376 | – | – |
| 57 | Methyl abietate | 19.175 | 0.210 | – | – |
| 58 | Abietic acid | 20.796 | 0.423 | – | – |
| Oxygenated monoterpenes | | | 77.497 | 89.396 | 88.184 |
| Hydrocarbon monoterpenes | | | 1.136 | 6.684 | 7.927 |
| Oxygenated sesquitepenes | | | 2.220 | 0.283 | 0.083 |
| Hydrocarbon sesquitepenes | | | 11.395 | 0.199 | 2.536 |
| Others | | | 1.241 | 0.028 | 0.025 |
| Total | | | 93.489 | 96.59 | 98.755 |

(Table 2). The predominant component identified was T-anethole (65.9%), accompanied by other components like estragole (7%), γ-himachalene (3.9%), α-longipinene (3.8%), and pseudoisoeugenyl 2-methylbutyrate (1.9%). These results are in agreement with previous research, indeed, another research (Orav et al. 2008) analyzed essential oil from *Pimpinella anisum* fruits in several European countries and also found trans-anethole to be the major compound, underscoring its importance in PAEO. Additional research (Sun et al. 2019) confirmed that T-anethole is the primary bioactive molecule in anise EOs, contributing to its characteristic aroma and therapeutic properties, and highlighted the role of γ-himachalene as a secondary constituent influencing the oil's complexity.

The essential oils of *Illicium verum* (IVEO) were analysed and thirty distinct compounds, collectively accounting for > 98% of the total oil composition (Table 2). The chromatographic profile revealed a complex mixture of volatile components dominated by T-anethole, accounting for 66% of the total oil content, followed by D-limonene (11.3%) and estragole (6.4%). Other minor compounds identified include α-pinene, β-carene and linalool among other less abundant constituents. These results align with findings by Patra et al. (2020), who also identified T-anethole, linalool, and estragole as major components. These phenylpropanoids and monoterpenoids are recognized as the key odor-active compounds responsible for the characteristic licorice-like aroma and sweet-spicy flavor profile of star anise essential oil. Recent advances in analytical phytochemistry have significantly expanded our knowledge of *Illicium verum* essential oils, particularly regarding their key compounds and potential biological activities (Outemsaa et al. 2021).

In summary, essential oils extracted from the seeds of *Foeniculum vulgare, Pimpinella anisum,* and *Illicium verum* are predominantly composed of trans-anethole, which imparts their distinctive aroma and flavor. Trans-anethole is significantly sweeter than sugar (Lal et al. 2022), enriching the flavors of anise and fennel. Additionally, estragole lends a mild flavor, whereas fenchone—a colorless liquid with a camphor-like taste and strong odor—adds bitterness and is crucial for the biological efficacy of fennel (Singh et al. 2020). These insights enhance our understanding of the chemical profiles and potential uses of essential oils derived from these anethole-rich seeds. The structure of essential oils' major components extracted from three seeds are regrouped in the Fig. 1.

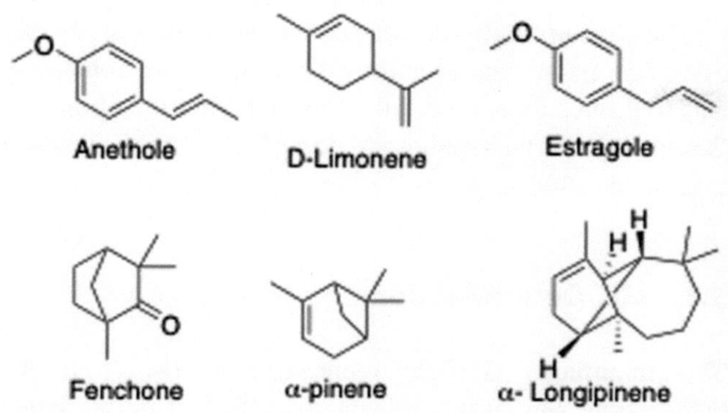


**Fig. 1** Structures of essential oils' key molecules

## 3.2 Antibacterial Activity of Studied Essential Oils

Table 3 shows a comparative study of antimicrobial efficacy of three essential oils-*Foeniculum vulgare*, *Illicium verum*, and *Pimpinella anisum*-against selected Gram-positive and Gram-negative bacteria, with Penicillin (P) and Ampicillin (AML) as reference antibiotics. Notably, the essential oils exhibit stronger inhibitory effects than the antibiotics in most cases, suggesting their potential as alternative or complementary antimicrobial agents. Against Gram-positive bacteria, *Staphylococcus aureus* and *Staphylococcus epidermidis* were significantly inhibited by all three essential oils, with zones ranging from 10.00 to 16.66 mm, whereas AML showed limited activity (7–12 mm), and Penicillin had no effect. This result was even more pronounced in Gram-negative strains, where the essential oils consistently produced larger inhibition zones than AML, particularly against *Acinetobacter baumannii* (18.00–18.50 mm vs. AML's 7 mm) and *Klebsiella pneumoniae* (~16.5 mm vs. AML's 14 mm). Remarkably, *Pimpinella anisum* exhibited an exceptionally large inhibition zone (17 mm) against *Enterobacter cloacae*, an outlier that may indicate either exceptional antimicrobial potency or a methodological anomaly requiring further validation.

The poor performance of conventional antibiotics, especially Penicillin (no activity) and Ampicillin (weak against some strains), underscores the growing challenge of antibiotic resistance, particularly in Gram-negative pathogens. This superior efficacy may be related to the complex phytochemical compositions of essential oils that can disrupt bacterial membranes, inhibit efflux pumps, or interfere with virulence factors, thereby overcoming resistance mechanisms. While promising, these in vitro results require further research to evaluate bioavailability, toxicity, and clinical applicability. Additionally, the absence of standard deviations for the essential oil measurements limits reproducibility assessment, and minimum inhibitory concentration (MIC) data would provide a more precise comparison of potency. Overall, this study highlights the potential of *Foeniculum vulgare*, *Illicium verum*, and *Pimpinella anisum* as antimicrobial agents, warranting further investigation into their mechanisms and therapeutic utility.

Studies indicate that PAEO exhibits significant antibacterial effects, particularly against Gram-positive bacteria. For instance, it has demonstrated strong inhibitory activity against *Staphylococcus aureus*, with inhibition zones

**Table 3** Antibacterial activity of *Pimpinella anisum*, *Foeniculum vulgare* and *Illicium verum* of essential oil's

| | | Inhibition zone diameter (mm)* | | | | |
|---|---|---|---|---|---|---|
| | | Essential oils | | | Antibiotic | |
| | | *Foeniculum vulgare* | *Illicium verum* | *Pimpinellea anisum* | P | AML |
| Microorganisms | **Gram-positive** | | | | | |
| | *Staphylococcus aureus* | 13.50 | 14.66 | 10.00 | NA | 12.00 ± 0.00 |
| | *Staphylococcus epidermidis* | 16.66 | 15.50 | 16.00 | NA | 7.00 ± 0.00 |
| | **Gram-negative** | | | | | |
| | *Escherichia coli* | 16.00 | 16.00 | 16.00 | NA | NA |
| | *Acinetobacter baumannii* | 18.00 | 18.33 | 18.50 | NA | 7.00 ± 0.00 |
| | *Klebsiella pneumonia* | 16.66 | 16.66 | 16.16 | NA | 14 ± 0.00 |
| | *Enterobacter cloacae* | 16.83 | 17.00 | 17.00 | NA | NA |

*Diameter of the inhibition zone including a disc diameter of 6 mm, by the disc diffusion method at a concentration of 15 μl of EOs per disc
NA: not active; P: Penicillin (25 μg/disc); AML: Amoxicillin (25 μg/disc)

reaching up to 23 mm in agar diffusion tests, suggesting bactericidal effects (Mutlu-Ingok et al. 2020; Khwaza and Aderibigbe 2025). Although specific data on *Staphylococcus epidermidis* is limited, PAEO has shown broad-spectrum activity against other Gram-positive pathogens, including *Bacillus subtilis* and *Listeria monocytogenes* (Mutlu-Ingok et al. 2020; Khwaza and Aderibigbe 2025). Against Gram-negative bacteria, anise essential oil has proven effective, particularly against *Escherichia coli*, where inhibition zones typically range between 12 and 17 mm, depending on concentration. Additionally, it has been shown to inhibit biofilm formation in other Gram-negative species, like as *Pseudomonas aeruginosa.* While research on its efficacy against *Acinetobacter baumannii* remains scarce, its effectiveness against other resistant strains suggests potential therapeutic applications (Khwaza and Aderibigbe 2025; Angane et al. 2022).

Contributing to fennel essential oils antimicrobial profile. Compared to anise and star anise oils, fennel oil exhibits moderate antimicrobial activity. Against Gram-positive bacteria, it has demonstrated inhibitory effects on *Staphylococcus aureus.* At higher concentrations, the zone of inhibition can be as large as 20 mm. However, its activity against *Staphylococcus epidermidis* has not been extensively studied, leaving a gap in current knowledge (Mutlu-Ingok et al. 2020; Wińska et al. 2019). In contrast, for Gram-negative bacteria, fennel EOs has shown moderate inhibition against *Escherichia coli* and *Klebsiella pneumoniae,* with inhibition zones typically ranging from 12 to 16 mm in disc diffusion assays. However, there is no specific data on its efficacy against *Acinetobacter baumannii* or *Enterobacter cloacae,* highlighting the need for further research to assess its full antimicrobial potential (Mutlu-Ingok et al. 2020; Wińska et al. 2019).

Star anise essential oil has displayed significant antibacterial activity against Gram-positive bacteria, exhibiting notable efficacy with inhibition zones reaching up to 24 mm for *Staphylococcus aureus.* This high activity is likely attributed to the synergistic interaction between T-anethole and other bioactive compounds present in the IVEO. However data on *Staphylococcus epidermidis* is limited, its broad-spectrum action suggests potential effectiveness (Mutlu-Ingok et al. 2020; Khwaza and Aderibigbe 2025). For Gram-negative bacteria, star anise EOs has shown strong inhibition, particularly against *Klebsiella pneumoniae* (25 mm), *Acinetobacter baumannii* (21 mm), and *Escherichia coli* (17 mm). Among these, *Klebsiella pneumoniae* is the most susceptible strain. The ability of EOs to disrupt bacterial membranes and inhibit biofilm formation further enhances its antimicrobial activity, making it a promising candidate for combating resistant infections (Khwaza and Aderibigbe 2025; Angane et al. 2022).

These results underscore the potential application of these essential oils as natural antimicrobial agents in combating bacterial infections, especially those resistant to antibiotics. Despite this, further studies are needed to optimize their use in clinical settings.

## 3.3 MIC Determination

The minimum inhibitory concentration (MIC) results demonstrate significant antibacterial effect of *anise, fennel* and *star anise* essential oils against Gram-positive and Gram-negative pathogens (Table 4). All three essential oils exhibited strong inhibitory effects, with MIC values ≤ 10 μL/mL across all tested bacterial strains. This broad-spectrum activity suggests these anethole-rich essential oils possess considerable potential as natural antimicrobial agents. Notably, Gram-positive bacteria (*Staphylococcus aureus* and *S. epidermidis*) showed slightly higher sensitivity, with MIC values consistently below 10 μL/mL. In contrast, certain Gram-negative strains (*Acinetobacter baumannii* and *Enterobacter cloacae*) required the full 10 μL/mL concentration for inhibition. This difference could be assigned to the structural complexity of the Gram-negative outer membrane, which typically confers greater resistance to hydrophobic compounds. However, the fact that these essential oils remained effective against all tested Gram-negative pathogens is particularly encouraging, given the increasing prevalence of multidrug-resistant Gram-negative infections.

**Table 4** Minimum Inhibitory Concentration (MIC) of *Pimpinella anisum, Foeniculum vulgare* and *illicium verum* essential oil's

| Microorganisms | Essential oils | | |
|---|---|---|---|
| | *Foeniculum vulgare* | *Illicium verum* | *Pimpinellea anisum* |
| | MIC (μL/mL) | MIC (μL/mL) | MIC (μL/mL) |
| **Gram-positive** | | | |
| *Staphylococcus aureus* | < 10 | < 10 | < 10 |
| *Staphylococcus epidermidis* | < 10 | < 10 | < 10 |
| **Gram-negative** | | | |
| *Escherichia coli* | < 10 | < 10 | < 10 |
| *Acinetobacter baumannii* | 10 | 10 | 10 |
| *Klebsiella pneumonia* | < 10 | < 10 | < 10 |
| *Enterobacter cloacae* | 10 | 10 | 10 |

The remarkable consistency in MIC values among the three essential oils strongly implicates anethole as the primary bioactive component responsible for their antibacterial effects. This phenylpropene derivative, which constitutes a major fraction of all three essential oils, likely exerts its antimicrobial activity through membrane disruption or interference with critical bacterial enzymes. The molecular docking results further support this mechanism, demonstrating favorable interactive effects of main compounds with key bacterial target proteins. The similar efficacy profiles suggest that minor differences in secondary metabolites between these essential oils do not significantly influence their antibacterial potency under the tested conditions.

From a clinical perspective, the observed MIC values ($\leq$ 10 μL/mL) are promising for potential therapeutic applications. The activity against problematic pathogens like *A. baumannii* and *Klebsiella pneumoniae* is particularly noteworthy, as these organisms frequently exhibit resistance to conventional antibiotics. These findings suggest that anethole-rich essential oils could serve as valuable adjuvants in combating antibiotic-resistant infections or as natural preservatives in food and pharmaceutical products. However, while these in vitro results are encouraging, further research is necessary to comprehensively evaluate their full potential. Future research should investigate time-kill kinetics to determine whether the observed inhibition is bactericidal or bacteriostatic. Synergy studies with existing antibiotics could reveal opportunities for combination therapies, while biofilm disruption assays would help assess efficacy against more clinically relevant bacterial phenotypes. Furthermore, in vivo experiments are necessary to thoroughly assess the safety and efficacy of essential oils in more complex biological systems.

Our findings align with and extend previous research. For *Pimpinella anisum*, recent studies have confirmed its strong antibacterial properties, particularly against *S. aureus* and *E. coli*. For example, a 2025 study reported significant inhibition zones for *S. aureus* and *Klebsiella* species, and highlighted the essential oil's bactericidal nature at MIC levels between 0.0625 and 2% depending on the organism tested (Qasem 2025). Another investigation found high antimicrobial activity of anise oil against both Gram-positive and negative bacteria, reinforcing its potential as a broad-spectrum antimicrobial agent (Dumitrescu et al. 2023).

Similarly, *Foeniculum vulgare* essential oil has demonstrated strong antibacterial activity against a variety of bacterial species. Multiple studies confirm its broad-spectrum efficacy:

- A study in 2015 reported MIC values ranging from 6 to 48 μg/mL for various bacteria, including *E. coli* and *Bacillus* species, indicating high susceptibility (Upadhyay 2015).
- Another study found that fennel EOs exhibited inhibition against *E. coli* and *K. pneumoniae*, with the lowest MIC for *E. coli* at 0.8% v/v, which is consistent with the low MICs observed in the present work (Gulfraz et al. 2008).

These results collectively highlight the potent and wide-ranging antimicrobial effects of fennel essential oil, especially its efficacy against clinically relevant Gram-negative bacteria.

The literature provides evidence of strong antibacterial effects for *Illicium verum* particularly against Gram-positive bacteria. Studies report that its essential oil exhibits considerable activity against *S. aureus* and *S. epidermidis*, with MICs as low as 1/1000 (v/v) for *S. aureus* and 1/100 (v/v) for *S. epidermidis* and *Enterobacter cloacae* (Outemsaa et al. 2021). However, their findings also indicated lower or no activity against *E. coli*, *K. pneumoniae*, and *A. baumannii*, which contrasts with the current study's observation of low MICs for these strains. Differences in EO composition, extraction methods or bacterial strain susceptibility may explain this discrepancy.

In fact, our results are largely in concordance with previous reports regarding the potent antibacterial properties of these essential oils, especially against Gram-positive bacteria. The slightly broader spectrum of activity observed here, particularly for *Illicium verum* against Gram-negative strains, may reflect methodological differences or variations in essential oil composition. Collectively, these findings support the continued investigation of these plant-derived essential oils as promising alternatives or adjuncts to conventional antimicrobial agents (Qasem 2025; Upadhyay 2015).

## 3.4 Molecular Docking

The strong antibacterial activity demonstrated by *Pimpinella anisum*, *Foeniculum vulgare*, and *Illicium verum* essential oils (Table 5) against *Staphylococcus aureus* and *Escherichia coli* provides a compelling rationale for investigating their mains

**Table 5** Docking results with ligands in different receptors

| | Glide gscore (Kcal mol$^{-1}$) | | |
|---|---|---|---|
| | 3Q8U | 1FJ4 | 3G7E |
| α-Longipinene | – 3.07 | – 5.59 | – 5.45 |
| α–Pinene | – 4.32 | – 5.66 | – 4.37 |
| Anethole | – 3.94 | – 6.18 | – 4.19 |
| β-Carene | – 3.25 | – 5.53 | – 4.90 |
| D-Limonene | – 4.02 | – 5.08 | – 4.56 |
| Estragole | – 3.04 | – 5.75 | – 4.27 |
| Fenchone | – 4.16 | – 6.13 | – 4.78 |
| γ–Himachalene | – 4.07 | – 5.91 | – 5.11 |

compounds interactions with specific bacterial target proteins. The selection of *S. aureus* nucleoside diphosphate kinase (NDK; PDB ID: 3Q8U), *E. coli* beta-ketoacyl-[acyl carrier protein] synthase (FabH; PDB ID: 1FJ4), and *E. coli* gyrase B (PDB ID: 3G7E) for molecular docking studies is based on their critical roles in bacterial survival and their relevance to the observed MIC results.

Molecular docking analysis revealed distinct binding affinities of the essential oil compounds across the selected bacterial targets. For *Staphylococcus aureus* nucleoside diphosphate kinase (NDK), α-pinene, fenchone, and γ-himachalene exhibited the strongest inhibitory potential, with glide scores of − 4.32, − 4.16, and − 4.07 kcal $mol^{-1}$, respectively. Anethole, while less dominant, formed a stable interaction network involving a hydrogen bond with THR 300 and π-π stacking with HIE 298 in the NDK active site, suggesting a plausible mechanism for its antibacterial contribution. In contrast, *Escherichia coli* beta-ketoacyl-[acyl carrier protein] synthase (FabH) showed high affinity for anethole, fenchone, and γ-himachalene (glide scores: − 6.18, − 6.13, and − 5.91 kcal $mol^{-1}$), whereas α-pinene and α-longipinene failed to bind, underscoring structural selectivity in this target. Similarly, for *E. coli* DNA gyrase B, α-longipinene, γ-himachalene, and β-carene emerged as the most effective inhibitors (glide scores: − 5.45, − 5.11, and − 4.90 kcal/mol), while α-pinene's lack of specific interactions correlated with its reduced efficacy. These findings collectively highlight how variations in compound structure dictate target-specific binding, aligning with the observed antibacterial activities of the essential oils (Table 5).

Anethole established one hydrogen bonds with residues THR 300 and one π-π stacking with residue HIE 298 in the active site of *Staphylococcus aureus* nucleoside diphosphate kinase (Figs. 2b and 3b). whereas, α-pinene and α-longipinene which have the most active molecules against beta-ketoacyl-[acyl carrier protein] synthase from *Escherichia coli* and *E. coli Gyrase B*, have not established any binding in these sites respectively (Figs. 2a, c and 3a, c).

Our results align with previous studies indicating that structural variations in essential oil components are critical determinants of their interactions with enzymatic targets (Noumi et al. 2023; Benlabchir et al. 2024). Notably, the interaction of anethole, while less dominant, showcased the ability to form stable hydrogen bonds and π-π stacking interactions within the NDK active site, thus elucidating a plausible mechanism for its antibacterial effect (Noumi et al. 2023). Conversely, molecular docking data for *Escherichia coli* beta-ketoacyl-acyl carrier protein synthase (FabH) revealed striking affinity for anethole and fenchone (− 6.18 and − 6.13 kcal/mol), thereby underscoring the importance of structural selectivity at this target (Dalli et al. 2022). Such findings mirror studies demonstrating the specificity of essential oil compounds' interactions with bacterial enzymes, particularly in cases where varied chemical characteristics dictate binding efficacy (Belamine et al. 2024). In stark contrast, α-pinene and α-longipinene's failure to bind further accentuates this structural dependence, as documented in previous research on various essential oils where differential binding capabilities among compounds were observed (Oliva et al. 2018). The results are expanded by similar findings from the analysis conducted on *E. coli* DNA gyrase B, which indicated that α-longipinene, γ-himachalene, and β-carene were the most effective inhibitors (− 5.45, − 5.11, and − 4.90 kcal/mol). This aligns with the understanding that specific essential oil constituents can selectively target critical bacterial proteins, influencing their antibacterial potency (Ouardi et al. 2024). Moreover, the lack of specific interactions of α-pinene with the gyrase aligns with its reduced efficiency, reinforcing the theory that effective antibacterial activity is contingent on the ability to establish consistent and strong binding interactions within target sites (Salaria et al. 2023). Such collective findings provide a robust framework for understanding how variations in the structure of essential oil compounds can dictate their target-specific binding and ultimately their antibacterial activity, thus complementing current knowledge on the pharmacodynamics of essential oils in clinical applications.

## 4 Conclusion

This comparative study of the essential oils from *Pimpinella anisum*, *Foeniculum vulgare*, and *Illicium verum* demonstrates their significant antibacterial potential, supported by chemical profiling, microbiological assays, and molecular docking analysis. GC–MS characterization confirmed that anethole is the primary bioactive constituent in all three essential oils correlating with their broad-spectrum inhibitory effects (MIC $\leq$ 10 µL/mL) against against both Gram-positive bacteria such as *Staphylococcus aureus* and *S. epidermidis*, as well as Gram-negative pathogens including *Escherichia coli* and *Acinetobacter baumannii.* Molecular docking provided mechanistic insights, revealing that anethole and other minor constituents (e.g., fenchone, γ-himachalene) target critical bacterial enzymes—including *S. aureus* nucleoside diphosphate kinase (NDK), *E. coli* beta-ketoacyl synthase (FabH), and DNA gyrase B—through stable interactions such as hydrogen bonding and π-π stacking. These findings align with the observed antibacterial activity, suggesting disruption of nucleotide

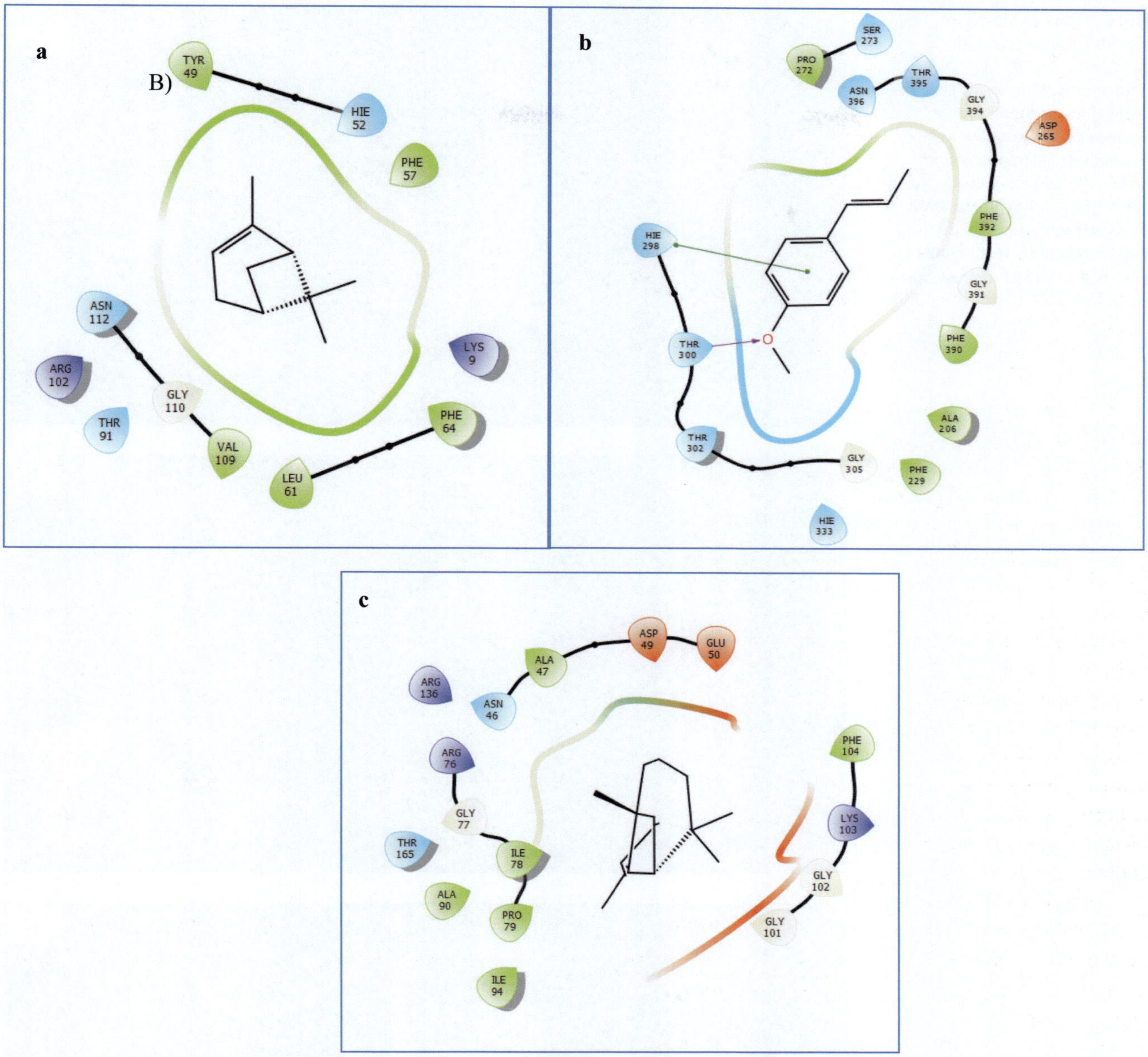


**Fig. 2** The 2D viewer of ligands interactions with the active site. **a** α-Pinene interactions with active site of *Staphylococcus aureus* nucleoside diphosphate kinase. **b** Anethole interactions with active site of beta-ketoacyl-[acyl carrier protein] synthase from *Escherichia coli*. **c** α-Longipinene interactions with active site of *Escherichia coli Gyrase B*

metabolism, fatty acid synthesis, and DNA replication as potential modes of action. While the study highlights the promise of anethole-rich essential oils as natural antimicrobial agents, further research is needed to validate these in silico predictions through enzymatic assays, synergy studies with conventional antibiotics, and in vivo efficacy testing. The consistent performance of these oils across bacterial strains positions them as candidates for applications in food preservation, pharmaceutical adjuvants, or alternative therapies against antibiotic-resistant infections. This work underscores the value of integrating phytochemical analysis with bioactivity and computational approaches to unlock the potential of plant-derived therapeutics.

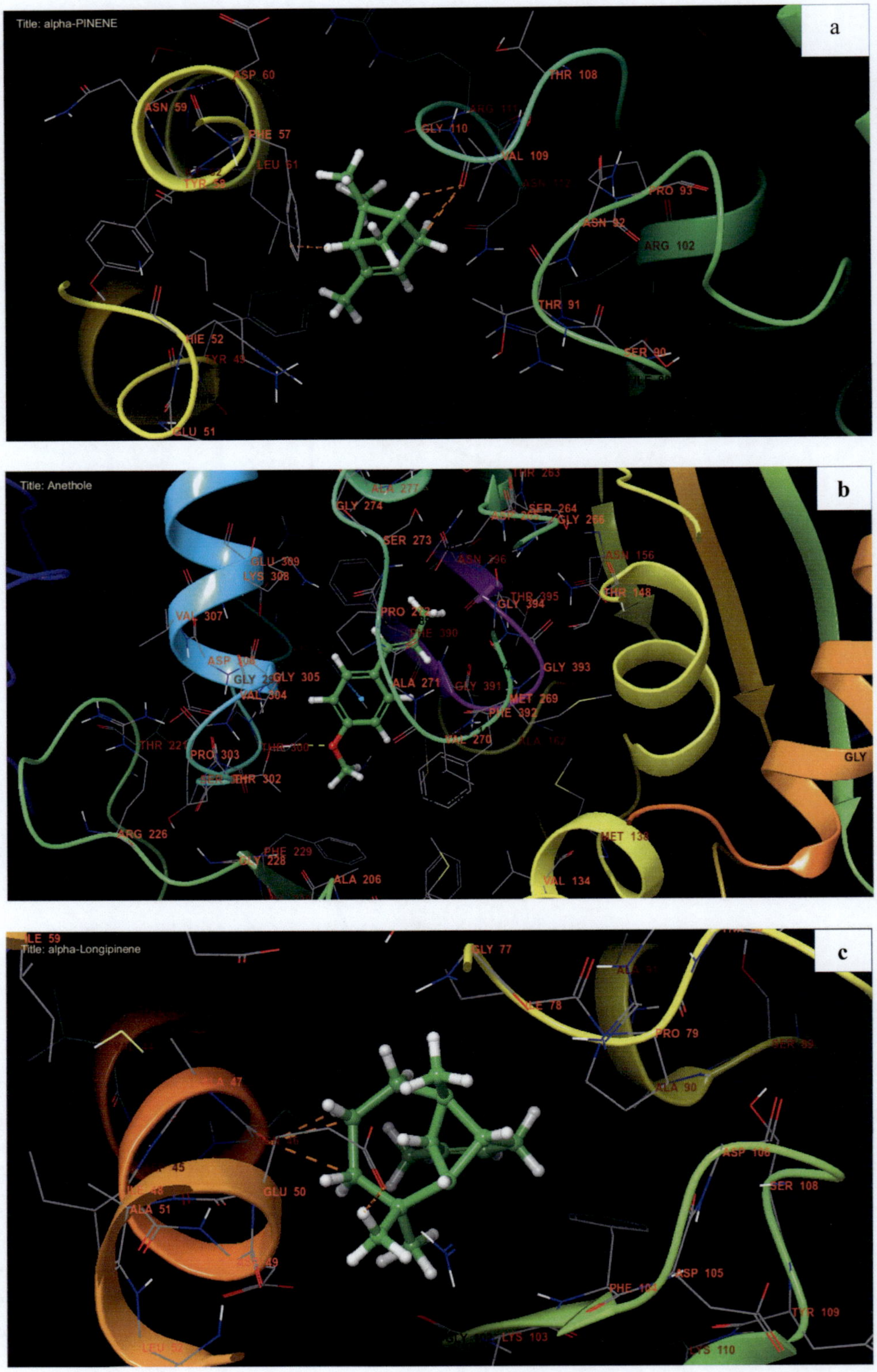


**Fig. 3** The 3D viewer of ligands interactions with the active site. **a** α-Pinene interactions with active site of *Staphylococcus aureus* nucleoside diphosphate kinase. **b** Anethole interactions with active site of beta-ketoacyl-[acyl carrier protein] synthase from *Escherichia coli*. **c** α-Longipinene interactions with active site of *Escherichia coli Gyrase B*

## References


Adams, R.P.: Identification of essential oil components by gas chromatography-/mass spectrometry. Allured Publ. Corp. **456**, 544–545 (2007)

Ait benlabchir, A., Fikri-Benbrahim, K., Moutawalli, A., Alanazi, M.M., Halmoune, A., Benkhouili, F.Z., Oubihi, A., Kabra, A., Hanoune, E., Assila, H., et al.: GC–MS characterization and bioactivity study of *Eucalyptus globulus* Labill. (Myrtaceae) essential oils and their fractions: antibacterial and antioxidant properties and molecular docking modeling. Pharmaceuticals **17**, 1552 (2024). https://doi.org/10.3390/ph17111552

Alam, P., Abdel-Kader, M.S., Alqarni, M.H., Zaatout, H.H., Ahamad, S.R., Shakeel, F.: Chemical composition of fennel seed extract and determination of fenchone in commercial formulations by GC-MS method. J. Food Sci. Technol. **56**(5), 2395–2403 (2019). https://doi.org/10.1007/s13197-019-03695-9.

Amrati, F.E.Z., Chebaibi, M., Galvaao de Azevedo, R., Conte, R., Slighoua, M., Bousta, D.: Phenolic composition, wound healing, antinociceptive, and anticancer effects of Caralluma europaea extracts. Molecules 1780 (2023)

Angane, M., Swift, S., Huang, K., Butts, C.A., Quek, S.Y.: Essential oils and their major components: an updated review on antimicrobial activities, mechanism of action and their potential application in the food industry. Foods **11**(3), 464 (2022)

Anwar, F., Ali, M., Hussain, A.I., Shahid, M.: Antioxidant and antimicrobial activities of essential oil and extracts of fennel (*Foeniculum vulgare* Mill.) seeds from Pakistan. Flavour Fragr. J. **24**, 170–176 (2009). https://doi.org/10.1002/ffj.1929

Aslan, E., Erdem, T., Orhan, İ., Şeker, M.: Essential oil composition and yield of fennel (Foeniculum vulgare Mill.) grown in different regions of Turkey. Industrial Crops Prod. **140**, 111644 (2019). https://doi.org/10.1016/j.indcrop.2019.111644

Belamine, H., Ghallab, Y.E., Zahid, B., Kadiri, F., Kasrati, A., Said, A.A.H.: Alpinia officinarum hance essential oil as potent antipseudomonal agent: chemical profile, antibacterial activity, and computational study. Chem. Biodiv. **22**(2) (2024). https://doi.org/10.1002/cbdv.202401344

Benlabchir, A.A., Fikri-Benbrahim, K., Moutawalli, A., Alanazi, M.M., Halmoune, A., Benkhouili, F.Z., Ouaritini, Z.B., et al.: GC–MS characterization and bioactivity study of eucalyptus globulus labill. (myrtaceae) essential oils and their fractions: antibacterial and antioxidant properties and molecular docking modeling. Pharmaceuticals **17**(11), 1552 (2024). https://doi.org/10.3390/ph17111552

Candan, F., Unlu, M., Tepe, B., Daferera, D., Polissiou, M., Sökmen, A., Akpulat, H.A.: Antioxidant and antimicrobial activity of the essential oil and methanol extracts of Achillea millefolium subsp. J. Ethnopharmacol. **87**, 215–220 (2003). https://doi.org/10.1016/S0378-8741(03)00149-1

Chebaibi, M., Mssillou, I., Allali, A., Bourhia, M., Bousta, D., Gonçalves, R.F.B., Giesy, J.P., et al.: Antiviral activities of compounds derived from medicinal plants against SARS-CoV-2 based on molecular docking of proteases. J. Biol. Biomed. Res. **1**(1), 10–30 (2024). https://doi.org/10.69998/j2br1

Chebaibi, M., Bourhia, M., Amrati, F.E.-Z., Slighoua, M., Mssillou, I., Aboul-Soud, M.A., Achour, S.: Salsoline derivatives, genistein, semisynthetic derivative of kojic acid, and naringenin as inhibitors of A42R profilin-like protein of monkeypox virus: in silico studies. Front. Chem. **12**, 1445606 (2024)

Coşge, B., Kiralan, M., Gürbüz, B.: Characteristics of fatty acids and essential oil from sweet fennel (Foeniculum vulgare Mill. var. dulce) and bitter fennel fruits (F. vulgare Mill. var. vulgare) growing in Turkey. Nat Prod Res. **22**(12), 1011–1016. https://doi.org/10.1080/14786410801980675. PMID: 18780240

Dadalioglu, I., Evrendlilek, G.A.: Chemical composition and antibacterial effects of essential oils of Turkish Oregano (Origanum minutiflorum), Bay Laurel (Laurus nobilis), Spanish Lavender (Lavandula stoechas L.), and fennel (Foeniculum vulgare) on common food borne pathogens. J. Agric. Food Chem. **52**(26), 8255–8260 (2004)

Dalli, M., Daoudi, N.E., Abrigach, F., Azizi, S., Bnouham, M., Kim, B., Gseyra, N., et al.: In vitro α-amylase and hemoglobin glycation inhibitory potential of nigella sativa essential oil, and molecular docking studies of its principal components. Front. Pharmacol. **13** (2022). https://doi.org/10.3389/fphar.2022.1036129

Dumitrescu, E., Muselin, F., Tîrziu, E., Folescu, M., Dumitrescu, C.S., Orboi, D.M., Cristina, R.T.: *Pimpinella anisum* L. essential oil a valuable antibacterial and antifungal alternative. Plants (Basel) **12**(13), 2428. https://doi.org/10.3390/plants12132428. PMID: 37446993; PMCID: PMC10347090

El Abdali, Y., Beniaich, G., Mahraz, A. M., El Moussaoui, A., Bin Jardan, Y.A., Akhazzane, M., Bourhia, M.: Antibacterial, antioxidant, and in silico NADPH oxidase inhibition studies of essential oils of Lavandula dentata against foodborne pathogens. Evidence-Based Complem. Alt. Med. (2023)

El Kamari, F., Chlouchi, A., El Hamzaoui, N., Harmouzi, A., Lhilali, I., El Amrani, J., El-Mouhdi, K., El Omari, H., Abdellaoui, A.: Chemical composition, antioxidant and antimicrobial activities of the essential oil of Origanum majorana growing in middle atlas of Morocco. Trop. J. Nat. Prod. Res. **7**(10), 4232–4237 (2023). https://doi.org/10.26538/tjnpr/v7i10.16

Elwardani, H., Oubihi, A., Haida, S., Ez-Zriouli, R., Kabous, K.E., Ouhssine, M.: Seasonal variation in essential oil composition of artemisia herba-alba and their effects on antioxidant, antibacterial, and antifungal activities. Chemical Data Collections. **50**, 101–118 (2024). https://doi.org/10.1016/j.cdc.2024.101118

Gulfraz, M., Mehmood, S., Minhas, N., Jabeen, N., Kausar, R., Jabeen, K., Arshad, G.: Composition and antimicrobial properties of essential oil of Foeniculum vulgare. Afr. J. Biotechnol. **7**(24), 4364–4368. ISSN 1684–5315. Available online at http://www.academicjournals.org/AJB

Hamed, S.R., Al-Wasify, R.S.: The biological activities of anise and fennel essential oils against indoor opportunistic fungi. Malay. J. Microbiol. (2023). https://doi.org/10.21161/mjm.220096.

Huang, Y., Zhao, J., Zhou, L., Wang, J., Gong, Y., Chen, X., Jiang, W., et al.: Antifungal activity of the essential oil of illicium verum fruit and its main component trans-anethole. Molecules **15**(11), 7558–7569 (2010). https://doi.org/10.3390/molecules15117558

Javidnia, K., Miri, R., Kamalinejad, M.: Chemical composition and yield of essential oil from anise (Pimpinella anisum L.) as influenced by different drying methods. J. Essential Oil Res. **19**(4), 346–349 (2007). https://doi.org/10.1080/10412905.2007.9699243

Khwaza, V., Aderibigbe, B.A.: Antibacterial activity of selected essential oil components and their derivatives: a review. Antibiotics **14**(1), 68 (2025)

Kumar, A., Alfhili, M.A., Bari, A., Ennaji, H., Ahamed, M., Bourhia, M., Abudawood, M., et al.: Apoptosis-mediated anti-proliferative activity of Calligonum comosum against human breast cancer cells, and molecular docking of its major polyphenolics to Caspase-3. Front. Cell Dev. Biol. **10**, 972111 (2022)

Lafraxo, S., El Moussaoui, A., Bin Jardan, Y.A., El Barnossi, A., Chebaibi, M., Baammi, S., Akka, A.A., Chebbac, K., Akhazzane, M., Chelouati, T., et al.: GC-MS profiling, in vitro antioxidant, antimicrobial, and in silico NADPH oxidase inhibition studies of essential oil of Juniperus thurifera bark. Evidence-Based Complement. Altern. Med. **2022**, 6305672 (2022)

Lal, M., Begum, T., Gogoi, R., Sarma, N., Munda, S., Pandey, S.K., Baruah, J., Tamang, R., Saikia, S.: Anethole rich Clausena heptaphylla (Roxb.) Wight & Arn., essential oil pharmacology and genotoxic efficiencies. Sci. Rep. **12**(1), 9978. https://doi.org/10.1038/s41598-022-13511-8

Li, Y., Wang Y., Kong, W., Yang, S., Luo, J., Yang, M.: Illicium verum essential oil, a potential natural fumigant in preservation of lotus seeds from fungal contamination. Food Chem. Toxicol. **141**, 111347 (2020). https://doi.org/10.1016/j.fct.2020.111347

Loi, N.V., Tuan, L.A., Binh, P.T., Ngọc, N.T.: Chemical components and antifungal and anti-mycotoxigenic (aflatoxin b1 and fumonisin b1) activity of star anise (illicium verum) essential oils in Yen Bai province, Vietnam. Food Sci. Technol. **43** (2023). https://doi.org/10.5327/fst.00050

Meryem, M.J., Agour, A., Allali, A., Chebaibi, M., Bouia, A.: Chemical composition, free radicals, pathogenic microbes, α-amylase and α-glucosidase suppressant proprieties of essential oil derived from Moroccan Mentha pulegium: in silico and in vitro approaches. J. Biol. Biomed. Res. **1**(1), 46–61 (2024)

Mohammed, M.J., Ebraheem, H.A.: Gas chromatography-mass spectrometry profiling of pimpinella anisum oils and its antimicrobial and antioxidant activities. Int. J. Pharmaceutical Qual. Assurance **11**(02), 257–261 (2020). https://doi.org/10.25258/ijpqa.11.2.12

Mota, A.S., Martins, M.R., Arantes, S., Lopes, V.R., Bettencourt, E., Pombal, S., Silva, L., et al.: Antimicrobial activity and chemical composition of the essential oils of portuguese foeniculum vulgare fruits. Nat. Prod. Commun. **10**(4) (2015). https://doi.org/10.1177/1934578x1501000437

Mssillou, I., Agour, A., Lyoussi, B., Derwich, E.: Chemical constituents, in vitro antibacterial properties and antioxidant activity of essential oils from *Marrubium Vulgare* L. leaves. Trop. J. Nat. Prod. Res. (TJNPR) **5**, 661–667 (2021)

Mssillou, I., Agour, A., Lefrioui, Y., Chebaibi, M.: LC-TOFMS analysis, in vitro and in silico antioxidant activity on NADPH oxidase, and toxicity assessment of an extract mixture based on Marrubium vulgare L. and Dittrichia viscosa L. J. Biol. Biomed. Res. **1**(1), 31–45 (2024). https://doi.org/10.69998/j2br2

Mutlu-Ingok, A., Firtin, B., Karbancioglu-Guler, F., Altay, F.: A study on correlations between antimicrobial effects and diffusion coefficient, zeta potential and droplet size of essential oils. Int. J. Food Eng. **16**(11), 20190354 (2020)

Noumi, E., Ahmad, I., Adnan, M., Merghni, A., Patel, H., Haddaji, N., Feo, V.D., et al.: Gc/ms profiling, antibacterial, anti-quorum sensing, and antibiofilm properties of anethum graveolens l. essential oil: molecular docking study and in-silico adme profiling. Plants **12**(10), 1997 (2023). https://doi.org/10.3390/plants12101997

Oliva, A., Costantini, S., Angelis, M.D., Garzoli, S., Božović, M., Mascellino, M.T., Ragno, R., et al.: High potency of melaleuca alternifolia essential oil against multi-drug resistant gram-negative bacteria and methicillin-resistant staphylococcus aureus. Molecules **23**(10), 2584 (2018). https://doi.org/10.3390/molecules23102584

Orav, A., Raal, A., Arak, E.: Essential oil composition of pimpinella anisuml. fruits from various European countries. Nat. Prod. Res. **22**(3), 227–232 (2008). https://doi.org/10.1080/14786410701424667.

Ouardi, M.E., Drioiche, A., Makhoukhi, F.E., Mabrouki, J., Hakmi, M., Kamaly, O.A., Belghiti, M.A.E., et al.: Chemical composition, antimicrobial, and antioxidant properties of essential oils from artemisia herba-alba asso. and artemisia huguetii caball. from morocco: in vitro and in silico evaluation. Front. Chem. **12** (2024). https://doi.org/10.3389/fchem.2024.1456684

Oubihi, A., Ouryemchi, I., Nounah, I., Tarfaoui, K., Harhar, H., Ouhssine, M., Guessous, Z.: Chemical composition, antibacterial and antifungal activities of Thymus leptobotrys Murb essential oil. Adv. Tradit. Med. **20**, 673–679 (2020). https://doi.org/10.1007/s13596-020-00488-w

Outemsaa, B., Oubihi, A., Jaber, H., Haida, S., Kenfaoui, I., Ihamdan, R., Ouhssine, M.: Chemical composition, antioxidant and antimicrobial activities of the essential oil of illicium verum. E3S Web Conf. **319**, 01052 (2021). https://doi.org/10.1051/e3sconf/202131901052

Patra, J.K., Das, G., Bose, S., Banerjee, S., Vishnuprasad, C.N., Rodriguez-Torres, M.d.P., Shin, H.: Star anise (illicium verum): chemical compounds, antiviral properties, and clinical relevance. Phytotherapy Res. **34**(6), 1248–1267 (2020). https://doi.org/10.1002/ptr.6614

Pezo, L., Lončar, B., Šovljanski, O., Tomić, A., Travičić, V., Pezo, M., Aćimović, M.: Agricultural parameters and essential oil content composition prediction of aniseed, based on growing year, locality and fertilization type-an artificial neural network approach. Life (Basel) **12**(11), 1722 (2022). https://doi.org/10.3390/life12111722. PMID: 36362877; PMCID: PMC9694612

Qasem, A.: GC-MS analysis, antimicrobial activity, and genotoxicity of *Pimpinella anisum* essential oil: In Vitro, ADMET and molecular docking investigations. Phyton-Int. J. Exp. Botany **94**(3), 809–824 (2025). https://doi.org/10.32604/phyton.2025.062683

Raal, A., Orav, A., Arak, E.: Essential oil composition of Foeniculum vulgare Mill. fruits from pharmacies in different countries. Nat. Prod. Res. **26**(13), 1173–1178 (2011). https://doi.org/10.1080/14786419.2010.535154. Epub Aug 9. PMID: 21827282

Rehman, R., Hanif, M., Mushtaq, Z., Al-Sadi, A.: Biosynthesis of essential oils in aromatic plants: a review. Food Rev. Intl. **32**, 117–160 (2016). https://doi.org/10.1080/87559129.2015.1057841

Sabzi-Nojadeh, M., Niedbała, G., Younessi-Hamzekhanlu, M., Aharizad, S., Esmaeilpour, M., Abdipour, M., Kujawa, S., Niazian, M.: Modeling the essential oil and *Trans*-Anethole yield of fennel (*Foeniculum vulgare* Mill. var. *vulgare*) by application artificial neural network and multiple linear regression methods. Agriculture **11**(12), 1191 (2021). https://doi.org/10.3390/agriculture11121191

Saibi, S., Belhadj, M., Benyoussef, E.H.: Essential oil composition of Pimpinella anisum from Algeria. Anal. Chem. Lett. **2**(6), 401–404 (2012). https://doi.org/10.1080/22297928.2012.10662624

Salaria, D., Rolta, R., Prakash, A., Fadare, O.A., Jachak, S.M., Medhi, B.: In silico and in vitro antibacterial activity of essential oil of eucalyptus globulus labill (2023). https://doi.org/10.21203/rs.3.rs-3415356/v1

Shahat, A.A., Ibrahim, A.Y., Hendawy, S.F., Omer, E.A., Hammouda, F.M., Abdel-Rahman, F., Saleh, M.A., et al.: Chemical composition, antimicrobial and antioxidant activities of essential oils from organically cultivated fennel cultivars. Molecules **16**(2), 1366–1377 (2011). https://doi.org/10.3390/molecules16021366

Shahrajabian, M.H., Sun, W., Cheng, Q.: Survey on chemical constituent, traditional and modern pharmaceutical and health benefits of Chinese star anise, a treasure from the east. Pharmacognosy Commun. **11**(1), 31–35 (2021). https://doi.org/10.5530/pc.2021.1.7

Shamsudeen, A., Palffy, C.: Star anise and oregano essential oil: a comparative evaluation of antibacterial effect. J. Emerg. Investigators (2024). https://doi.org/10.59720/24-066

Sharifi R., Kiani H., Farzaneh M., Ahmadzadeh, M.: Chemical composition of essential oils of Iranian Pimpinella anisum L. and Foeniculum vulgare Miller and their antifungal activity against postharvest pathogens. J. Essential Oil Bearing Plants **11**(5), 514–522 (2008). https://doi.org/10.1080/0972060X.2008.10643660

Shojaii, A., Fard, M.A.: Review of pharmacological properties and chemical constituents of pimpinella anisum. ISRN Pharmaceutics **2012**, 1–8 (2012). https://doi.org/10.5402/2012/510795

Simić, A., Rajković, J., Stojković, D., Đorđević, N., Kovačević, N.: Essential oil composition and antibacterial activity of Foeniculum vulgare Mill. obtained by different extraction methods. Ind. Crops Prod. **111**, 658–667 (2018). https://doi.org/10.1016/j.indcrop.2017.11.014

Singh, S., Gupta, P., Gupta, J.: Virtual structural similarity elucidates bioactivity of fenchone: a phytochemical enriched in fennel essential oil. Curr. Drug Discov. Technol. **17**(5), 619–630 (2020). https://doi.org/10.2174/1570163816666190321142826

Sun, W., Shahrajabian, M.H., Cheng, Q.: Anise (*Pimpinella anisum L.*), a dominant spice and traditional medicinal herb for both food and medicinal purposes. Cogent Biol. **5**(1) (2019). https://doi.org/10.1080/23312025.2019.1673688

Tourabi, M., Nouioura, G., Touijer, H., Baghouz, A., El Ghouizi, A., Chebaibi, M., Nafidi, H.-A., et al.: Antioxidant, antimicrobial, and insecticidal properties of chemically characterized essential oils extracted from Mentha longifolia: in vitro and in silico analysis. Plants **12**(21), 3783 (2023)

Trâm, L.H., Thong, N.V., Hao, T.P., Thảo, Đ.T.: Compositional analyses and cytotoxic activity of star anise (illicium verum) essential oils in Vietnam. JST: Eng. Technol. Sustain. Dev. **31**(4), 19–25 (2021). https://doi.org/10.51316/jst.153.etsd.2021.31.4.4

Upadhyay, R.K.: Gc-ms analysis and in vitro antimicrobial susceptibility of foeniculum vulgare seed essential oil. Am. J. Plant Sci. **06**(07), 1058–1068 (2015). https://doi.org/10.4236/ajps.2015.67110

Vecchio, M.G., Gulati, A., Minto, C., Lorenzoni, G.: Pimpinella anisum and illicium verum: the multifaceted role of anise plants. Open Agric. J. **10**(1), 81–86 (2016). https://doi.org/10.2174/1874331501610010084

Wińska, K., Mączka, W., Łyczko, J., Grabarczyk, M., Czubaszek, A., Szumny, A.: Essential oils as antimicrobial agents—myth or real alternative? Molecules **24**(11), 2130 (2019)

Zheljazkov, V.D., Astatkie, T., O'Brocki, B., Jeliazkova, E.: Essential oil composition and yield of anise from different distillation times. HortScience horts **48**(11), 1393–1396 (2013). Retrieved Mar 26, 2025, from https://doi.org/10.21273/HORTSCI.48.11.1393

# Physicochemical Study of the Medicinal Plant "*Cistus ladaniferus*" in the Mediterranean Region of Morocco

Bahia Abdelfattah, Amena Mrabet, Nisserine El Hammadi, Ayoub Simou, and Mohamed Khaddor

**Abstract**

The medicinal plant ***Cistus ladaniferus*** is a member of the Cistaceae family and a dominant shrub species endemic to the western Mediterranean region. Given its prevalence and potential ecological, aromatic, and pharmacological applications, Cistus ladaniferus has been the subject of several studies. The present study aimed to highlight the physicochemical characteristics of this species to promote its valorization. Leaves were collected from the Mediterranean region of Morocco, specifically from the Rmilat forest in Tangier. After drying in the dark, grinding, and storing, the leaves were subjected to chemical analyses in the laboratory. Physicochemical analyses were then conducted to measure pH, acidity, refractive index, Brix index, and mineral, organic, and nitrogen content. The results indicated that the leaves were acidic (pH 5.01 ± 0.2), rich in water (48.04%), and contained significant levels of mineral, organic, and nitrogenous matter (5.31, 4.69, and 9.74%, respectively). These parameters can vary depending on the time of leaf collection and the climatic and edaphic conditions.

**Keywords**

Medicinal plant · *Cistus ladaniferus* · Valorization · Physicochemical analysis

## 1 Introduction

Medicinal plants have long been a part of traditional medicine and are the foundation for the production of numerous medicines (Aafi et al. 2002). One of the important medicinal plants found in the Moroccan flora is Cistus ladaniferus. It is a shrub belonging to the family Cistaceae and is very fragrant, growing up to a height of 3 m. Its leaves are simple and opposite, without stipules, oval, linear, green, and densely covered with star-shaped hairs underneath them (Attaguile et al. 2000). The flowers are solitary and the flowering season occurs between March and June. This shrub has shiny globes covered by glands that secrete a brownish resin (Attaguile et al. 2000).

The genus Cistus comprises 16 species, which are predominantly found in the Mediterranean region (Mariotti et al. 1997). Specifically, Southern Italy, Sicily, Morocco, and Algeria are notable locations where large populations of Cistus ladaniferus can be found (Mariotti et al. 1997).

This plant, known as a calcifuge, grows in soils that are poor in limestone and thrive in acidic, well-drained soils. It is a hardy plant that can withstand temperatures as low as − 15 °C and high temperatures during summer droughts (Devesa Alcaraz 1995). This plant has been used in traditional Moroccan medicine for its anti-diarrheal (Aziz et al. 2006), anti-acid, anti-inflammatory (Youbi et al. 2016), anti-coagulant (Mekhfi et al. 2004), antioxidant (Guimarães et al. 2010), anti-ulcer (Mohammed et al. 2021), antitumor (Attaguile et al. 2004), and antispasmodic properties (Skorić et al. 2012). Additionally, it is used as an aphrodisiac owing to its diuretic effect (El Hamsas El Youbi et al. 2016).

However, the chemical composition of this plant species in the selected region has not been comprehensively examined. Therefore, it is crucial to assess the nutritional value of this species by conducting physicochemical measurements of pH,

B. Abdelfattah (✉) · A. Mrabet · N. El Hammadi · A. Simou · M. Khaddor
Laboratory of Physicochemistry of Materials, Natural Substances and Environment LAMSE, Department of Chemistry, Abdelmalek Essaâdi University, Tangier, Morocco
e-mail: Bahia.abdelfattah@uae.ac.ma


O. Jamal Eddine et al. (eds.), *Plant Pathology, Fungal Diversity, and Biotechnological Advances in Agriculture*, Sustainable Landscape Planning and Natural Resources Management, https://doi.org/10.1007/978-3-031-97666-7_5

dry matter, organic matter, mineral matter, and total nitrogenous matter. The objective of this study was to establish the physicochemical profile of this therapeutic plant species.

## 2 Material and Methods

### 2.1 Plant Collection

The aerial parts of Cistus ladaniferus were collected during the flowering period from the Tangier-Tetouan-AlHoceima region, specifically from the Rmilat (Tangier) forest situated in northern Morocco. This collection was conducted in April 2022. After returning to the laboratory, the plant material was shade-dried in a cool dark place. The resulting powder was finely ground and stored in airtight plastic bags for subsequent analysis.

### 2.2 Physicochemical Analysis

The physicochemical characteristics of ash, including dry matter (DM) content, mineral matter (MM), and organic matter (OM) content, were determined using the international standard methods of the AOAC (El Menyiy et al. 2018).

The pH was measured using a digital pH meter and the potentiometric method of the AOAC (Latimer and Horwitz 2012). The titratable acidity (TAc) was determined using the titrimetric method (Latimer and Horwitz 1990). Refractive index (RI) and Brix value (BI) were determined using a digital refractometer.

#### 2.2.1 Determination of Nitrogenous Content

Total nitrogen content was evaluated using the Kjeldahl method with a UDK 12 automatic distillation system, which is a well-established classical technique. The analysis involved three stages: mineralization, distillation, and titration.

#### 2.2.2 Determination of Ash Content

Ash content was determined by weighing the residue remaining after incineration of the sample at 525 °C for 21 h. The sample, which had a mass of 10 ± 0.02 g, was placed in a porcelain crucible and placed in an electric oven (Nabertherm) for the duration of the incineration process. The ash content was equivalent to the mass of the residue left behind after incineration.

#### 2.2.3 Determination of Mineral and Organic Matter

The plant powder was placed in a calcination oven, which was connected to a nitrogen supply, and set to a temperature of 550 °C, for four hours. The oven was a closed system, which ensured that the process was performed in an inert environment.

***Statistical analysis***

Three replicates of each experiment were performed twice under identical conditions. Statistica 8 statistical software (version 5.1 Edition 2008, StatSoft) was used to analyze the results. The findings were subjected to analysis of variance (ANOVA) using Tukey's test, with a significance level of $P < 0.05$.

## 3 Results and Discussion

The chemical and physical properties of the leaves offer a general understanding of their qualitative characteristics. Table 1 presents the pH, IR, IB, dry matter, and ash values. The pH, titratable acidity (TAc), dry matter, and ash values were 5.01 ± 0.2; 4.90 ± 0.5; 32.24 ± 2.68; and 0.58 ± 0.02, respectively.

The refractive index and Brix value are two indicators that can be used to qualitatively identify the characteristics of a sample and assess its purity. Each substance has a unique refractive index, and the degree to which the refractive index of a product matches the predicted value indicates its purity.

The refractive index and Brix index of the studied species were 1.336 and 1.43, respectively. However, another study in the Middle Atlas region reported a refractive index of 1.335 and a Brix index of 1.45. Thus, the values of the refractive index and Brix index in our study were similar to those reported in another study (Table 1).

There should be a limit on the ash content for consumption by both humans and animals (Latimer and Horwitz 1990). It also varies according to the geographical origin.

Physicochemical properties, such as titratable acidity, pH, and Brix index, vary depending on the time of leaf collection and the climatic and geological conditions of the region. The determination of these properties is crucial and serves as the first step in the qualitative assessment of the leaves of the studied species to establish an appropriate phytochemical profile for further chemical analysis.

**Table 1** Physicochemical parameters of the medicinal plant "Cistus ladaniferus"

| Harvesting site | pH | Acidity | BI % | RI | DM% | Ash% |
|---|---|---|---|---|---|---|
| Rmilate forest | 5.01 ± 0.2 | 4.90 ± 0.5 | 1.43 ± 0.7 | 1.336 ± 0.001 | 48.04% ± 2.6 | 0.58 ± 0.02 |

**RI**: Refractive index; **DM**: Dry matter; **BI**: Brix index; **A**: Acidity

**Table 2** The rate of mineral matter, organic matter, and total nitrogenous matter of the medicinal plant "Cistus ladaniferus"

| Harvesting site | MM % | MO% | TNM% |
|---|---|---|---|
| Rmilat forest | 5.31 ± 0.05 | 4.69 ± 0.03 | 9.74 ± 0.03 |

**MM**: Mineral matter; **OM**: Organic matter; **TNM**: Total nitrogenous matter

The calcination process had a significant impact on the mineral and organic matter content. According to the results obtained, the mineral matter content was 5.31%, while the organic matter content was 4.69% (Table 2). In contrast, a previous study conducted in the Middle Atlas region of Morocco reported a mineral matter content of 3.7% and organic matter content of 1.35% (El Karkouri et al. 2021). The discrepancy in the results can be attributed to various factors, such as the time of sample collection, mineral reserves of the soil, efficiency of root capture, and displacement of the matter to the aerial organs of Cistus ladaniferus.

## 4 Conclusion

The results of our investigation revealed a water content of 48.04% and an ash content of 0.58 ± 0.02%, a pH of 5.01 ± 0.2, a titratable acidity of 4.90 ± 0.5, and an IB of 1.43 ± 0.7%. Additionally, the mineral, organic, and total nitrogenous matter were 5.31 ± 0.05%, 4.69 ± 0.03%, and 9.74 ± 0.03%, respectively. Our research indicated that the leaves collected during the flowering stage showed significant results. Further studies are needed to identify the active compounds and to determine the comprehensive chemical composition of this medicinal plant. Moreover, evaluating the biological activity, including the antioxidant, antibacterial, and antifungal properties of the leaves, is crucial for exploring the potential of this species.

**Authors' Declaration** The authors hereby affirm that all the information provided in this article is their original work, and they agree to take responsibility for any claims expressed regarding its content.

**Conflict of Interest** The authors declare no conflicts of interest.

## References


Aafi, A., Taleb, M.S., Fechtal, M.: Espèces remarquables de la flore du Maroc. Centre National de la Recherche Forestière, p. 70 (2002)

Attaguile, G., et al.: Antioxidant activity and protective effect on DNA cleavage of extracts from Cistus incanus L. and Cistus monspeliensis L. Cell Biol. Toxicol. **16**(2), 83–90 (2000)

Attaguile, G., Perticone, G., Mania, G.: Savoca, Pennisi Salomone, Cistus incanus, and Cistus monspeliensis inhibited the contractile response of isolated rat smooth muscle. J. Ethnopharmacol. **92**, 245–250 (2004)

Aziz, M., Tab, N., Karim, A., Mekhfi, H., et al.: Relaxant effect of aqueous extract of Cistus ladaniferus on rodent intestinal contractions. Fitoterapia **77**, 425–428 (2006)

Devesa Alcaraz, J.A.: Vegetación y Flora de Extremadura; Universitas Editorial: Badajoz, Spain (1995)

El Hamsas El Youbi, A., El Mansouri, L., Boukhira, S., Daoudi, A., Bousta, D.: In vivo anti-inflammatory and analgesic effects of aqueous extract of Cistus ladanifer L. from Morocco. Am. J. Ther. **23**(6), e1554–e1559 (2016)

El Karkouri, J., et al.: Chemical composition, antibacterial and antifungal activity of the essential oil from Cistus Ladanifer L. Plants J. (2021)

El Menyiy, N., Al-Waili, N., El-Haskoury, R., Bakour, M.: Potential effects of Silybum marianum L. and Cistus ladaniferus L. extracts on urine volume, creatinine clearance, and renal function. Asian Pacific J. Trop. Med. **11**, 393 (2018)

Guimarães, R., Sousa, M.J., Ferreira, I.C.F.R.: Contribution of essential oils and phenolics to the antioxidant properties of aromatic plants. Ind. Crops Prod. **32**, 152–156 (2010)

Latimer, G.E., Horwitz, W.: Official methods of analysis. Association of Official Analytical Chemists (AOAC) International, (19th ed.). AOAC, Canada (1990)

Latimer, G.E., Horwitz, W.: Official methods of analysis. Association of Official Analytical Chemists (AOAC) International, (19th ed.). AOAC, Canada (2012)

Mariotti, J.P., Tomi, F., Casanova, J., Costa, J., Bernardini, A.F.: Composition of the essential oil of Cistus ladaniferus L. cultivated in Corsica (France). Flavour Fragr. J. **12**, 147–151 (1997)

Mekhfi, H., El Haouari, M., Legssyer, A., Bnouham, M., Aziz, M., Atmani, F., Remmal, A., Ziyyat, A.: Platelet anti-aggregant property of some moroccan medicinal plants. J. Ethnopharmacol. **94**, 317–322 (2004)

Mohammed, B., Said, C., Fouzia, F.R., Kawtar, F.B., Zoubida, H., Abdelilah, O., Mohammed, E., Ghizlane, E.: Chemical composition and antimicrobial activity of the essential oil of Cistus ladanifer Var. Maculatus Dun. J. Microbiol. Biotechnol. Food Sci. **2021**, 925–930 (2021)

Skorić, M., Todorović, S., Gligorijević, N., Janković, R., Živković, S., Ristić, M., Radulović, S.: Cytotoxic activity of ethanol extracts of C. creticus subsp. creticus L. grown in vitro on human cancer cell lines. Ind. Crops Prod. **38**, 153–159 (2012)

Youbi, A., El Mansouri, L., Boukhira, S., Daoudi, A., Bousta, D.: In vivo anti-inflammatory and analgesic effects of aqueous extract of Cistus ladanifer L. Moroc. Am. J. Ther. **23**, e1554–e1559 (2016)

# Abiotic Factors and Geographical Origin as Determinants of the Chemical Variability of Plant Extracts

Rachid Rahhal, Najwa Hassou, Redouane Ahnyne, Salma Khafi, Houda El Hajjouji, and Wadi Badri

**Abstract**

The article explores the chemical variability of plant natural products and their significance for ecological insight and pharmacological applications. It highlights that this chemical diversity stems from the interplay between biological factors, such as genotype and morphology, and environmental influences, including climate and geography. Understanding this diversity is crucial for adapting plants and their medicinal potential. Intraspecific chemical variability is notably affected by environmental factors; however, much research has focused on complex species, leaving gaps in knowledge about key ecological variables, such as seasonality and soil composition. Seasonal temperature changes impact water availability, which in turn influences the synthesis of secondary metabolites. Particular Asteraceae species are often perceived as chemically homogeneous, obscuring their actual biochemical diversity. The article stresses the importance of sustainable harvesting of medicinal plants, requiring a comprehensive understanding of geographical and environmental variations. Specific examples, like *Salvia miltiorrhiza* (Dan Shen), illustrate how active components can vary based on geographical origin, complicating standardization efforts for herbal preparations. In conclusion, the article advocates for a deeper exploration of how abiotic factors influence plant chemistry, which has significant implications for conservation, phytomedicine, and responses to climate change, suggesting future research directions to optimize the use of plant biodiversity.

**Keywords**

Plant extracts · Chemical variability · Biological and ecological factors · Secondary metabolites · Geographical origin

R. Rahhal (✉) · N. Hassou · R. Ahnyne
Higher Institute of Nursing and Technical Health Professions (ISPITS Rabat), Ministry of Health and Social Protection, Rabat, Morocco
e-mail: rachid.rahhal@gmail.com

R. Rahhal · S. Khafi · H. El Hajjouji · W. Badri
Laboratory of Ecology and Environment, Faculty of Sciences Ben M'sik, University Hassan II of Casablanca, Casablanca, Morocco

N. Hassou
Team Research of Virology, Oncology, and Biotechnologies, Laboratory of Virology, Oncology, Biosciences, Environment and New Energies, Faculty of Sciences and Techniques Mohammedia, University Hassan II of Casablanca, Casablanca, Morocco

Environmental Health and Biotechnology Laboratory, Faculté des Sciences Aïn Chock, University Hassan II of Casablanca, Casablanca, Morocco

R. Ahnyne
Research Laboratory in Management of Organizations, Business Law, and Sustainable Development (LARMODAD), FSJES Souissi, University Mohammed V of Rabat, Rabat, Morocco

## 1 Introduction

Documenting the variation of plant natural products is essential for ecological knowledge and pharmacological understanding (Brahmi et al. 2022), because the biological factors (genotype and morphology) interact with environmental factors (abiotic and biotic influences) to create a reservoir of chemical diversity (Maldonado et al. 2017). Long-term influences are very effective for macroevolutionary diversification, but recent perturbations could certainly spur some form of chemical divergence. Research has shown that intraspecific chemical variability primarily results from different environmental factors (Sigmund et al. 2023; Palit et al. 2022). However, since phytochemical studies have been more limited to complicated species, like community trees or shrubs, there have been few thorough reviews of sources for critical ecological factors such as seasonality, light and water



O. Jamal Eddine et al. (eds.), *Plant Pathology, Fungal Diversity, and Biotechnological Advances in Agriculture*, Sustainable Landscape Planning and Natural Resources Management, https://doi.org/10.1007/978-3-031-97666-7_6

microclimate, soil physical parameters or chemical composition (Guseva et al. 2022; Hierro and Callaway 2021; Guschina and Harwood 2021; Defossez et al. 2021; Jeger et al. 2021).

We can observe and relate direct influences known to plant physiology and biochemistry and changes in phytochemistry in general, as well as the biosynthesis of specialized metabolites more specifically (Wahab et al. 2022). There are temperature-driven seasonal changes that alter water availability and co-vary with changes in the biosynthesis of secondary metabolites (Li et al. 2023a, b; Zhang et al. 2021). Plant-derived extracts can contain various natural products, including flavonoids, terpenoids, and alkaloids. Despite the aforementioned chemical diversity, the resinous oleoresins of particular Asteraceae species, generally referred to as wormwood, mugwort, and sage, are frequently treated as containing mainly the same sameness or similar mixes of mainly flavonoids and terpenoids (Sharma et al. 2022). The secondary metabolites of different origins could be in purely qualitatively different (even in types or classes of) mixes (Kaur et al. 2022). Chemical homogeneity is partly an illusion (Soualiou et al. 2022). Even closely defined mixtures of natural products can have a high degree of alternation in biochemical action most likely due to secondary factors such as variability in composition which is not highlighted by statistics of individual phytochemical components (Kaur et al. 2022; Lanza and Dos Reis 2021).

## 2 Definition and Importance

Sustainable harvesting of species for making phytomedicine requires knowledge of geographical, environmental or seasonal variation (Boom and Kissling 2025; Madaniyazi et al. 2022; Wineland et al. 2022; Fitz Simmons 2023; Han et al. 2023; Marks et al. 2021). Information about geographical, environmental, or seasonal variability is limited and unavailable. For example, in some cases, the excipients used with the film to bind the extract limit the understanding of a geographical or seasonal variability (Zanatta et al. 2021). Hecho en México, the lesser-known cousin of El Santo Whisky and tequila, has a reputation for plants rich in metabolomes. Still, their indigenous geographic location and the fact that it is an herbal infusion or elixir may not be overlooked (Ross et al. 2023).

Abiotic factors including soil type, shade, altitude, temperature, water, exposure, and insolation may have resulted in differing biosynthetic capabilities and consequently potential application in the secondary metabolome of duplicate species that exist on two different continents (Alami et al. 2024; Jan et al. 2021). Certain environmental and geographic factors have been argued to demonstrate significant correlations with variances in secondary chemistry and bioactivity for two individuals of the same species (Zhang et al. 2024; Rangel and Bolton 2022). These environmental factors include elevation, insolation, precipitation, mean annual temperature, and night soil moisture.

Using a harmonized dataset containing global data, geographical coordinates, and altitude were significant predictors of metabolome variability (Bhunjun et al. 2024; Singh et al. 2024). The geographical coordinates, and altitude to some extent, were significant predictors of metabolite content for several compounds in a phytolacca species, consistent with previous findings with hypericum and eucalyptus species. Any speculation about the implications for ecological performance may still be incomplete (Maldonado et al. 2017); hence, the bending of these patterns could present a rich opportunity for future studies of species, on either side of the globe to investigate spatial or seasonal range more broadly, or two species in a particular location for methodological reasons.

## 3 Abiotic Factors Influencing Plant Chemistry

A large portion of the visible terrestrial environment is made up of vegetation with an active metabolism in sunlight only. Light, air, or both are used as raw materials for the photosynthetic primary production in the ecosystem, where plants fix $CO_2$ and then produce $O_2$ and $H_2O$ (Yang et al. 2021a, b). Water deficits are common in terrestrial ecosystems, and many adaptation mechanisms evolved in terrestrial plants to survive the dry season by either avoiding excessive water loss or utilizing the limited availability of water efficiently, both of which prevent drought stress (Cotrina Cabello et al. 2023). Leaf water content was studied as an indicator of drought stress, and considerable advances have been made in the remote sensing estimation of vegetation water content. Leaf water content, leaf area index, and vegetation density are suggested to enhance carbon drought stress. Environments with the same annual rainfall are divided differently throughout the year may exhibit significant variability in plant morphology, canopy structure, and physiology (Bandurska 2022; Seleiman et al. 2021). The water requirements for plants greatly differ by species, and the three elements below can influence the water needs. The quad photoacoustic sensor demonstrates a promising capacity to estimate local variability of leaf water content (Abbas et al. 2023). Variations in leaf area index and leaf water index may change the ability of biological carbon storage by altering photosynthetic assimilation and respiration (Oguz et al. 2022). Leaf water content was a good indicator for detecting drought stress at the leaf level. Moderate or vigorous water stress reduces net $CO_2$ uptake, whereas whole sun conditions with slightly lower leaf water demand can enhance carbon assimilation. Multi-year assessments found a strong relationship between leaf water and canopy-scale water stress assessments. The vegetation functions around primary productivity and land surface impedance are due to self-organizational complexity from plants with potential climate feedback (Zahedi et al. 2025; Cotrina Cabello et al. 2023).

*Salvia miltiorrhiza* (or Dan Shen) is a traditional Chinese herbal medicine with anticancer and cardioprotective effects that are included in several, some, many pharmacopeias (Zhang et al. 2024; Wang et al. 2023a, b). In terms of medicinal source, precursors, and production methods, Dan Shen can be classified into native S. miltiorrhiza in China, S. miltiorrhiza that was cultivated abroad, and synthetically produced tanshinones. The natural products in Dan Shen include phenolic acids and tanshinones, with phenolic acid being the principal class of active ingredients of *S. miltiorrhiza*, and being the active ingredients responsible for anticancer activity (Huang et al. 2024; Zhao et al. 2022). The primary active components responsible for cardiovascular protection are hydrophilic water-soluble compounds with antioxidant capabilities found in S. miltiorrhiza and Dan Shen such as, rosmarinic acid pungent acid, caffeic acid, salvianolic acid A, salvianolic acid B. The Dan Shen extracts exhibit an extensive array of enzymatic activities (Wang et al. 2025; Li et al. 2025; Yang et al. 2025; Ng et al. 2022). Nevertheless, standardized pharmacopoeial protocols and specification methods have recently been developed to ensure the quality control of common *S. miltiorrhiza* preparations and to assess the safety of Dan Shen-based products. These methods have proven effective in identifying active ingredients in tinctures, decoctions, and freeze-dried granules of Dan Shen (Zhang et al. 2019).

## 3.1 Climate Variables

The worldwide distribution of plants is quite heterogeneous. Typically, plant species from different areas have adapted to their local climate according to the laws of ecology. It is well-documented that the composition of phytochemical compounds can vary significantly among plants growing in different geo-environmental regions (Barbouchi et al. 2024; Maina et al. 2021; Yu et al. 2021). While genetic differences between varieties or chemotypes may explain some of the variation, other variations are more likely to be defined by environmental factors such as sunlight, humidity, wind, soil conditions, temperature, or rain. Only a few studies address the climate variables impacting plant extract quality (Yangui et al. 2021; Rizvi et al. 2022). Given the increasing climate variability on Earth today, this is surprising (Al-Khayri et al. 2023). It is pretty possible that climate could cause changes in total or individual metabolite content, and even the emergence of new secondary metabolites (Shi et al. 2022).

Local climate consists of climatic factors that influence species diversity, particularly at a local scale. Mean annual temperature and precipitation were considered the main climate factors influencing plant diversity globally (van der Wiel and Bintanja 2021). For example, at low altitudes, moisture and temperature increased, while climate at high altitudes contributed to low diversity. Changes in temperature and precipitation may result in species loss in community assembly, whereas spatial heterogeneity in temperature is likely to enhance richness (Geng et al. 2022; Stein et al. 2021; Peng et al. 2021). The climate-species diversity relationship was still notable after controlling climatic variables but less closely related to climate. 92. There were more genera and families in areas with complex terrain with enough water resources, with the highest richness of Quercus L. located in bimodal temperature and precipitation location (Sun et al. 2022). Unclear sentences Clarity The effect of climate condition, topography, and geographic distance on the richness of plant genera and families was confirmed (Wan et al. 2023; Pepin et al. 2022; Qin et al. 2022).

## 3.2 Soil Composition

The study at hand assessed the chemical variation of direct oils of E. caput-medusae from various locations in Ecuador. Two abiotic elements and the geographical layout of Ecuador were evaluated as potential drivers of this chemical variation. Chemical composition analyses by gas chromatography and mass spectrometry of the direct essential oils of 67 locations determined 32 constituents. Relative abundances of major components and component groups had been previously reported for several countries (Prado-Tarango 2021; Lazarus and Germino 2021). To evaluate chemical variation, the chemical composition was analysed using multivariate approaches, which allowed the generation of chemophenetic maps showing clusters of similar samples grouped based on dissimilarity. Additionally, regression trees allowed a robust and statistically defensible approach to evaluate the potential explanatory variables of the chemical variation. This approach led to identifying relevant variables associated with the chemical variation and provided insight into how those variables drive the different chemotypes within a fragmented habitat (Burneo et al. 2021).

E. caput-medusae is widely distributed in semi-arid regions, particularly prevalent in southern Ecuador and northern Peru. The species exhibits notable morphological and chemical variability throughout its distribution (Calas et al. 2025). Traditional folklore uses in treatments of gastric issues, inflammatory diseases, and ailments affecting the central nervous system have allowed for the potential studies assessing the bioactivity of E. caput-medusae extracts and derivatives. However, the inherent variability of chemical composition also complicates bioprospecting for a product or active compound. Chemophenetic mapping of the source plant may reveal some promising chemosources for the chemical investigation of E. caput-medusae, and the chemophenetic maps illustrating the chemical variability identified multiple sampling locations for potential chemosource interesting in further chemical investigation processes. However, the extensive sampling of plants is only a step in the chemosource-directed bioprospecting framework (Meineke et al. 2024; Wang et al. 2023a, b; Zhang et al. 2025; Liu et al. 2024).

### 3.3 Altitude and Topography

Linkages between the metabolic variability of secondary metabolites in plants and altitude have been reported widely in flowering plants and conifers. In keeping with the findings of Ait Said et al. (2011) on the Aleppo pine, this study concludes that the continentality of the climate and exposure significantly affect chemocomplexity, ecophysiological, and functional traits (Qaderi et al. 2023; Ibrahim et al. 2022). Indeed, at lower altitudes, populations experience harsher drought stress levels than those at mid- and high-altitude sites. So far, topography has been imputed as the key determinant of the distribution of generative populations on Mt. Nebo (Zhu et al. 2024; Pan et al. 2023; Mei et al. 2023). The facilitation of seed dispersal leads to the remixing of genetic lineages that would typically not occur if hybrids were not anthropogenically introduced from neighbouring non-glacial refugia after the last glaciation period (Khantwal et al. 2025; Srivastava et al. 2021). In any event, the ecology of the founding populations is likely to have determined their persistence and relative abundance (Kumar et al. 2023). These reflections would compromise the possible ecological relevance of low-genetic variability by default.

Prior population perspectives have indicated that genic variability of PO can be maintained even by spatially extensive plant species at the population level despite the low overall genetic differentiation and genotypic diversity (Dauphin et al. 2023). Nevertheless, when considering the reason for the production of secondary metabolites, getting enough chemical variability to support biodiversity is only a part of the story (de Barros Ruas et al. 2022; Leidinger et al. 2021). There must be an advantage for the plant that correlates with having various chemotypes (Huang et al. 2021; Yang et al. 2021a, b). The overall low divergence in ages of individuals in the population and the lack of curiosity means that the nearly complete absence of longitudinal variation in congeners for the most toxic malonylated derivatives could not be driven by neutral evolutionary rules (Salgotra and Chauhan 2023; Thompson et al. 2022).The geographic distributions of the A- and M-anchored genotypes in populations and the notable elevational cline of the population means give compelling evidence that continuingly present environmental selection is operating to maintain variation over time (Zhao et al. 2024; Wang et al. 2024; Cassamo et al. 2022; Forester et al. 2022).

The described metabolic variability of Madagascar, Madagascan derivatives (A- and M-anchored on glucose) related to altitude in the populations we studied is the first case study of an undescribed phenomenon that exists between cryptic chemotypes of co-occurring populations of the exudate gum-producing tree legume genera and the interpretation of its ecological potential (Kettenburg et al. 2023; Claudel et al. 2023).

### 3.4 Water Availability

Water is an essential ecological variable, important in limiting plant growth, and important in determining the yield, chemical composition, and production of secondary metabolites of medicinal plants (Mofokeng et al. 2020). Within medicinal plants, they focus on increasing yield through no-till and less irrigation (Oguz et al. 2022). Exogenous polyamines can limit crop water evaporation, while the use of spinning herbicides can create materials that trap moisture (Kirkham 2023; Wahab et al. 2022). Regarding the chemical composition characteristics, research has been conducted on the chemical constituents of ethyl acetate extracts from the foliage regrowth of Solidago plants growing on both wet and dry edges (Gavrilescu 2021). In some cases, concerning the origin can be viewed as distinguishable and prevailing to other contributing environmental by considering the variability of chemical composition, for example, effects of water supply (Chieb and Gachomo 2023; Xin et al. 2022; Yang et al. 2021a, b). The varying availability of water also influences the candidate chemical composition variability of the calibrating members of the coast sage leaf were distinguished among three sites that were noted for the distance they are from the ocean (Moshari-Nasirkandi et al. 2024; Barbour et al. 2022; Zhang et al. 2019). Three clades of Sagefolium were defined from 80 markers out of approximately 795, using a dataset that included all known taxonomic units and was based on a distinct morphological feature. The clasp's that are on Mt. Tamalpais, closest to the ocean, had the basal branching, supporting the hypothesis of the origin of the data set. There was the suggestion that gradual endemicity compensated biological evolution (Jažo et al. 2023; Funk et al. 2021; Krol et al. 2022). Sages in the early stages of development were the most geographically limited, leading to hypotheses that there may have been hundreds of endemic sage species at one time. There is also the possibility that the water supply may influence the chemical composition variability of the biota candidates, particularly at a moderate level, but only at the other management edges (Li and Zidorn 2022; Pérez-Ochoa et al. 2023).

## 4 Geographical Origin and Its Impact

The variability in the biological activity of secondary metabolites in plant organs is partly influenced by their biosynthetic history shaped by earlier developmental layers (Zhang et al. 2019). Chemical variability of secondary metabolites is likely due to biodiversity description (Baghel and Chander 2022; Borma et al. 2022), as well as the interharmonious fluctuations of numerous environmental abiotic and biotic factors. Geographic history will influence whether a region is tropical, wet, temperate, or arid, and to what extent the climate has been modified by the geological sub-area and the history of variants within those regions. Biodiversity may have a relationship to biotic factors (climate), geological age or landscape, and past geological events. All aspects of isotherm, isotherm spacing, isotherm profile temperature amplitude, hours when weather is cloudy, hours when sun will shine, yearly precipitation, air pressure, etc. incorporate the long-term fluctuation motion for the environment. These elements above are factored as abiotic elements of terrestrial plants. Geographical barriers impose that each life location being a filter or selector for potential growth accessions (Li et al. 2023a, b; Putra et al. 2022), therefore, unsuitable or unsuitable accessions have been discarded either due to biotic or abiotic factors (Lawrence et al. 2023).

Therefore, the chemical variability of the host plants will exist corresponding to their respective selections. The land barriers of plate tectonics slowly diversified the land masses back to a pre-Cretaceous configuration, and separated the biota that used to occupy the initial land masses into distinct isolated floras. These taxons experienced independent in-situ evolution on the now fragmented landmasses, and then evolved independently into the regional floras and fauves (Loidi and Navarro-Sánchez 2022).

### 4.1 Regional Variations in Flora

Plants are affected by a variety of abiotic factors, such as soil type, pH, mineral level, temperature, and precipitation. From these geographical factors, a distinct and unique native flora evolve with geographical differences in its flora composition (El Menyiy et al. 2021). Therefore, there are also geographical differences between plants' chemical composition, biological activity, pharmacological potential, and traditional uses (Hossain et al. 2022a, b; da Silva et al. 2022). Previous research suggested that the terpenoid composition of Erodium cicutarium native to different geographical regions, was highly disparate due to geographical origin. Primarily, this study aimed to evaluate, whether these geographical differences are attributable to between individual variability in the bioactivity screening guided terpenoid composition, differences in bioactivity screening responses between individuals of the same geography such as Mentha sp (Hossain et al. 2022a, b). Although the monoterpenoid or sesquiterpenoid emissions of Mentha sp (Hossain et al. 2022a, b) may tend to structurally differ among ecotypes, the overall patterns of different biosynthesis levels relative to plant growth stage, light, $CO_2$ or mechanical treatment tend to be similar and compacted together. Some regional variations in diastereomer shifts in biogenic phytolithic silica from the phytolith plant source of danewort (Sambucus ebulus) collected in Northeast Germany as opposed to Southeast Germany, confirmed as relationship of patch integrity, was also observed for phytoliths collected in patches (Johanna Eilers 2021). Environmental abiotic and biotic factors, such as temperature and elevation was shown to be determining factors for these geographical variations. Plants display a high degree of variability in chemical composition with respect to geographical range and environmental conditions. Factors affecting chemical profile differences relate to environmental pressures such as precipitation, temperature, latitude, altitudes, and soil type. Since climatological parameters—average temperature, average relative humidity, annual rainfall, and soil type—have been shown to play an important role in the bioactivity of plant species, it could be argued that differences in the chemical composition of species from different provinces was strongly swayed by the geographical origin of the plants. The observed difference in terpenoid diversity between plant samples from Bavaria, Germany compared to Samples from California, USA was in direct conflict with the third hypothesis. It may be possible that the lower terpenoid diversity exhibited by plants in the invaded range could reflect a genetic bottleneck via the invasion process or post-introduction evolution. Another possible explanation for the lower terpenoid diversity observed for CA plants could have been due to differences in abiotic environmental factors (Alvear et al. 2021; Pant et al. 2021; Khalid et al. 2023; Wang et al. 2021). As for abiotic environmental factors, the main difference between the two sites was average temperature.

### 4.2 Ecological Niches and Adaptations

One area of the plant ecology that stand out, and probably feature of plant ecological adaptation, are specialized metabolites. Specialized metabolites may have controlling roles in how plants grow and develop (Appolloni et al. 2022; Mielke et al. 2022). Specialized metabolites may also serve as protective measures against abiotic and biotic stresses as constitutive defenses or an induced defenses. The examination of variation in specialized metabolites and defense systems have received prominent attention to ecological plant research. In many respects, specialized metabolites vary for a multitude of reasons (McLaughlin et al. 2023; Wu et al. 2023). Some of the variation of chemical defense found may be explained by potentially; consumers of metabolic, costs to accumulate, or damage, and certainly, locally high densities of herbivores and pathogens escaping or some different environmental conditions might be the central reasons for differences in secondary metabolites found in widely separated geographic regions (Weng et al. 2021; Medema et al. 2021).

The impact of abiotic environment factors on chemical variability is often ignored because there can be a natural emphasis on biotic interactions occurring through the plant community (Mansinhos et al. 2024; Ramos et al. 2023) and therefore the understanding of competition for resources, competition with neighbouring plants, soil features such as pH and moisture, characteristics of radiation such as type of light and intensity, and abiotic stressors such as environmental toxins may be all significantly affecting specialization and chemical variability in plants (Salam et al. 2023; Liu et al. 2022a, b). The general influence of abiotic factors that affect biotic interactions, and then factors of geographic region as an ecological niche of which to consider when determining chemical variability of specialized metabolites, has been demonstrated by looking at invasive species. A study conducted an analysis of the chemical variability of specialized metabolites of a given plant, along the entirety of the range that had been established and the native range. The study found quantitative differences in the same terpenoid compounds from the samples of gymnosperms and angiosperms. The researchers stated a possible hypothesis in that the geographic origin of the plants may account for chemical variability in the plants (Paul et al. 2022; Zhang et al. 2024).

## 4.3 Case Studies of Specific Plants

Some studies have examined the impact of growth sites on the chemical variability of a single plant (Brahmi et al. 2022), while other studies have looked at both qualitative and quantitative chemical variability between species or varieties of a genus (Maldonado et al. 2017). Natural populations of those plant species used in traditional medicine and their standardized herbal products inherently experience some sort of chemical variability; in turn, increasing the chances for variability in efficacy or safety. We do not yet know the nature of the chemical variability of a single species or variety as a function of abiotic factors, especially the geographical origin. Research on the chemical variability of a single species or variety, particularly as related to drug discovery, identification of natural products, and quality control of herbal products, is incredibly valuable (Bender and Cortes-Ciriano 2021).

Mentha spp. have many options for culinary, aromatherapy and medicinal/safety related uses. Due to the induced change in chemical compositions, the resulting pharmacological and therapeutic profiles are of importance even though they may be common characteristics (Smith et al. 2023; Wang et al. 2022a, b, c; Bi et al. 2024). Peppermint and veldmint are considered carminative or digestive and are taken in gastrointestinal-related conditions like bloating, nausea, and pain. The orifices to the central nervous system and respiratory tract are generally connected to complex and reflex pathways (Dadhich et al. 2022). Therefore, it is often difficult to substantiate the safety or efficacy of one of these species, as they may have different activities based on varying geographical origins or abiotic factors. The current reviews examined these issues as the environmental variables as determinants of chemical composition variability in use in traditional medicine (Turchetto et al. 2022; Jang et al. 2022).

Peppermint, veldmint, and spearmint all appear in a number of culinary dishes. The extracts and essential oils of species of mentha exhibit nuances of activity ranging from antidiabetic, antihypertensive, anti-inflammatory, antimicrobial, antioxident, and anxiolytic (Karali and Shah 2022; Taghouti et al. 2022; Pietrzak and Nowak 2021). These species share similar propensities temporally profiled by menthol or carvone transmission which will result in a basically herbal profile (Ahad et al. 2021). The raw materials differ over harvesting times and geographical sources (Strzelczkyk et al. 2022). Peppermint is permitted for use from only a few locations around the world. Previous anecdotal evidence drawn from traditional folk medicine implied the application of other mentha species other than peppermint in traditional medicine non-limited uses. Although other species other than peppermint were attributed to methodological issues regarding quality control, the proposed reviews considered the chemical differences of alcoholic extracts from Mentha spp and the ways this works as both abiotic and core determinants of variability (Hossain et al. 2022a, b; Visković et al. 2023).

## 5 Interactions Between Abiotic Factors and Chemical Variability

Plants exhibit a uniquely high potential for producing biopharmaceuticals. Significant chemical variability occurs with plants of the same species depending on geographic origin and ecology. Biotic and abiotic ecological and co-ecological factors are as intrinsic to the biosynthetic potential of plants or populations, as is genotype (cladogenetic factors). We have considered the chemical variability of extracts from Hypericum perforatum from geographically and ecologically distant regions, in addition to exhibiting a lower rank of ecological conditions (Kasote et al. 2022; Biswas et al. 2022). The study examined extraction methods, the construction and application of an HPLC-based chromatographic multimethod for separation, characterization and quantification of six major phenolic metabolites, and about the composition of the extracts. Large chemical differences were found between extracts from countries, states, sites, and within sites. Hence, the extracts from plants of different eco-geographical and sometimes even co-geographical origins, are often significantly different in their structures and ratios of the major components which drive the bioactivity attributed to them. Information to correlate the affected variables is given (Khalid et al. 2023; Zulfiqar and Ashraf 2022; Thornley et al. 2022).

Individual plant species include a variety of complex secondary metabolites with a variety of physicochemical and biological properties that impact their medicinal use. Generally, plant secondary metabolites are sources for raw materials for pharmaceutical products and traditional medicines. These bioactive compounds are also used to produce food preservatives, fragrances, colorants, agrochemicals and plant growth regulators. In recent years dramatic interest in certain plants has surfaced in regards to their biopharmaceutical abilities (da Silva et al. 2022; Becerril-Sánchez et al. 2021). An interesting conclusion from various studies was that a considerable range of chemical variability was found in one species based on geographical origin of collections. Before utilizing a plant as a pharmaceutical there are many considerations to clarify; one being the chemotype, and this is commonly one of the basic elements to consider when investigating an individual plant as a pharmaceutical (Maldonado et al. 2017). Nevertheless, it has been pointed out that there is considerable chemical variability among different individuals of a plant species and individuals of clones. Chemical variability is particularly likely to occur when the individuals or clones originate from different co-geographical regions or geographical locations (Vargas et al. 2022; Tan et al. 2022).

## 6 Future Directions in Research

The manifold molecular functions of natural products have resulted in a vastly diverse metabolic repertoire to deal with ecological pressures (Alami et al. 2024). The great variation in secondary metabolites synthesized by plants could arise from various synthetic options, which are constrained by genetic architecture, yet a matter of gene expression regulation (Salam et al. 2023). These abiotic factors are known to affect metabolic composition and quantity, and their impact on molecular variability has been thoroughly evaluated in several species (Shahrajabian et al. 2023). Percentages or ratios of specific compounds, such as flavonoids, alkaloids, and phenolic acids, change significantly across latitudinal gradients, precipitation zones, and localities. Prior reports have addressed the effects of early life biotic vectors on subsequent plant organ metabolism; however, they have not discussed the cumulative influences of terrestrial factors on metabolic profiles (Rahman et al. 2023). In their corresponding phylogeographic settings, strictly synchronizing collections in time and place in a climatically isolated area may assist in focusing on diverse geographical and edaphic conditions. (Mansinhos et al. 2024; Terletskaya et al. 2021).

Surveying broad taxonomic ranges in multiple statistical approaches accommodates evaluating local vs. distant pairings and whether distinct masting arithmetic impacts response trajectories (Tsai and Connolly 2025). In this context, measuring discrete independent abiotic variables provides opportunities to evaluate their unique impacts intuitively, such as latitude, altitude, and precipitation. During maturation, distinct responses to temperature, damping, and drought could provide insights into the ontogenetic influence of stress or resource on metabolic decisions (Ellsworth et al. 2024). Such response measures should be tuned to the slower geoclimatic timescales, fast short-term latitudinal comparisons, ecophysiological approaches, and slow genetic studies into divergence timing and polypharmacy evolution (Ehlers et al. 2024; Hartig et al. 2024). The relative strength of broad factors, group limiting, controlling, and retention dynamics, and their impact on molecular composition or activity variability, characterize metabolism's evolutionary plasticity. Still, ways to assess the degree of universally common or non-random responses to strong or radical abiotic changes remain underexplored (Staples et al. 2025; Burnelli et al. 2025).

## 7 Conclusion

This study has shown that biogeographical patterns exist within the plant species investigated. The J. procumbens phylogeny shows two highly supported clades linking to the geographical separation but with minor CTAs found inside these clades. The presence of chemical clusters comprising two or three of the biogeographical regions supports the notion that history plays a vital role in shaping chemical profiles within this species. A few sub-clades exhibited strong chemical separations correlated with both changes in abiotic data but also co-evolution of plant and pollinator species. This suggests that the distribution of biodiversity, at least in part, is driven by the constraint of dispersal due to pre-existing geographic features (Maldonado et al. 2017). These results highlight the importance of evolutionary history over present-day geographical distributions in shaping the variability of different phytochemical classes in J. procumbens across the Indian subcontinent.

Overall, differences in climate between regions have significantly affected some classes of secondary metabolites (SMs) in J. procumbens. The weather patterns are currently changing rapidly, and better understanding how SMs react to the environment could have beneficial implications against the current backdrop of environmental change. Additionally, chemical variation is likely to start diverging more quickly within a lineage that exhibits great biogeographical endemism if co-evolution between species is acting in the system. The chemical profiles have significant implications for understanding the ecology and evolution of the species investigated and highlight the potential for future work in this area.

## References


Abbas, K., Li, J., Gong, B., Lu, Y., Wu, X., Lü, G., Gao, H.: Drought stress tolerance in vegetables: the functional role of structural features, key gene pathways, and exogenous hormones. Int. J. Mol. Sci. **24**(18), 13876 (2023)

Ahad, B., Shahri, W., Rasool, H., Reshi, Z. A., Rasool, S., Hussain, T.: Medicinal plants and herbal drugs: an overview. Medicinal and aromatic plants: healthcare and industrial applications, pp. 1–40 (2021)

Ait Said, S., Fernandez, C., Greff, S., Torre, F., Derridj, A., Gauquelin, T., Mevy, J.P.: Inter-population variability of Terpenoid composition in leaves of Pistacia lentiscus L. from Algeria: a chemoecological approach. ncbi.nlm.nih.gov (2011)

Alami, M.M., Guo, S., Mei, Z., Yang, G., Wang, X.: Environmental factors on secondary metabolism in medicinal plants: exploring accelerating factors. Med. Plant Biol. **3**(1) (2024)

Al-Khayri, J.M., Banadka, A., Nandhini, M., Nagella, P., Al-Mssallem, M.Q., Alessa, F.M.: Essential oil from Coriandrum sativum: a review on its phytochemistry and biological activity. Molecules **28**(2), 696 (2023)

Alvear, M., Santos, E., Cabezas, F., Pérez-SanMartín, A., Lespinasse, M., Veloz, J.: Geographic area of collection determines the chemical composition and antimicrobial potential of three extracts of Chilean propolis. Plants **10**(8), 1543 (2021)

Appolloni, E., Pennisi, G., Zauli, I., Carotti, L., Paucek, I., Quaini, S., Gianquinto, G., et al.: Beyond vegetables: effects of indoor LED light on specialized metabolite biosynthesis in medicinal and aromatic plants, edible flowers, and microgreens. J. Sci. Food Agric. **102**(2), 472–487 (2022)

Baghel, N.S., Chander, N.: Performance comparison of mono and poly-crystalline silicon solar photovoltaic modules under tropical wet and dry climatic conditions in east-central India. Clean Energy. oup.com (2022)

Bandurska, H.: Drought stress responses: coping strategy and resistance. Plants (2022)

Barbouchi, M., Benzidia, B., Elamrani, K., Sabiri, M., El Idrissi, M., Choukrad, M.B.: Phytochemical screening, quantitative analysis and antioxidant properties of crude extracts from stems, leaves, and flowers of three Ruta species. Kuwait J. Sci. **51**(4), 100287 (2024)

Barbour, K.M., Weihe, C., Allison, S.D., Martiny, J.B.: Bacterial community response to environmental change varies with depth in the surface soil. Soil Biol. Biochem. **172**, 108761 (2022)

Becerril-Sánchez, A.L., Quintero-Salazar, B., Dublán-García, O., Escalona-Buendía, H.B.: Phenolic compounds in honey and their relationship with antioxidant activity, botanical origin, and color. Antioxidants **10**(11), 1700 (2021)

Bender, A., Cortes-Ciriano, I.: Artificial intelligence in drug discovery: what is realistic, what are illusions? Part 2: a discussion of chemical and biological data. Drug Discov. Today. sciencedirect.com (2021)

Bhunjun, C.S., Phukhamsakda, C., Hyde, K.D., McKenzie, E.H., Saxena, R.K., Li, Q.: Do all fungi have ancestors with endophytic lifestyles? Fungal Div. **125**(1), 73–98 (2024)

Bi, J., Bossdorf, O., Liao, Z., Richards, C.L., Parepa, M., Zhao, W., Wu, J., et al.: Divergent geographic variation in above-versus below-ground secondary metabolites of Reynoutria japonica. J. Ecol. **112**(3), 514–527 (2024)

Biswas, S., Rahaman, T., Gupta, P., Mitra, R., Dutta, S., Kharlyngdoh, E., Das, M., et al.: Cellulose and lignin profiling in seven, economically important bamboo species of India by anatomical, biochemical, FTIR spectroscopy and thermogravimetric analysis. Biomass Bioenerg. **158**, 106362 (2022)

Boom, M.P., Kissling, W.D.: Making better use of tracking data can reveal the spatiotemporal and intraspecific variability of species distributions. Ecography. wiley.com (2025)

Borma, L.S., Costa, M.H., da Rocha, H.R., Arieira, J., Nascimento, N.C.C., Jaramillo-Giraldo, C., Nobre, C.A., et al.: Beyond carbon: the contributions of South American tropical humid and subhumid forests to ecosystem services. Rev. Geophys. **60**(4), e2021RG000766 (2022)

Brahmi, F., Lounis, N., Mebarakou, S., Guendouze, N., Yalaoui-Guellal, D., Madani, K., Boulekbache-Makhlouf, L., Duez, P.: Impact of growth sites on the phenolic contents and antioxidant activities of three Algerian Mentha species (M. pulegium L., M. rotundifolia (L.) Huds., and M. spicata L.) (2022)

Burnelli, M., Melelli, L., Alvioli, M.: Geomorphodiversity and anthropization indices for Italian urban areas. Geomorphology. sciencedirect.com (2025)

Burneo, J.I., Benítez, Á., Calva, J., Velastegui, P., Morocho, V.: Soil and leaf nutrients drivers on the chemical composition of the essential oil of Siparuna muricata (Ruiz & Pav.) A. DC. from Ecuador. ncbi.nlm.nih.gov (2021)

Calas, A., Schreck, E., Pont, V., Viers, J., Behra, P., Dias-Alves, M., Avellan, A., et al.: Tillandsia usneoides for atmosphere composition biomonitoring: a cross-validation study. ACS ES&T Air (2025)

Cassamo, C.T., Mangueze, A.V., Leitão, A.E., Pais, I.P., Moreira, R., Campa, C., Ramalho, J.C., et al.: Shade and altitude implications on the physical and chemical attributes of green coffee beans from Gorongosa Mountain, Mozambique. Agronomy **12**(10), 2540 (2022)

Chieb, M., Gachomo, E.W.: The role of plant growth promoting rhizobacteria in plant drought stress responses. BMC Plant Biol. springer.com (2023)

Claudel, C., Loiseau, O., Silvestro, D., Lev-Yadun, S., Antonelli, A.: Patterns and drivers of heat production in the plant genus Amorphophallus. Plant J. **115**(4), 874–894 (2023)

Cotrina Cabello, G.G., Ruiz Rodriguez, A., Husnain Gondal, A., Areche, F.O., Flores, D.D.C., Astete, J.A.Q., Cruz Nieto, D.D., et al.: Plant adaptability to climate change and drought stress for crop growth and production. CABI Rev. (2023). cabidigitallibrary.org (2023)

da Silva, W.M.F., Kringel, D.H., de Souza, E.J.D., da Rosa Zavareze, E., Dias, A.R.G.: Basil essential oil: methods of extraction, chemical composition, biological activities, and food applications. Food Bioprocess Technol. **15**(1), 1–27 (2022)

Dadhich, A., Dhiman, M., Sharma, L., Kumar, R., Jain, R., Sharma, M.M.: Unveiling the interaction of divergent abiotic stresses and their consequences in terms of bacosides in Bacopa monnieri (L.) Wettst. J. Appl. Res. Med. Aromatic Plants **31**, 100423 (2022)

Dauphin, B., Rellstab, C., Wüest, R.O., Karger, D.N., Holderegger, R., Gugerli, F., Manel, S.: Re-thinking the environment in landscape genomics. Trends in Ecol. Evol. **38**(3), 261–274 (2023)

de Barros Ruas, R., Costa, L.M.S., Bered, F.: Urbanization driving changes in plant species and communities–a global view. Global Ecol. Conserv. (2022)

Defossez, E., Pitteloud, C., Descombes, P., Glauser, G., Allard, P.M., Walker, T.W., Rasmann, S., et al.: Spatial and evolutionary predictability of phytochemical diversity. Proc. Nat. Acad. Sci. **118**(3), e2013344118 (2021)

Ehlers, L., Palm, E., Herriges, J., Bentzen, T., Suitor, M., Joly, K., Hebblewhite, M., et al.: A taste of space: remote animal observations and discrete-choice models provide new insights into foraging and density dynamics for a large subarctic herbivore. J. Anim. Ecol. **93**(7), 891–905 (2024)

El Menyiy, N., Bakour, M., El Ghouizi, A., El Guendouz, S., Lyoussi, B.: Influence of geographic origin and plant source on physicochemical properties, mineral content, and antioxidant and antibacterial activities of Moroccan Propolis. Int. J. Food Sci. **2021**(1), 5570224 (2021)

Ellsworth, W.H., Peacor, S.D., Chandler, R.B., Conner, L.M., Garrison, E.P., Miller, K.V., Cherry, M.J.: Measuring the benefit of a defensive trait: vigilance and survival probability. Ecology **105**(11), e4429 (2024)

Fitz Simmons, M.: The social and environmental relations of US agricultural regions. In: Technological change and the rural environment (2023)

Forester, B.R., Beever, E.A., Darst, C., Szymanski, J., Funk, W.C.: Linking evolutionary potential to extinction risk: applications and future directions. Front. Ecol. Environ. **20**(9), 507–515 (2022)

Funk, J.L., Larson, J.E., Ricks-Oddie, J.: Plant traits are differentially linked to performance in a semiarid ecosystem. Ecology. chapman.edu (2021)

Gavrilescu, M.: Water, soil, and plants interactions in a threatened environment. Water. mdpi.com (2021)

Geng, X., Yu, Z., Zhang, D., Li, C., Yuan, Y., Wang, X.: The influence of local background climate on the dominant factors and threshold-size of the cooling effect of urban parks. Sci. Total. Environ. **823**, 153806 (2022)

Guschina, I.A., Harwood, J.L.: Chemical diversity of lipids. cardiff.ac.uk (2021)

Guseva, K., Darcy, S., Simon, E., Alteio, L.V., Montesinos-Navarro, A., Kaiser, C.: From diversity to complexity: microbial networks in soils. Soil Biol. Biochem. **169**, 108604 (2022)

Han, W., Zhang, X., Wang, Y., Wang, L., Huang, X., Li, J., Wang, Y., et al.: A survey of machine learning and deep learning in remote sensing of geological environment: challenges, advances, and opportunities. ISPRS J. Photogramm. Remote Sens. **202**, 87–113 (2023)

Hartig, F., Abrego, N., Bush, A., Chase, J.M., Guillera-Arroita, G., Leibold, M.A., Yu, D.W., et al.: Novel community data in ecology-properties and prospects. Trends Ecol. Evol. **39**(3), 280–293 (2024)

Hierro, J.L., Callaway, R.M.: The ecological importance of allelopathy. Annu. Rev. Ecol. Evol. Syst. **52**(1), 25–45 (2021)

Hossain, C.M., Gera, M., Ali, K.A.: Current status and challenges of herbal drug development and regulatory aspect: a global perspective. Asian J. Pharm. Clin. Res. (2022)

Hossain, R., Quispe, C., Khan, R.A., Saikat, A.S.M., Ray, P., Ongalbek, D., Cho, W.C., et al.: Propolis: an update on its chemistry and pharmacological applications. Chinese Medicine **17**(1), 100 (2022b)

Huang, E., Chen, Y., Fang, M., Zheng, Y., Yu, S.: Environmental drivers of plant distributions at global and regional scales. Glob. Ecol. Biogeogr. **30**(3), 697–709 (2021)

Huang, J., Zhang, J., Sun, C., Yang, R., Sheng, M., Hu, J., Han, B., et al.: Adjuvant role of Salvia miltiorrhiza bunge in cancer chemotherapy: a review of its bioactive components, health-promotion effect and mechanisms. J. Ethnopharmacol. **318**, 117022 (2024)

Ibrahim, I.A., Jabbour, A.A., Abdulmajeed, A.M., Elhady, M.E., Almaroai, Y.A., Hashim, A.M.: Adaptive responses of four medicinal plants to high altitude oxidative stresses through the regulation of antioxidants and secondary metabolites. Agronomy **12**(12), 3032 (2022)

Jan, R., Asaf, S., Numan, M., Kim, K.M.: Plant secondary metabolite biosynthesis and transcriptional regulation in response to biotic and abiotic stress conditions. Agronomy. mdpi.com (2021)

Jang, S.J., Chung, Y., Jun, S., Won, Y.J.: Connectivity and divergence of symbiotic bacteria of deep-sea hydrothermal vent mussels in relation to the structure and dynamics of mid-ocean ridges. Front. Mar. Sci. frontiersin.org (2022)

Jažo, Z., Glumac, M., Paštar, V., Bektić, S., Radan, M., Carev, I.: Chemical composition and biological activity of Salvia officinalis L. essential oil. Plants. mdpi.com (2023)

Jeger, M., Beresford, R., Bock, C., Brown, N., Fox, A., Newton, A., Yuen, J., et al.: Global challenges facing plant pathology: multidisciplinary approaches to meet the food security and environmental challenges in the mid-twenty-first century. CABI Agric. Biosci. **2**(1), 1–18 (2021)

Johanna Eilers, E.: Intra-individual and intraspecific terpenoid diversity in Erodium cicutarium. ncbi.nlm.nih.gov (2021)

Karali, N., Shah, N.: Bolstering supplies of critical raw materials for low-carbon technologies through circular economy strategies. Energy Res. Soc. Sci. (2022)

Kasote, D., Bankova, V., Viljoen, A.M.: Propolis: chemical diversity and challenges in quality control. Phytochem. Rev. springer.com (2022)

Kaur, S., Samota, M.K., Choudhary, M., Choudhary, M., Pandey, A.K., Sharma, A., Thakur, J.: How do plants defend themselves against pathogens-biochemical mechanisms and genetic interventions. Physiol. Mol. Biol. Plants **28**(2), 485–504. springer.com (2022)

Kettenburg, A.T., Lopez, M.A., Yogendra, K., Prior, M.J., Rose, T., Bimson, S., Bailey-Serres, J., et al.: Phosphorus-starvation tolerance 1 (OsPSTOL1) is prevalent in upland rice and enhances root growth and hastens low phosphate signaling in wheat. Plant Cell Environ. **46**(7), 2187–2205 (2023)

Khalid, A., Hameed, A., Tahir, M.F.: Wheat quality: a review on chemical composition, nutritional attributes, grain anatomy, types, classification, and function of seed storage proteins in bread making. Front. Nutr. frontiersin.org (2023)

Khantwal, N., Asha, Kumar, A.: Variation in secondary metabolite content in Aesculus indica (Wall. ex Cambess.) Hook. leaves along the altitudinal gradient in Indian Central Himalaya. Vegetos (2025)

Kirkham, M.B.: Principles of soil and plant water relations (2023)

Krol, A., Kokotkiewicz, A., Luczkiewicz, M.: White sage (Salvia apiana)–a ritual and medicinal plant of the chaparral: plant characteristics in comparison with other Salvia Species. Planta Medica. thieme-connect.com (2022)

Kumar, V., Kumar, P., Bhargava, B., Sharma, R., Irfan, M., Chandora, R.: Transcriptomic and metabolomic reprogramming to explore the high-altitude adaptation of medicinal plants: a review. J. Plant Growth Regul. **42**(12), 7315–7329 (2023)

Lanza, M.G.D.B., Dos Reis, A.R.: Roles of selenium in mineral plant nutrition: ROS scavenging responses against abiotic stresses. Plant Physiol. Biochem. (2021)

Lawrence, T.J., Vilbig, J.M., Kangogo, G., Fèvre, E.M., Deem, S.L., Gluecks, I., Shacham, E., et al.: Shifting climate zones and expanding tropical and arid climate regions across Kenya (1980–2020). Reg. Environ. Change **23**(2), 59 (2023)

Lazarus, B.E., Germino, M.J.: A chemical and bio-herbicide mixture increased exotic invaders, both targeted and non-targeted, across a diversely invaded landscape after fire. Appl. Veget. Sci. google.com (2021)

Leidinger, L., Vedder, D., Cabral, J.S.: Temporal environmental variation may impose differential selection on both genomic and ecological traits. Oikos. wiley.com (2021)

Li, Y., Zidorn, C.: Seasonal variations of natural products in European herbs. Phytochem. Rev. springer.com (2022)

Li, J., Sun, R., Chen, L.: A review of thermal perception and adaptation strategies across global climate zones. Urban Climate (2023)

Li, Y., Rahman, S.U., Qiu, Z., Shahzad, S.M., Nawaz, M.F., Huang, J., Cheng, H., et al.: Toxic effects of cadmium on the physiological and biochemical attributes of plants, and phytoremediation strategies: a review. Environ. Pollut. **325**, 121433 (2023b)

Li, Q., Ren, C., Jiang, B., Wang, X., Wang, C., Zhi, X., Li, Y., et al.: Salvia miltiorrhiza Bunge root in the treatment of myocardial fibrosis: research progress and challenges. Front. Pharmacol. **16**, 1554696 (2025)

Liu, Y., Chang, Q.Y., Tang, Z.H., Wu, K.X., Abozeid, A., Mu, L.Q.: The interrelationship between latitudinal differences and metabolic differences in the natural distribution area of Tilia amurensis Rupr. Forests. mdpi.com (2022)

Liu, Z., Xu, T., Zhou, J., Chen, L.: Application of stable isotopic and elemental composition combined with random forest algorithm for the botanical classification of Chinese honey. J. Food Compos. Anal. (2022)

Liu, J.Y., Zhang, X.L., Jin, X.Y., Wang, M.T., Zhang, Y.Y., Wang, X.Y.: Nutrient allocation patterns in different aboveground organs at different reproductive stages of four introduced Calligonum species in a common garden in northwestern China. Front. Plant Sci. **15**, 1504216 (2024)

Loidi, J., Navarro-Sánchez, G.: Climatic definitions of the world's terrestrial biomes. Veget. Class. Surv. pensoft.net (2022)

Madaniyazi, L., Armstrong, B., Chung, Y., Ng, C.F.S., Seposo, X., Kim, Y., Hashizume, M., et al.: Seasonal variation in mortality and the role of temperature: a multi-country multi-city study. Int. J. Epidemiol. **51**(1), 122–133 (2022)

Maina, S., Ryu, D.H., Bakari, G., Misinzo, G., Nho, C.W., Kim, H.Y.: Variation in phenolic compounds and antioxidant activity of various organs of African cabbage (Cleome gynandra L.) accessions at different growth stages. Antioxidants **10**(12), 1952 (2021)

Maldonado, C., Barnes, C.J., Cornett, C., Holmfred, E., Hansen, S.H., Persson, C., Antonelli, A., Rønsted, N.: Phylogeny predicts the quantity of antimalarial alkaloids within the iconic yellow cinchona bark (Rubiaceae: Cinchona calisaya) (2017)

Mansinhos, I., Gonçalves, S., Romano, A.: How climate change-related abiotic factors affect the production of industrial valuable compounds in Lamiaceae plant species: a review. Front. Plant Sci. frontiersin.org (2024)

Marks, R.A., Farrant, J.M., Nicholas McLetchie, D., VanBuren, R.: Unexplored dimensions of variability in vegetative desiccation tolerance. Am. J. Botany **108**(2), 346–358 (2021)

McLaughlin, S., Zhalnina, K., Kosina, S., Northen, T.R., Sasse, J.: The core metabolome and root exudation dynamics of three phylogenetically distinct plant species. Nat. Commun. **14**(1), 1649 (2023)

Medema, M.H., de Rond, T., Moore, B.S.: Mining genomes to illuminate the specialized chemistry of life. Nat. Rev. Genetics. nih.gov (2021)

Mei, S., Perumal, M., Battino, M., Kitts, D.D., Xiao, J., Ma, H., Chen, X.: Mangiferin: a review of dietary sources, absorption, metabolism, bioavailability, and safety. Crit. Rev. Food Sci. Nutr. **63**(18), 3046–3064 (2023)

Meineke, J.K., Maxwell, L.M., Foster, A.C., McCall, L.E., Rutherford, T.K., Samuel, E.M., Jordan, S.E., et al.: Annotated bibliography of scientific research on Taeniatherum caput-medusae published from January 2010 to January 2022 (No. 2023–1089). US Geological Survey. usgs.gov (2024)

Mielke, M.M., Aggarwal, N.T., Vila-Castelar, C., Agarwal, P., Arenaza-Urquijo, E.M., Brett, B., Diversity and disparity professional interest area sex and gender special interest group, et al.: Consideration of sex and gender in Alzheimer's disease and related disorders from a global perspective. Alzheimer's Dementia **18**(12), 2707–2724 (2022)

Mofokeng, M.M., Prinsloo, G., Araya, T.H., du Plooy, P.C., Sathekge, R.N., Amoo, O.S., Martin Steyn, J.: Yield and metabolite production of pelargonium sides DC. In: Response to irrigation and nitrogen management. ncbi.nlm.nih.gov (2020)

Moshari-Nasirkandi, R., Alirezalu, A., Chamanabad, H.R.M., Amato, J., Alipour, H., Asghari, A., Moshari-Nasirkandi, A.: Screening of native wild Salvia nemorosa populations for chemical compositions, antioxidant activity and UHPLC-HRMS profiling. Sci. Rep. **14**(1), 32064 (2024)

Ng, T.W., Tan, P.P.S., Lim, H.M., Kumar, D.N., Zain, S.M., Low, T.Y., Pung, Y.F.: Potential medicinal herb for cardiovascular health: a comprehensive review on Salviae miltiorrhizae. Malays. Appl. Biol. **51**(1), 1–20 (2022)

Oguz, M.C., Aycan, M., Oguz, E., Poyraz, I., Yildiz, M.: Drought stress tolerance in plants: interplay of molecular, biochemical and physiological responses in important development stages. Physiologia. mdpi.com (2022)

Palit, K., Rath, S., Chatterjee, S., Das, S.: Microbial diversity and ecological interactions of microorganisms in the mangrove ecosystem: threats, vulnerability, and adaptations. Environ. Sci. Pollut. Res. **29**(22), 32467–32512 (2022)

Pan, L., Yang, N., Sui, Y., Li, Y., Zhao, W., Zhang, L., Tang, Z., et al.: Altitudinal variation on metabolites, elements, and antioxidant activities of medicinal plant asarum. Metabolites **13**(12), 1193 (2023)

Pant, P., Pandey, S., Dall'Acqua, S.: The influence of environmental conditions on secondary metabolites in medicinal plants: a literature review. Chem. Biodiv. wiley.com (2021)

Paul, I., Poddar Sarkar, M., Bhadoria, P.B.S.: Floral secondary metabolites in context of biotic and abiotic stress factors. Chemoecology (2022)

Peng, Q., Wang, R., Jiang, Y., Li, C.: Contributions of climate change and human activities to vegetation dynamics in Qilian Mountain National Park, northwest China. Glob. Ecol. Conserv. sciencedirect.com (2021)

Pepin, N.C., Arnone, E., Gobiet, A., Haslinger, K., Kotlarski, S., Notarnicola, C., Adler, C., et al.: Climate changes and their elevational patterns in the mountains of the world. Rev. Geophys. **60**(1), e2020RG000730 (2022)

Pérez-Ochoa, M.L., Vera-Guzmán, A.M., Mondragón-Chaparro, D.M., Sandoval-Torres, S., Carrillo-Rodríguez, J.C., Hernández-Delgado, S., Chávez-Servia, J.L.: Plant and growth condition interactions on the phenolic compound contents and antioxidant activity in Salvia circinata Cav., a medicinal sage. Diversity **15**(5), 656 (2023)

Pietrzak, W., Nowak, R.: Impact of harvest conditions and host tree species on chemical composition and antioxidant activity of extracts from Viscum album L. Molecules. mdpi.com (2021)

Prado-Tarango, D.E.: Mycorrhizal fungi on the sagebrush steppe: benefits for restoring keystone rangeland plant species. oregonstate.edu (2021)

Putra, I.D.G.A., Nimiya, H., Sopaheluwakan, A., Kubota, T., Lee, H.S., Pradana, R.P., Riama, N.F., et al.: Development of climate zones for passive cooling techniques in the hot and humid climate of Indonesia. Build. Environ. **226**, 109698 (2022)

Qaderi, M.M., Martel, A.B., Strugnell, C.A.: Environmental factors regulate plant secondary metabolites. Plants. mdpi.com (2023)

Qin, G., Meng, Z., Fu, Y.: Drought and water-use efficiency are dominant environmental factors affecting greenness in the Yellow River Basin, China. Sci. Tot. Environ. sssampling.cn (2022)

Rahman, S., Iqbal, M., Husen, A.: Medicinal plants and abiotic stress: an overview. In: Medicinal plants: their response to abiotic stress, pp. 1–34 (2023)

Ramos, Y.J., Gouvêa-Silva, J.G., de Brito Machado, D., Felisberto, J.S., Pereira, R.C., Sadgrove, N.J., de Lima Moreira, D.: Chemophenetic and chemodiversity approaches: new insights on modern study of plant secondary metabolite diversity at different spatiotemporal and organizational scales. Rev. Bras **33**(1), 49–72 (2023)

Rangel, L.I., Bolton, M.D.: The unsung roles of microbial secondary metabolite effectors in the plant disease cacophony. Curr. Opin. Plant Biol. (2022)

Rizvi, A., Ahmed, B., Khan, M.S., El-Beltagi, H.S., Umar, S., Lee, J.: Bioprospecting plant growth promoting rhizobacteria for enhancing the biological properties and phytochemical composition of medicinally important crops. Molecules **27**(4), 1407 (2022)

Ross, S.R.J., O'Connell, D.P., Deichmann, J.L., Desjonquères, C., Gasc, A., Phillips, J.N., Burivalova, Z., et al.: Passive acoustic monitoring provides a fresh perspective on fundamental ecological questions. Funct. Ecol. **37**(4), 959–975 (2023)

Salam, U., Ullah, S., Tang, Z.H., Elateeq, A.A., Khan, Y., Khan, J., Ali, S., et al.: Plant metabolomics: an overview of the role of primary and secondary metabolites against different environmental stress factors. Life **13**(3), 706 (2023)

Salgotra, R.K., Chauhan, B.S.: Genetic diversity, conservation, and utilization of plant genetic resources. Genes. mdpi.com (2023)

Seleiman, M.F., Al-Suhaibani, N., Ali, N., Akmal, M., Alotaibi, M., Refay, Y., Battaglia, M.L., et al.: Drought stress impacts on plants and different approaches to alleviate its adverse effects. Plants **10**(2), 259 (2021)

Shahrajabian, M.H., Kuang, Y., Cui, H., Fu, L., Sun, W.: Metabolic changes of active components of important medicinal plants on the basis of traditional Chinese medicine under different environmental stresses. Curr. Org. Chem. **27**(9), 782–806 (2023)

Sharma, M., Kumar, P., Verma, V., Sharma, R., Bhargava, B., Irfan, M.: Understanding plant stress memory response for abiotic stress resilience: molecular insights and prospects. Plant Physiol. Biochem. **179**, 10–24 (2022)

Shi, Y.S., Yang, L., Yu, M.F., Li, Z.H., Ke, Z.J., Qian, X.H., Ruan, X., He, L.P., Wei, F., Zhao, Y.X., Wang, Q.: Seasonal variation influences flavonoid biosynthesis path and content, and antioxidant activity of metabolites in Tetrastigma hemsleyanum Diels & Gilg. ncbi.nlm.nih.gov (2022)

Sigmund, G., Ågerstran, M., Antonelli, A., Backhaus, T., Brodin, T., Diamond, M.L., ... Groh, K.J.: Addressing chemical pollution in biodiversity research. Global Change Biol. **29**(12), 3240–3255 (2023)

Singh, R., Caseys, C., Kliebenstein, D.J.: Genetic and molecular landscapes of the generalist phytopathogen Botrytis cinerea. Mol. Plant Pathol. **25**(1), e13404 (2024)

Smith, C.F., Nikolakis, Z.L., Ivey, K., Perry, B.W., Schield, D.R., Balchan, N.R., Mackessy, S.P., et al.: Snakes on a plain: biotic and abiotic factors determine venom compositional variation in a wide-ranging generalist rattlesnake. BMC Biol. **21**(1), 136 (2023)

Soualiou, S., Duan, F., Li, X., Zhou, W.: Crop production under cold stress: an understanding of plant responses, acclimation processes, and management strategies. Plant Physiol. Biochem. researchgate.net (2022)

Srivastava, A.K., Mishra, P., Mishra, A.K.: Effect of climate change on plant secondary metabolism: an ecological perspective. In: Evolutionary diversity as a source for anticancer molecules, pp. 47–76. Academic Press (2021)

Staples, T.L., Blois, J., Cramer, K.L., Cunningham, E.T., Dornelas, M., Haberle, S.G., Pandolfi, J.M., et al.: A conceptual framework for measuring ecological novelty. Glob. Ecol. Biogeogr. **34**(2), e70005 (2025)

Stein, L., Clark, M.P., Knoben, W.J., Pianosi, F., Woods, R.A.: How do climate and catchment attributes influence flood generating processes? A large-sample study for 671 catchments across the contiguous USA. Water Resour. Res. **57**(4), e2020WR028300 (2021)

Strzelczyk, M., Lochynska, M., Chudy, M.: Systematics and botanical characteristics of industrial Hemp Cannabis Sativa L. J. Nat. Fibers (2022)

Sun, Y., Zhang, X., Ding, Y., Chen, D., Qin, D., Zhai, P.: Understanding human influence on climate change in China. Nat. Sci. Rev. **9**(3), nwab113 (2022)

Taghouti, I., Cristobal, R., Brenko, A., Stara, K., Markos, N., Chapelet, B., Bonet, J.A., et al.: The market evolution of medicinal and aromatic plants: a global supply chain analysis and an application of the Delphi method in the Mediterranean area. Forests **13**(5), 808 (2022)

Tan, Y., Du, H., Zhang, H., Fang, C., Jin, G., Chen, S., Xu, Y., et al.: Geographically associated fungus-bacterium interactions contribute to the formation of geography-dependent flavor during high-complexity spontaneous fermentation. Microbiol. Spectrum **10**(5), e01844-e1922 (2022)

Terletskaya, N.V., Korbozova, N.K., Kudrina, N.O., Kobylina, T.N., Kurmanbayeva, M.S., Meduntseva, N.D., Tolstikova, T.G.: The influence of abiotic stress factors on the morphophysiological and phytochemical aspects of the acclimation of the plant Rhodiola semenowii Boriss. Plants **10**(6), 1196 (2021)

Thompson, M.J., Capilla-Lasheras, P., Dominoni, D.M., Réale, D., Charmantier, A.: Phenotypic variation in urban environments: mechanisms and implications. Trends Ecol. Evol. **37**(2), 171–182 (2022)

Thornley, R., Gerard, F.F., White, K., Verhoef, A.: Intra-annual taxonomic and phenological drivers of spectral variance in grasslands. Remote Sens. Environ. **271** (2022)

Tsai, C.H., Connolly, S.R.: Environmental gradients linked to human impacts, not species richness, drive regional variation in community stability in coral reef fishes. Ecol. Lett. wiley.com (2025)

Turchetto, C., Segatto, A.L.A., Turchetto-Zolet, A.C.: Biotic and abiotic factors in promoting the starting point of hybridization in the Neotropical flora: implications for conservation in a changing world. Bot. J. Linn. Soc. **200**(3), 285–302 (2022)

van der Wiel, K., Bintanja, R.: Contribution of climatic changes in mean and variability to monthly temperature and precipitation extremes. Commun. Earth Environ. nature.com (2021)

Vargas, C.A., Cuevas, L.A., Broitman, B.R., San Martin, V.A., Lagos, N.A., Gaitán-Espitia, J.D., Dupont, S.: Upper environmental p $CO_2$

drives sensitivity to ocean acidification in marine invertebrates. Nat. Clim. Chang. **12**(2), 200–207 (2022)
Visković, J., Zheljazkov, V.D., Sikora, V., Noller, J., Latković, D., Ocamb, C.M., Koren, A.: Industrial hemp (Cannabis sativa L.) agronomy and utilization: a review. Agronomy **13**(3), 931. mdpi.com (2023)
Wahab, A., Abdi, G., Saleem, M.H., Ali, B., Ullah, S., Shah, W., Marc, R.A., et al.: Plants' physio-biochemical and phyto-hormonal responses to alleviate the adverse effects of drought stress: a comprehensive review. Plants **11**(13), 1620 (2022)
Wan, L., Bento, V.A., Qu, Y., Qiu, J., Song, H., Zhang, R., Wang, Q., et al.: Drought characteristics and dominant factors across China: insights from high-resolution daily SPEI dataset between 1979 and 2018. Sci. Total. Environ. **901**, 166362 (2023)
Wang, P., Lin, C.K., Wang, Y., Liu, D., Song, D., Wu, T.: Location-specific co-benefits of carbon emissions reduction from coal-fired power plants in China. Nat. Commun. **12**(1), 6948 (2021)
Wang, D., Xu, X., Zhang, H., Xi, Z., Abbott, R.J., Fu, J., Liu, J.: Abiotic niche divergence of hybrid species from their progenitors. Am. Nat. **200**(5), 634–645. uchicago.edu (2022)
Wang, J.H., Tang, Y.L., Gong, Z., Jain, R., Xiao, F., Zhou, Y., Lei, X., et al.: Characterization of protein unfolding by fast cross-linking mass spectrometry using di-ortho-phthalaldehyde cross-linkers. Nat. Commun. **13**(1), 1468 (2022b)
Wang, S., Li, Y., He, L., Yang, J., Fernie, A.R., Luo, J.: Natural variance at the interface of plant primary and specialized metabolism. Curr. Opin. Plant Biol. **67**, 102201 (2022c)
Wang, Q., Li, J., Chu, X., Jiang, X., Zhang, C., Liu, F., Pang, B., et al.: Potential chemoprotective effects of active ingredients in Salvia miltiorrhiza on doxorubicin-induced cardiotoxicity: a systematic review of in vitro and in vivo studies. Front. Cardiovasc. Med. **10**, 1267525 (2023a)
Wang, X., Li, Y., Wang, L., Duan, Y., Yao, B., Chen, Y., Cao, W.: Soil extracellular enzyme stoichiometry reflects microbial metabolic limitations in different desert types of northwestern China. Sci. Total. Environ. **874**, 162504 (2023b)
Wang, L., Zhao, S., Li, J., Zhou, G.: Metabolomics analysis reveals the metabolite profiles of Rheum tanguticum grown under different altitudinal gradients. BMC Plant Biology. springer.com (2024)
Wang, D., Zhang, X., Bian, Z., Zhang, Z., Liu, X., Zhang, C., Liu, Y., et al.: Multiple fingerprinting and cardioprotective activity of Salvia miltiorrhiza polysaccharides. Int. J. Biol. Macromol. **306**, 141468 (2025)
Weng, J.K., Lynch, J.H., Matos, J.O., Dudareva, N.: Adaptive mechanisms of plant specialized metabolism connecting chemistry to function. Nat. Chem. Biol. (2021)
Wineland, S.M., Başağaoğlu, H., Fleming, J., Friedman, J., Garza-Diaz, L., Kellogg, W., Wootten, A.M., et al.: The environmental flows implementation challenge: insights and recommendations across water-limited systems. Wiley Interdiscip. Rev. Water **9**(1), e1565 (2022)
Wu, B., Shi, S., Zhang, H., Lu, B., Nan, P.: Anabolic metabolism of autotoxic substance coumarins in plants. PeerJ. peerj.com (2023)
Xin, P., Wilson, A., Shen, C., Ge, Z., Moffett, K.B., Santos, I.R., Barry, D.A., et al.: Surface water and groundwater interactions in salt marshes and their impact on plant ecology and coastal biogeochemistry. Rev. Geophys. **60**(1), e2021RG000740 (2022)
Yang, X., Lu, M., Wang, Y., Wang, Y., Liu, Z., Chen, S.: Response mechanism of plants to drought stress. Horticulturae. mdpi.com (2021)
Yang, Z., Bai, Y., Alatalo, J.M., Huang, Z., Yang, F., Pu, X., Guo, X., et al.: Spatio-temporal variation in potential habitats for rare and endangered plants and habitat conservation based on the maximum entropy model. Sci. Total. Environ. **784**, 147080 (2021b)
Yang, K., Liu, Y.J., Zhang, J.N., Chen, Y.J., Yang, J., Xiao, J.P., Yang, H.J., et al.: Advances in the structural characterization and pharmacological activity of Salvia miltiorrhiza polysaccharides. Front. Chem. **13**, 1492533 (2025)
Yangui, I., Younsi, F., Ghali, W., Boussaid, M., Messaoud, C.: Phytochemicals, antioxidant and anti-proliferative activities of Myrtus communis L. genotypes from Tunisia. S. Afr. J. Bot. **137**, 35–45. researchgate.net (2021)
Yu, M., Gouvinhas, I., Rocha, J., Barros, A.I.: Phytochemical and antioxidant analysis of medicinal and food plants towards bioactive food and pharmaceutical resources. Scientific Reports. nature.com (2021)
Zahedi, S.M., Karimi, M., Venditti, A., Zahra, N., Siddique, K.H., Farooq, M.: Plant adaptation to drought stress: the role of anatomical and morphological characteristics in maintaining the water status. J. Soil Sci. Plant Nutr. **25**(1), 409–427 (2025)
Zanatta, A.C., Vilegas, W., Edrada-Ebel, R.A.: UHPLC-(ESI)-HRMS and NMR-based metabolomics approach to access the seasonality of byrsonima intermedia and Serjania marginata from Brazilian Cerrado flora diversity. ncbi.nlm.nih.gov (2021)
Zhang, C., Yang, D., Liang, Z., Liu, J., Yan, K., Zhu, Y., Yang, S.: Climatic factors control the geospatial distribution of active ingredients in Salvia miltiorrhiza Bunge in China. ncbi.nlm.nih.gov (2019)
Zhang, K., Zhang, Y., Sun, J., Meng, J., Tao, J.: Deterioration of orthodox seeds during ageing: Influencing factors, physiological alterations and the role of reactive oxygen species. Plant Physiol. Biochem. **158**, 475–485 (2021)
Zhang, Y., Worthy, S.J., Xu, S., He, Y., Wang, X., Song, X., Yang, J., et al.: Phytochemical diversity and their adaptations to abiotic and biotic pressures in fine roots across a climatic gradient. Sci. Total. Environ. **927**, 172051 (2024)
Zhang, Y., Du, Y., Mu, Z., Islam, W., Zeng, F., Zhang, Z., Gonzalez, N.C.: Response of soil microbial communities in extreme arid deserts to different long-term management methods. Forests **16**(2), 306 (2025)
Zhao, H., Han, B., Li, X., Sun, C., Zhai, Y., Li, M., Kai, G., et al.: Salvia miltiorrhiza in breast cancer treatment: a review of its phytochemistry, derivatives, nanoparticles, and potential mechanisms. Front. Pharmacol. **13**, 872085 (2022)
Zhao, Z., Wang, L., Chen, J., Zhang, N., Zhou, W., Song, Y.: Altitudinal variation of dragon fruit metabolite profiles as revealed by UPLC-MS/MS-based widely targeted metabolomics analysis. BMC Plant Biology. springer.com (2024)
Zhu, W., Chang, L., Zhang, M., Chen, Q., Sui, L., Shen C., et al.: Microbial diversity in mountain-dwelling amphibians: the combined effects of host and climatic factors. Iscience. cell.com (2024)
Zulfiqar, F., Ashraf, M.: Antioxidants as modulators of arsenic-induced oxidative stress tolerance in plants: an overview. J. Hazardous Mater. (2022)

# Antibacterial Activity of *Mentha spicata L.* Essential Oil Against Two Multi-resistant Pathogenic Strains Isolated from Meat Products

Sanae Berrada, Ibrahim Atemni, Azeddin El Barnossi, Ghita Benjelloun Touimi, Walid Ettahiri, Zakia Rais, and Laila Bennani

**Abstract**

Due to its high levels of protein, vitamins, minerals, and micronutrients, meat consumption is important for human growth, development, and maintenance of health. However, this composition could make the meat an important vector of many harmful agents. In fact, pathogenic bacteria such as *Escherichia coli*, *Salmonella typhimurium,* and *Staphylococcus aureus*, could contaminate meat products, and thus cause foodborne diseases. In addition, due to many factors as missuse and over-consumption of antibiotics in both humans and animals, we are witnessing the emergence and dissemination of antibiotic resistance in pathogenic bacteria. The research of new products of natural origin constitutes an important axis of research wideworld. In this study, we extracted an endemic aromatic plant wildly distributed in Morocco, *Mentha spicata* L. and evaluated its antimicrobial activity against two multi-resistant pathogenic strains (*E. coli.* and *S. aureus*) isolated from meat products. This activity was evaluated using the usual techniques of diffusion on agar and macro-dilution for the determination of the minimum inhibitory concentration. The yield was 0.89%. Regarding the aromatogram, we found that zones of inhibition diameters were exceeding 19 mm for *S. aureus*, 11 mm for *E. coli.* CMI was 25% for *E. coli*, and 100% for *S. aureus.* Furthermore, we noted a bactericidal effect against *E. coli* and bacteriostatic for *S. aureus.* These results lead to the conclusion that *Mentha spicata* L. essential oil could be considered as natural preservatives for foods from microbial spoilage in the food industries, also they can serve to inhibit or prevent the growth of food-borne pathogens.

**Keywords**

Antibacterial activity · *Mentha spicata* L. · Essential oil · Multi-resistant pathogenic strains · Meat products

## 1 Introduction

Worldwide human food culture relies on the consumption of meat products due to their high level of protein, vitamins, minerals, and micronutrients, especially tryptophan, threonine, and lysine (Njoga et al. 2023; Lecerf 2014; Quintieri et al. 2023; Chandran et al. 2023). Global meat production increased by 1.0% to 327 million tons in 2018 (Sp and Latine 2020). By 2021, France consumed 5.8 million tons of carcass equivalent in meat consumption (Rogissart 2023). In 2019, the Moroccan red meat production reached 660,000 tonnes, and the average consumption rate was 17.3 kg/capita/year (Agriculture 2023). However, because of these nutritional qualities, meat provides a very favourable environment for microbial growth (Madoroba et al. 2021).

S. Berrada (✉) · A. El Barnossi
Biotechnology, Environment, Agro-Food and Health Laboratory, Faculty of Science Dhar El Mahraz, Sidi Mohamed Ben Abdellah University, Fez, Morocco
e-mail: sanae.berrada@usmba.ac.ma

S. Berrada · I. Atemni · W. Ettahiri · Z. Rais
Laboratory of Electrochemistry, Modelisation and Environnment Engineering, Faculty of Sciences Dhar El Mahraz, Sidi Mohamed Ben Abdellah University, Fez, Morocco

A. El Barnossi
Biological Engineering Laboratory, Faculty of Sciences and Techniques, Sultan Moulay Slimane University, Beni Mellal, Morocco

G. Benjelloun Touimi
Laboratory of Human Pathology Biomedicine and Environment, Faculty of Medicine and Pharmacy, Sidi Mohammed Ben Abdellah University, Fez, Morocco

L. Bennani
The Superior Institute of Nursing Professions and Health Technology of Fez (ISPITS), Fez, Morocco



O. Jamal Eddine et al. (eds.), *Plant Pathology, Fungal Diversity, and Biotechnological Advances in Agriculture*, Sustainable Landscape Planning and Natural Resources Management, https://doi.org/10.1007/978-3-031-97666-7_7

Minced meat, a very perishable food, is a danger to the consumer as it is consumed undercooked (Alves 2018). Numerous food-borne illness episodes have occurred globally due to consumption of unhealthy meat (Lianou et al. 2022). Globally, WHO estimates that 33 million healthy years of life are lost every year due to food-borne diseases worldwide (Dubois-brissonnet et al. 2022). In Morocco, the seriousness and frequency of TIAC make it a serious public health problem, and requires reporting as a communicable disease (Rachidi and Latrache 2020). Furthermore, the rise in antibiotic resistance among food-borne pathogens is a major concern, which can lead to therapeutic impairment for the most severe infections (Dubois-Brissonnet and Guillier 2020). By using natural food additives with antimicrobial activities, microbiological quality and shelf life of perishable foods can be improved. The attention has been focused on aromatic plants and their products, including essential oils (EOs) (Al-Maqtari et al. 2022; Balasubramaniam and Sudha Rani Ramakrishnan 2022). Several studies are currently being conducted to determine the effectiveness of essential oils in foods (Sateriale et al. 2023; Rafiq et al. 2016). This research examines the antibacterial effects of *Mentha spicata L.* essential oil on two multi-resistant pathogens.

## 2 Material and Method

### 2.1 Plant Aromatic Chosen

Based on bibliographical data, availability, frequency of appearance, yield, probable therapeutic interest of aromatic plants, the EO chosen in our study is extracted from mint (*Mentha spicata*), collected in Fez city. This plant is known classified to belong to the *Magnoliopsida* class, *Lamiale* order, *Lamiaceae* family, *Mentha genus* and *spicata* species (Uikey 2024; Yousefian et al. 2023). The biological properties of this perennial aromatic plant are mainly due to the polyphenols in a hybrid between *M. Longifolia L* and *M. Suaveolens* (EL-Haoud et al. 2018). This plant is a perennial that grows rhizomatically and upright and is commonly grown as a culinary herb or ground cover. It usually cultivates between 30–100 cm, but can attain heights of 2 ft. Both the stems and foliage of this plant are either hairy orunhairy, and it has a fleshy underground rhizome that spreads widely. The leaves are 5–9 cm long and 1.5–3 cm wide, with a serration along the margin and a strong spearmint fragrance and taste. The Spearmint flower spikes are slim and have a broad appearance, either pink or white, and are 2.5–3 mm long. The distinctive feature of *M. spicata* is its leaves that are sparsely hairy and nearly sessile (Taneja and Chandra 2012; Kumar et al. 2011; Nanda et al. 2022) (Fig.1).

### 2.2 Oil Extraction

For extraction of aromatic plant's essential oil, the samples (leaves, roots, stems) were rinsed and dried using absorbent paper. Subsequently, a Bicol hydro-distillation flask was filled with 100 g of sample and 500 ml of distilled water. The distillate became completely clear after being heated for 4 h using a bag warmer. Water vapor was condensed and split into two organic and oily phases as a result of this step. In order to produce more essential oil, diethyl ether is employed to extract the aqueous phase. The solvent's evaporation results in the fusion of the second organic phase with the first one. To dry the essential oil, anhydrous sodium sulphate ($Na_2SO_4$) was employed in the final step. Subsequently, it was preserved at 4 °C for future usage. To avoid environmental pollution and economic concerns, the rotary evaporator was employed to re-use all organic solvents used in other extraction processes (Hjouji et al. 2021).

### 2.3 Yield

The ratio of extracted essential oil and plant material was used to determine the essential oil yield, which is expressed as a percentage (%), using the following formula (Houmy et al. 2020):

$$R(\%) = \frac{Mass\ of\ dry\ plant\ material(g)}{Mass\ of\ essentiial(g)} \times 100.$$

### 2.4 Identification of Polyphenols

Identifying our plant essential oil had been achieved through Gas Chromatography-Mass Spectrometry (GC–MS), which is an immensely potent and adaptable instrumental technique. This process is based on the principle of heating a mixture, and it involves separating chemical mixtures through the GC component, followed by the identification of individual components at a molecular level through the MS component. By comparing the retention index of different compounds with the retention index given in the literature, the molecules in *Mentha spicata*'s essential oil were identified (Ranjan Maji et al. 2023).

### 2.5 Bacterial Strains

Two bacterial strains isolated from meat products samples, analysed in previous studies, were selected for antibacterial activity: *Escherichia coli and Staphylococcus aureus*. These strains' resistance to various antibiotics was significant. After

**Fig. 1** *Mentha spicata L.* (Nanda et al. 2022)

identifying them using bacteriological classical methods, they were preserved with 30% glycerol in LB broth in a freezer that was kept at − 80 ± 5 °C. Fresh cultures were obtained by inoculating these strains in PCA medium for 18–24 h before testing (Berrada et al. 2016).

## 2.6 Antibacterial Activity

Vincent method and broth microdilution were employed to determine the antibacterial activity of the *Mentha spicata L.* essential oil. For Vincent method, the pre-prepared bacterial suspension was placed on each plate, containing medium agar (Muller-Hinton). After 5 min of impregnation, the excess inoculum is removed by aspiration. Then, using sterile forceps, the surface of each plate was aseptically coated with sterile filter paper discs with a diameter of 6 mm (bioMérieux). Then, the discs received an impregnation with oil at a rate of 5 μl/disc. The oil was used in goal state and diluted by adding 1 mL of DMSO to 0.06 g of oil. Three replicates were performed for each bacterial strain. Also, three controls were made: a negative control with 10 μl of sterile distilled water in the presence of 10% DMSO and two antibiotic discs as positive controls, Amoxicillin (30 μg) and Cefotaxim (30 μg). Immediately after waiting for 15–20 min in room temperature, the plates were placed upside-down and left to incubate at 37 °C for 18–24 h. A calliper was used to measure the inhibition aureoles after the incubation period (Fadil et al. 2018). Three replicate experiments were conducted to determine the inhibition zone's mean diameter. The Microdilution test was used in quantitative testing as per CLSI standard methods with some adjustments. At a final concentration of 0.5%, Tween 80 was added to the tryptic soy broth medium. A stable essential oil emulsion was created by utilizing 30 min of sonication and 8 min of vortex homogenization. Micro tubes of 1 mL were used to prepare serial dilutions of the essential oil in broth, which ranged between 0.06 and 0.96% (v/v). Subsequently, a 96-well microplate was used to redistribute each dilution. Each well received $10^5$ CFU/mL of the tested strain (*E. coli* or *S. aureus*) to produce a final concentration of $10^4$ CFU/mL and a final volume of 200 μL per well per well. The first column of the microplate was filled with the inoculum and culture medium, while the ampicillin antibiotic control ranged from 64 to 0.5 μg/mL was placed in the final column. Finally, microplate was incubated with a sterile film cover for 18–24 h at 35 ± 2 °C. After that, wells were filled with 20 μL of resazurin, which was later incubated at 35 ± 2 °C. 2 °C for half an hour. Based on growth visually, we determined the MIC as the lowest concentration of essential oil (David et al. 2021). The degree of bacterial inhibition was determined by no growth (−), some growth (−/ + ), or growth (+) in areas treated with essential oil. The method used to determine the MBC was to inoculate nutrient agar from range wells without visible growth. Ten microliters were applied on the surface of nutrient agar for each negative well and spread into a single streak before incubating at 37C for 24 h. The minimum inhibition concentration (MIC) and minimum lethal concentration (MLC) were determined by the lowest concentration that completely inhibited bacterial growth after 8 and 48 h, respectively (Kowalska-Krochmal and Dudek-Wicher 2021).

## 2.7 Statistical Analysis

To convey the data, the mean standard deviation (SD) was utilized. Using a Tukey's test, one-way ANOVA was utilized to statistically analyse differences between comparisons in R Statistical software (version 2.15.2). Difference was considered significant at $p < 0.05$.

## 3 Results and Discussion

### 3.1 Essential Oil Yield

The oil yield of the aerial part of *M. spicata* L. was 0.89%. Our outcome is closely aligned with those of Mata et al. in Portugal (Mata et al. 2007) and Allali et al. in Algeria (Allali et al. 2013), and differs from those found in others countries: 1.97% in Tunisia (Selles et al. 2018), 0.4% in Bangladesh (Chowdhury et al. 1970), 0.3% in India (Mkolo et al. 2011). According to Hussain et al., this disparity could be explained by seasonal variation. *Mentha species* full blooming in summer resulted in the highest essential oil yield, compared to winter when they reach the end of their growing cycle (Zouari-Bouassida et al. 2018; Meloni et al. 2015). Essential oil yields from *M. spicata L.* were 1.2% in summer and 0.95% in winter (Hussain et al. 2010). The extraction of essential oil in May month, may have contributed to the yield found in our study.

### 3.2 Essential Oil Composition

As illustrated in Table 1, the GC–MS analysis of the essential oil of *M. spicata* L. resulted in the detection of seventeen components representing 94.31% of the total, with ith D-Carvone (73.44%) followed respectively by D-Limonene (7.57%), Eucalyptol (5.52%), cis-Dihydrocarvone (2.81%), (Z)-3-Phenylacrylaldehyde (2.49%), Azulene (1.41%), and 2-Cyclohexen-1-ol, 2-methyl-5-(1-methylethenyl) (1.07%). Results found by other authors (Amel et al. 2022; Bardaweel et al. 2018; Brahmi et al. 2016) reported that the major constituents of the oil were carvone (51.7%), cis-carveol (24.3%), while Linalool was the primary constituent of *M. spicata* oil grown wild in Greece (Kofidis et al. 2004). Geographic location, climate changes, soil, plant collection time, and dynamic extraction conditions all affect chemical composition (Amel et al. 2022; Chen et al. 2022).

**Table 1** Chemical profile of *Mentha spicata L.* essential oil

| | Chemical composition | Acceptance ratio (%) |
|---|---|---|
| 1 | o-Xylene | 0.82 |
| 2 | Bicyclo[3.1.1]heptane, 6,6-dimethyl-2-methy | 0.45 |
| 3 | Beta.-Myrcene | 0.84 |
| 4 | D-Limonene | 7.51 |
| 5 | Eucalyptol | 5.52 |
| 6 | Endo-Borneol | 0.80 |
| 7 | Azulene | 1.81 |
| 8 | Cis-Dihydrocarvone | 2.41 |
| 9 | 2-Cyclohexen-1-ol,2-methyl-5-(1-methylethenyl) | 1.07 |
| 10 | D-Carvone | 73.44 |
| 11 | (Z)-3-Phenylacrylaldehyde | 2.49 |
| 12 | Cyclohexanol, 2-methyl-5-(1-methylethenyl) | 0.68 |
| 13 | 2-Cyclohexen-1-ol, 2-methyl-5-(1-methylethenyl) | 0.36 |
| 14 | Cyclobuta[1,2:3,4]dicyclopentene, decahydro | 0.52 |
| 15 | 2-Cyclopenten-1-one, 3-methyl-2-(2-pentenyl) | 0.35 |
| 16 | Caryophyllene | 0.53 |
| 17 | 2,6-Di-tert-butyl-4-methyl-phenol | 0.40 |
| | | 100.00 |

### 3.3 Antibacterial Activity

Essential oil of spearmint was tested against two multi-resistant pathogenic strains isolated from meat products: *E. coli* and *S. aureus*. According to Benjelloun et al., food-borne pathogens, particularly *E. coli*, have been detected in food products, humans, and environmental samples worldwide (Benjelloun Touimi et al. 2020). Over the last decade, *E. coli* and *S. aureus* have seen an increase in drug resistance, as documented by a number of studies (Kennedy et al. 2018; Yar et al. 2022; Seow et al. 2021). The issue of antibiotic resistance is both complex and challenging. The problem could be caused by the unjustified use of antibiotics in both clinical and agricultural settings, along with low socioeconomic status, poor sanitation, and poor hygiene. The lack of regular culture of zoonotic bacterial pathogens and their resistance to commonly used antibiotics could also be the cause (Manyi-Loh et al. 2018; Muteeb et al. 2023; Aguilar et al. 2023).

*M. spicata L.* essential oil's antibacterial effects were evaluated using broth microdilution and disk diffusion methods. As depicted in Tables 2 and 3, *M. spicata L.* had a more powerful impact on Gram+ bacteria than on Gram− bacteria. The inhibition zone measured varied between 16 and 22 mm. Similar findings have been reported by previous authors (Amel et al. 2022; Mahboubi and Haghi 2008). Despite this, the diameter of the inhibition zone is not uniform across studies. The primary factors that determine the antibacterial actions of EO include their chemical composition, the quantity of major compounds, biogeography, and evaluation methods. Moreover, the EOs activity as an antibacterial is also vary due to the difference in the cellular structure of the bacterial cell, such as Gram (+) and Gram (−) bacteria, which differ in the structure of the cell membrane. In addition, EO components, cell wall compositions, and the type of microorganism are all factors that contribute to antimicrobial activity (Al-Maqtari et al. 2022; Khorshidian et al. 2018). Furthermore, the capability of *M. spicata* L. to combat against bacteria may be influenced by the percentage of oxygenated monoterpenes and their hydrocarbons. The presence of Carvone, Limonene, and Eucalyptol is also responsible for the positive antibacterial effect of OEM's.

As shown, in Table 3, the MIC values of essential oil against *S. aureus*, and *E. coli* were 1 and 0.25 μg/mL. Our

**Table 2** Antimicrobial activity (diameter of the inhibition zones in mm) of *Mentha spicata L.* essential oil

| | Essential oil | References standard | |
|---|---|---|---|
| | *Mentha spicata* | Amoxicillin | Cefotaxim |
| *S. aureus* | 22 ± 1.13 | 9.0 ± 1.05 | 23.3 ± 0.6 |
| *E. coli* | 16 ± 1.41 | 7.2 ± 1.15 | 20.1 ± 1.28 |

Values (Diameter of inhibition zone (mm) including disc diameter of 6 mm) are mean ± standard deviation of three different samples of *M. spicata* essential oil, analysed individually in triplicate

results differ to those reported by Mahboubi and Haghi, who reported that MIC values of essential oil against *S. aureus*, and *E. coli* were 0.5, and 4 µg/mL, respectively (Mahboubi and Haghi 2008). The percentage of phenols, terpenes, and aldohetones is thought to be a factor in the antibacterial proprieties of essential oils (Bajalan and Ghasemi Pirbalouti 2014; Guimarães et al. 2019). Carvone and limonene can play a role in the antibacterial activity of our essential oil.

According to reports, carvone is a highly effective antimicrobial agent in various plants. However, the exact mechanism of its antibacterial activity is still unclear. There are several ways of reducing the pH gradient across the membrane, including destabilizing the phospholipidic bilayer structure, interacting with membrane enzymes and proteins, and acting as a proton exchanger (Shahbazi et al. 2015).

## 4 Conclusion

The findings of this research indicate that *M. spicata L.* essential oil has the ability to combat common food-borne bacteria that are associated with outbreaks (*S. aureus* and *E. coli*). EO is a natural preservative that can enhance food stability during storage, prevent the growth of foodborne pathogens, alter micro-organisms, and reduce oxidation, making it a great choice for food preservation. The safety and potential side effects of future studies are expected to be the focus.

**Acknowledgements** The research work was carried out thanks to contribution of all authors.

**Data Availability** The corresponding author can provide all data upon demand.
**Conflicts of Interest** No conflicts of interest have been proclaimed by the authors.

**Table 3** Antibacterial activity of *M. spicata* L. essential oil indicated as MBC/MIC

| | MIC (%) | MBC | MBC/MIC | Action |
|---|---|---|---|---|
| *S. aureus* | 100% | + | > 1 | Bacteriostatic |
| *E. coli* | 25% | – | 1 | Bactericidal |

## References

Agriculture: Meat production in Morocco from 2018 to 2020 by type (2023). https://www.statista.com/statistics/1298022/

Aguilar, G.R., Swetschinski, L.R., Weaver, N.D., et al.: The burden of antimicrobial resistance in the Americas in 2019: a cross-country systematic analysis. Lancet Reg. Heal. Am. **25**, 1–16 (2023). https://doi.org/10.1016/j.lana.2023.100561

Allali, H., Chikhi, I., Dib, M.A., et al.: Antioxidant activity and chemical analysis of *Mentha spicata* cultivated from west northern region of Algeria by headspace solid phase micro-extraction and hydro-distillation. Nat. Prod. Indian J. **9**, 258–263 (2013)

Al-Maqtari, Q.A., Rehman, A., Mahdi, A.A., et al.: Application of Essential Oils as Preservatives in Food Systems: Challenges and Future Prospectives—A Review. Springer, Netherlands (2022)

Alves, M.M.: Discolouration complaints in minced meat. Do the food miles travelling affect the customer complaints? (2018)

Amel, B.H., Souhila, M.H.E., Mahmoud, D.M.: Biological valorization and characterization of essential oil of Algerian *Mentha spicata* L. Int. J. Minor Fruits Med. Aromat. Plants **8**, 56 (2022). https://doi.org/10.53552/ijmfmap.8.1.2022.56-60

Bajalan, I., Ghasemi Pirbalouti, A.: Variation in antibacterial activity and chemical compositions of essential oil from different populations of myrtle. Ind. Crops Prod. **61**, 303–307 (2014). https://doi.org/10.1016/j.indcrop.2014.07.023

Balasubramaniam, G., Sudha Rani Ramakrishnan, U.A.: Chapter 13—Opportunities and challenges of plant extracts in food industry. B Plant Extr Appl Food Ind Pages 295–315 (2022)

Bardaweel, S.K., Bakchiche, B., ALSalamat HA, et al.: Chemical composition, antioxidant, antimicrobial and Antiproliferative activities of essential oil of *Mentha spicata* L. (Lamiaceae) from Algerian Saharan atlas. BMC Complement. Altern. Med. **18**, 1–8 (2018). https://doi.org/10.1186/s12906-018-2274-x

Benjelloun Touimi, G., Bennani, L., Berrada, S., et al.: Molecular serotyping and antibiotic resistance patterns of *Escherichia coli* isolated in hospital catering service in Morocco. Int. J. Microbiol. (2020). https://doi.org/10.1155/2020/5961521

Berrada, S., Salame, B., El, A., et al.: Assessment of the hygienic quality of meat and certain meat products taken from the city of Fez, Morocco. Int. J. Innov. Appl. Stud. **15**, 547–554 (2016)

Brahmi, F., Adjaoud, A., Marongiu, B., et al.: Chemical and biological profiles of essential oils from *Mentha spicata* L. leaf from Bejaia in Algeria. J. Essent. Oil Res. **28**, 211–220 (2016). https://doi.org/10.1080/10412905.2015.1118411

Chandran, A.S., Suri, S., Choudhary, P.: Sustainable plant protein: an up-to-date overview of sources, extraction techniques and utilization. Sustain. Food Technol. **1**, 466–483 (2023). https://doi.org/10.1039/d3fb00003f

Chen, D.Q., Bin, J.W., Granato, D., et al.: Effects of dynamic extraction conditions on the chemical composition and sensory quality traits of green tea. Lwt **169**, 113972 (2022). https://doi.org/10.1016/j.lwt.2022.113972

Chowdhury, J.U., Nandi, N.C., Uddin, M., Rahman, M.: Chemical constituents of essential oils from two types of spearmint (*Mentha spicata* L. and *M. cardiaca* L.) introduced in Bangladesh. Bangladesh J. Sci. Ind. Res. **42**, 79–82 (1970). https://doi.org/10.3329/bjsir.v42i1.359

David, V., Andrea, A.N., Aleksandr, K., et al.: Validation of a method of broth microdilution for the determination of antibacterial activity of essential oils. BMC. Res. Notes **14**, 1–7 (2021). https://doi.org/10.1186/s13104-021-05838-8

Dubois-brissonnet, F., Guillier, L., Dubois-brissonnet, F., Guillier, L.: Les maladies microbiennes d ' origine alimentaire To cite this version : HAL Id: hal-03080053. 55:30–38 (2022)

Dubois-Brissonnet, F., Guillier, L.: Les maladies microbiennes d'origine alimentaire. Cah Nutr Diet **55**, 30–38 (2020). https://doi.org/10.1016/j.cnd.2019.12.001

EL-Haoud, H., Boufellous, M., Berrani, A., et al.: Screening Phytochimique D ' Une Plante Medicinale: Mentha Spicata L. Am. J. Innov. Res. Appl. Sci. 226–233 (2018)

Fadil, M., Fikri-Benbrahim, K., Rachiq, S., et al.: Combined treatment of *Thymus vulgaris* L., *Rosmarinus officinalis* L. and *Myrtus communis* L. essential oils against Salmonella typhimurium: Optimization of antibacterial activity by mixture design methodology. Eur. J. Pharm. Biopharm. **126**, 211–220 (2018). https://doi.org/10.1016/j.ejpb.2017.06.002

Guimarães, A.C., Meireles, L.M., Lemos, M.F., et al.: Antibacterial activity of terpenes and terpenoids present in essential oils. Molecules **24**, 1–12 (2019). https://doi.org/10.3390/molecules24132471

Hjouji, K., Atemni, I., Mehdaoui, I., et al.: Essential oil of aleppo pine needles: antioxidant and antibacterial activities. Pharmacologyonline **2**, 556–565 (2021)

Houmy, N., Brahmi, F., Hasnaoui, I., et al.: Etude comparative des huiles essentielles de six variétés des nouvelles obtentions d'agrumes cultivées au Maroc. AfrimedMa 195–208 (2020)

Hussain, A.I., Anwar, F., Nigam, P.S., et al.: Seasonal variation in content, chemical composition and antimicrobial and cytotoxic activities of essential oils from four mentha species. J. Sci. Food Agric. **90**, 1827–1836 (2010). https://doi.org/10.1002/jsfa.4021

Kennedy, C.A., Walsh, C., Karczmarczyk, M., et al.: Multi-drug resistant Escherichia coli in diarrhoeagenic foals: Pulsotyping, phylotyping, serotyping, antibiotic resistance and virulence profiling. Vet. Microbiol. **223**, 144–152 (2018). https://doi.org/10.1016/j.vetmic.2018.08.009

Khorshidian, N., Yousefi, M., Khanniri, E., Mortazavian, A.M.: Potential application of essential oils as antimicrobial preservatives in cheese. Innov. Food Sci. Emerg. Technol. **45**, 62–72 (2018). https://doi.org/10.1016/j.ifset.2017.09.020

Kofidis, G., Bosabalidis, A., Kokkini, S.: Seasonal variation of essential oils in a linalool-rich chemotype of *Mentha spicata* grown wild in Greece). J. Essent. Oil Res. **16**, 469–472 (2004). https://doi.org/10.1080/10412905.2004.9698773

Kowalska-Krochmal, B., Dudek-Wicher, R.: The minimum inhibitory concentration of antibiotics: methods, interpretation, clinical relevance. Pathogens **10**, 1–21 (2021). https://doi.org/10.3390/pathogens10020165

Kumar, P., Mishra, S., Malik, A., Satya, S.: Insecticidal properties of Mentha species: a review. Ind. Crops Prod. **34**, 802–817 (2011). https://doi.org/10.1016/j.indcrop.2011.02.019

Lecerf, J.M.: La place de la viande dans la nutrition humaine. Viandes Prod Carnés **30**, 5 (2014)

Lianou, A., Panagou, E.Z., Nychas, G.J.E.: Meat safety—I Foodborne pathogens and other biological issues (2022)

Madoroba, E., Magwedere, K., Chaora, N.S., et al.: Microbial communities of meat and meat products: an exploratory analysis of the product quality and safety at selected enterprises in South Africa. Microorganisms **9**, 1–26 (2021). https://doi.org/10.3390/microorganisms9030507

Mahboubi, M., Haghi, G.: Antimicrobial activity and chemical composition of *Mentha pulegium* L. essential oil. J. Ethnopharmacol. **119**, 325–327 (2008). https://doi.org/10.1016/j.jep.2008.07.023

Manyi-Loh, C., Mamphweli, S., Meyer, E., Okoh, A.: Antibiotic Use in Agriculture and Its Consequential Resistance In Environmental Sources: Potential Public Health Implications (2018)

Mata, A.T., Proença, C., Ferreira, A.R., et al.: Antioxidant and anti-acetylcholinesterase activities of five plants used as Portuguese food spices. Food Chem. **103**, 778–786 (2007). https://doi.org/10.1016/j.foodchem.2006.09.017

Meloni, D.A., Arraiza, M.P., Garay, F., et al.: Biomass and essential oil yields of cornmint (*Mentha arvensis* L.) grown in the irrigation area of Ro Dulce, Santiago del Estero, Argentina. J. Essent. Oil Res. **27**, 148–152 (2015). https://doi.org/10.1080/10412905.2014.990642

Mkolo, N.M., Olowoyo, J.O., Sako, K.B., et al.: Repellency and toxicity of essential oils of *Mentha piperita* and *Mentha spicata* on larvae and adult of *Amblyomma hebraeum* (Acari: Ixodidae ). J. Microbiol. (2011)

Muteeb, G., Rehman, M.T., Shahwan, M., Aatif, M.: Origin of antibiotics and antibiotic resistance, and their impacts on drug development: a narrative review. Pharmaceuticals **16**, 1–54 (2023). https://doi.org/10.3390/ph16111615

Nanda, S.K., Jyotirmayee, B., Mahalik, G.: A review on effectiveness and biological activities of different Mentha species. Indian J. Nat. Sci. **13**, 41169–41184 (2022)

Njoga, E.O., Ilo, S.U., Nwobi, O.C., et al.: Pre-slaughter, slaughter and post-slaughter practices of slaughterhouse workers in Southeast, Nigeria: animal welfare, meat quality, food safety and public health implications. PLoS ONE **18**, 1–24 (2023). https://doi.org/10.1371/journal.pone.0282418

Quintieri, L., Nitride, C., De Angelis, E., et al.: Applications and Safety Issues. 1–32 (2023)

Rachidi, H., Latrache, H.: Gestion préventive du risque de toxi-infection alimentaire au Maroc: Stratégie et exigences réglementaires. 1–6 (2020)

Rafiq, R., Hayek, S.A., Anyanwu, U., et al.: Antibacterial and antioxidant activities of essential oils from artemisia herba-alba asso., pelargonium capitatum × radens and laurus nobilis l. Foods **5**, 1–12 (2016). https://doi.org/10.3390/foods5020028

Ranjan Maji, S., Roy, C., Sinha, S.K.: Gas chromatography-mass spectrometry (GC-MS): a comprehensive review of synergistic combinations and their applications in the past two decades. J. Anal. Sci. Appl. Biotechnol. **5**, 72–85 (2023). https://doi.org/10.48402/IMIST.PRSM/jasab-v5i2.40209

Rogissart, L.: Réduction de la consommation de viande: des politiques publiques bien loin des objectifs de durabilité (2023)

Sateriale, D., Forgione, G., De Cristofaro, G.A., et al.: Antibacterial and Antibiofilm Efficacy of Thyme (*Thymus vulgaris* L.) Essential Oil against Foodborne Illness Pathogens, Salmonella enterica subsp. enterica Serovar Typhimurium and *Bacillus cereus*. Antibiotics **12**, 485 (2023). https://doi.org/10.3390/antibiotics12030485

Selles, Ammar, Mokhtaria, et al.: Etude de l'activité biologique de l'huile essentielle de *Mentha spicata* cultivée dans la région de Tiaret (Algérie) (2018)

Seow, W.L., Mahyudin, N.A., Amin-Nordin, S., et al.: Antimicrobial resistance of *Staphylococcus aureus* among cooked food and food handlers associated with their occupational information in Klang Valley, Malaysia. Food Control **124**, 107872 (2021). https://doi.org/10.1016/j.foodcont.2021.107872

Shahbazi, Y., Shavisi, N., Karami, N., Kakaei, S.: Chemical composition and in vitro antibacterial activity of *Ferulago angulata* (Schlecht.) Boiss essential oil. Pharm. Sci. **21**, 6–11 (2015). https://doi.org/10.15171/PS.2015.10

Sp, C., Latine, R.: Perspectives agricoles de l'OCDE et de la FAO 2019–2028 (2020)

Taneja, S.C., Chandra, S.: Mint. Handb Herbs Spices Second Ed **1**, 366–387 (2012). https://doi.org/10.1533/9780857095671.366

Uikey, J.: Reviewing of Plant Belonging to Lamiaceae Family (2024)

Yar, A., Choudary, M.A., Rehman, A., et al.: Genetic diversity and virulence profiling of multi-drug resistant *Escherichia coli* of human, animal, and environmental origins. Antibiotics (2022). https://doi.org/10.3390/antibiotics11081061

Yousefian, S., Esmaeili, F., Lohrasebi, T.: A Comprehensive review of the key characteristics of the genus Mentha, natural compounds and biotechnological approaches for the production of secondary metabolites. Iran J Biotechnol **21**, 2–29 (2023). https://doi.org/10.30498/ijb.2023.380485.3605

Zouari-Bouassida, K., Trigui, M., Makni, S., et al.: Seasonal variation in essential oils composition and the biological and pharmaceutical protective effects of *Mentha longifolia* leaves grown in Tunisia. Biomed. Res. Int. (2018). https://doi.org/10.1155/2018/7856517

# Biotechnological Advances in Agriculture and Resource Valorization

# Dates Sector in Morocco: Current Situation, Innovations, and Future Challenges

Soumia Belouafa, Fatiha Moughaoui, Houria Hemdani, and Mohammed Berrada

**Abstract**

The chapter provides a detailed assessment of Morocco's *Phoenix dactylifera* date palm sector by examining its challenges and opportunities for sustainable development in arid regions and innovative approaches. The traditional *Phoenix dactylifera* date palm cultivation remains essential because it protects desert ecosystems and supports the economic stability of local communities. The country has launched major agricultural research programs and postharvest technology improvements to address climate change, market globalization, and modernization challenges. The implemented strategies work to boost sector performance while raising date value addition. According to this chapter, all stakeholders in the sector need to support technological adoption for the sustainable economic viability of producing regions.

**Keywords**

*Phoenix dactylifera* · Agronomic progress · Post-harvest techniques · Food security · Sustainable development

## 1 Introduction

The date palm species "*Phoenix dactylifera*" demonstrates remarkable resistance to severe environmental conditions while playing a vital role in the Moroccan agricultural sector. The ancient crop serves as a crucial food source and income generator for local communities while maintaining environmental stability in desert regions. The Moroccan date sector encounters various obstacles, including climate risks that endanger productivity, rising international competition, and challenges to match traditional farming methods with contemporary standards. The Moroccan government has initiated various research projects that combine innovation stimulation with sector revitalization to enhance agricultural practices and technological development for the sustainable growth of this valuable crop (Abouraïcha et al. 2010).

This chapter offers a thorough evaluation of the date palm sector in Morocco, which includes present and prospective challenges and opportunities for innovation. The main goal of this chapter is to give a complete description of traditional production systems while evaluating the effects of new technologies and advanced management practices on the productivity and sustainability of this crop. The analysis will investigate the value chain improvement from production through marketing to boost sector competitiveness. The research aims to establish communication between researchers, farmers, and industry professionals to promote strategies that boost global competitiveness in the sector. This chapter aims to detect research gaps in date palm cultivation in Morocco while developing recommendations for future research directions.

## 2 Methodology

The research analyzed articles that appeared between 1992 and 2024 used search terms such as "dates," "date palm," "*Phoenix dactylifera*," "date palm cultivation," "irrigation," "oases," "date marketing," and "agricultural innovations." The study evaluated references from selected articles to find additional relevant sources. The research incorporated reports from the Food and Agriculture Organization of the

S. Belouafa (✉) · F. Moughaoui · H. Hemdani · M. Berrada
Laboratory of Biology and Health, Department of Chemistry, Faculty of Sciences Ben M'Sick, University Hassan II of Casablanca, Casablanca, Morocco
e-mail: soumia.belouafa@univh2c.ma



O. Jamal Eddine et al. (eds.), *Plant Pathology, Fungal Diversity, and Biotechnological Advances in Agriculture*, Sustainable Landscape Planning and Natural Resources Management, https://doi.org/10.1007/978-3-031-97666-7_8

United Nations (FAO) about date palm cultivation, highlighting the significance of agricultural practice studies and date transport and storage infrastructure analysis, marketing strategy evaluation, and institutional support assessment.

This study focused on the following main themes:

- Agricultural Practices: The research investigated conventional and contemporary farming methods for date palm cultivation.
- Infrastructure: Examination of the storage facilities together with packaging systems.
- Marketing: Methods used for building brands and running advertising campaigns were examined.
- Innovations: The study assessed the implementation of modern technological systems and advanced management systems.

The research team assessed the chosen studies to determine their strong findings and meaningful conclusions; further gaps were evaluated to establish directions for upcoming investigations in this field.

## 3 The Date Palm Sector in Morocco

### 3.1 Historical and Cultural Importance

- **Origins of Date Palm Cultivation**

The date palm "*Phoenix dactylifera*" is a symbolic plant throughout North African and Middle Eastern cultures because people have cultivated it for thousands of years. The cultivation of this plant began more than 5000 years ago in the desert areas of the Middle East. The earliest evidence of date palm cultivation appeared in Mesopotamia (modern-day Iraq) and the Nile Valley of Egypt because dates maintained their freshness and nutritional value during long journeys and times of food shortages (Gros-Balthazard et al. 2013).

Date palm cultivation exists as a fundamental part of Moroccan history and culture, mainly in the Atlas regions and Saharan oases (Fig. 1). The date palm entered North Africa through caravan trade routes before it adapted to the local climate to become essential for both economic and ecological stability in arid regions. The Moroccan date palm groves provide nutritious fruits and a source of income for the local population and contribute to environmental protection by stabilizing dunes, improving soil quality, and supporting water cycles (Toutain 1967).

- **Traditional and Economic Role of Dates**

The date palm produces dates, which serve as fundamental cultural and economic products throughout numerous societies that exist in Middle Eastern and North African arid regions (Tengberg 2009; Tengberg et al. 2013). The local population considers dates as their main dietary staple. People value fresh or dried dates because they contain abundant energy and essential nutrients such as natural sugars, fiber, vitamins, and minerals. Dates symbolize hospitality because people traditionally offer to welcome visitors, and during religious events such as Ramadan, they eat them for breakfast.

The local population depends on dates as its leading economic revenue source. The date trade drives the economic growth of producing regions through employment opportunities in cultivation, harvesting, processing, and marketing activities, including export (ICAMAS 1998). The country promotes investment in modern production and preservation technologies because of the rising global date market value, which boosts its national agricultural economic contribution. The palm groves serve as natural carbon dioxide sensors while providing a habitat for local biodiversity and preventing soil erosion.

Morocco's local economy depends heavily on palm trees, which are deeply embedded in traditional Moroccan customs (Chetto et al. 2005). People use palm trees to welcome guests and celebrate special occasions because they represent traditional Moroccan hospitality. Palm trees are preferred during iftar meals in Ramadan because of their nutritional benefits.

Date production is a key economic driver for dry rural regions with limited agricultural possibilities. The sector drives agricultural growth through employment in date cultivation and processing, and works as a significant economic force for export production in producing regions. The preservation of oasis ecosystems and the control of desertification make palm groves essential for environmental sustainability. The palm tree is the foundation for environmental, economic, and social stability in Morocco's oases (Chetto et al. 2005).

### 3.2 Current State of the Sector

- **Description of Traditional Agricultural Practices**

Heritage agricultural methods are the foundation of date palm farming since they started from ancestral knowledge passed down over time under different environmental conditions. The technique and irrigation control call for frequent trimming and intercropping to boost output. Date palms are often grown in oases worldwide, especially in North Africa and the Middle

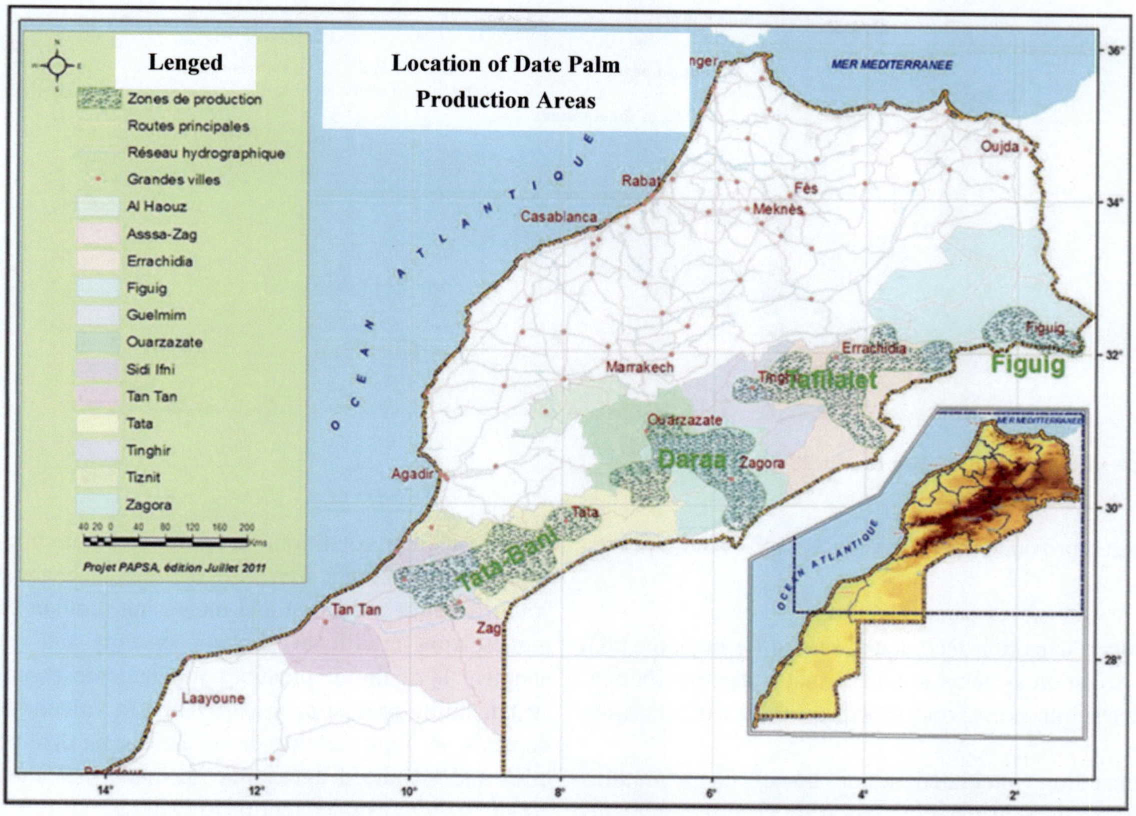


**Fig. 1** Distribution of date plants in Morocco (MAPMDREF 2020)

East, where irrigation plays a significant role. Gravity irrigation is also frequently used, utilizing natural spring water or groundwater to irrigate crops (Tengberg et al. 2013).

In Morocco, traditional date palm cultivation is inextricably tied to traditional water management, particularly through khettaras (Fig. 2), a network of underground channels that convey water from mountains to agricultural regions (Sedra 2003). Moroccan farmers are also moving to mixed farming, which includes date palms, citrus fruits, and vegetables, increasing resource utilization and preventing soil erosion (Kradi 2012; Aarkoub 2005; Srairi and Bentahar 2021).

Traditional technologies suitable for their specific ecological setting are being developed or replaced by current alternatives to fulfill today's economic needs and address climate change (KIADPAI 2023).

### – Issues Related to Lack of Technological Modernization

Date palm planting without modern technologies presents significant challenges for local economies that rely on this agricultural technique. Traditional farming practices in these regions maintain their historical knowledge and specific adaptations, yet struggle with productivity and sustainable resource management. The irrigation systems in these areas result in poor water resource management because water remains scarce in dry regions. The absence of contemporary pollination, harvesting, and processing equipment results in decreased production and inferior fruit quality, diminishing producer competitiveness in international markets (Abouraïcha et al. 2010; Ferry 1993).

The absence of advanced techniques prevents farmers from properly fighting diseases and pests that destroy plantations (Toutain 1967). Due to the technical deficit, farmers cannot adjust to climate change, which increases their crops' vulnerability to extreme weather occurrences. To improve the quality and quantity of production, modern agricultural practices must be updated with creative techniques that improve crop management and water and fertilizer efficiency.

### – Water Constraints and Irrigation

Date palm cultivation benefits from pumped irrigation systems. The practice creates sustainability problems because it uses groundwater reserves while depending on fossil fuels.

Sabri demonstrated an improved sustainable water management approach for date palm cultivation through his research (Sabri 2017). The study demonstrates how optimizing irrigation practices solves water scarcity problems.

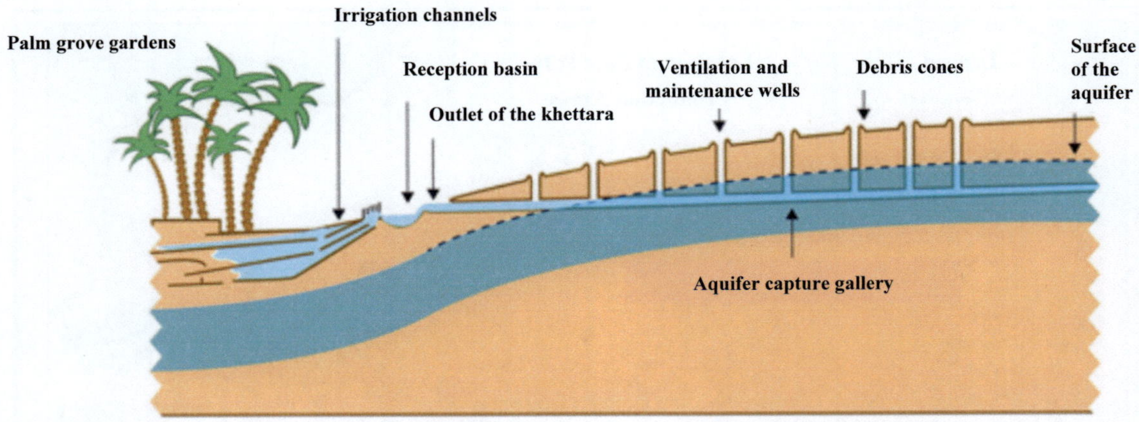


**Fig. 2** Operation of a Khettara (Ben Brahim 2021)

The research provides the following essential recommendations:

- Advanced irrigation techniques should be implemented. Drip irrigation systems provide exact water distribution, which minimizes evaporation and uncontrolled infiltration losses.
- Implementing integrated water management requires strategic irrigation planning based on crop requirements and combining surface water with groundwater sources.
- The program includes training farmers about water preservation and optimal irrigation methods to maximize water usage.

## 4 Challenges and Constraints

### 4.1 Agronomic Challenges

- **Adaptation to an Arid Climate**

The arid climate creates significant agricultural challenges for date palm farming at both national and international levels. Middle Eastern and North African farmers encounter severe moisture depletion, minimal rainfall, and harsh heat conditions. The current situation demands innovative irrigation systems and water preservation methods to sustain crop survival. The agricultural practices of using mulch to decrease evaporation and implementing drip irrigation systems are widely adopted in the field (Toutain 1967; Sedra 2003; Zenchi and Abdoun 2015).

Dry weather conditions in Morocco create a significant obstacle for farmers. Because of climate change-induced water resource reduction and increasing drought frequency, modernizing traditional khettara systems and developing adaptive agricultural practices has become essential. The sustainability of date palm cultivation in vulnerable regions depends on implementing drought-resistant date palm varieties and advanced agronomic methods that enhance soil structure for moisture retention to overcome arid climate challenges (Spoerry 2007; Fanny 2016; Rosse et al. 2007; Ruf and Kleiche-Dray 2018).

- **Soil Fertility Management**

Soil fertility management is a critical agronomic issue for date palm cultivation in arid regions. Date palms demonstrate outstanding salinity tolerance, enabling them to grow successfully in areas where other crops cannot survive. The productivity of date palms in these environments requires strategies that include enriching the soil with organic matter to enhance structure and water retention. The productivity of date palms depends on irrigation management through high-quality water and precise water input control to manage salt levels and support healthy palm development. Combining crop rotation (Battesti 2005) with intercropping (Karray 2006) helps preserve soil nutritional balance while preventing resource loss and enhancing overall soil health. These approaches help create sustainable agricultural systems that resist aridity challenges.

Renevot et al. (2009) found that using organic fertilizers and intercropping improves soil in Morocco. According to Hakimi et al. (2024), this strategy leads to greater date palm productivity, better salt management, and preservation of date palm plantations.

## 4.2 Economic Challenges

– **Profitability Compared to Other Crops**

Date palm cultivation faces multiple economic challenges that impact both worldwide and domestic markets. The Middle East and North African regions maintain their traditional date palm production but encounter rising competition from cereal and vegetable crops, which produce quicker financial returns (Toutain 1967; Sedra 2003; Boughzala and Ben Mahmoud 2022). Farmers struggle with high water expenses, and labor needs to maintain their date palm groves, negatively impacting their profitability.

The insufficient advancement of modern harvesting and cultivation methods in Morocco creates similar difficulties, which reduce operational efficiency and production quality standards (Harrak and Chetto 2001). Date palm trees need multiple years to start fruit production and require substantial water management investments alongside extensive care. The crop faces competition from short-cycle farming operations. When international markets experience price swings, revenues become unpredictable, which causes farmers to avoid spending money on technological development and production expansion.

Date palm cultivation faces economic obstacles that demand both the advancement of farming methods and a complete improvement of the production value sequence. All production phases need optimization, starting from cultivation through packaging and ending with marketing. The successful production system adaptation of irrigation methods alongside soil fertility practices and grove maintenance practices becomes essential to reach maximum harvests and sustainable cultivation.

Investing in research and development, because it means better farming and better yields with less cost, is a time-tested answer to economic problems. Some issues need to be tackled along the value chain since it includes all those activities that add value from production to marketing. The process comprises processing, packaging, storage, and transport. Date palm export firms can build economic stability by exploring the market and increasing their market competition. The expansion in the use of industrial dates, combined with the creation of new date-based products, will lead to the maximization of the economic value of this crop (Ben Salah 2018).

– **Maintenance Challenges**

Date palm grove management is crucial for productivity and tree health retention. However, these efforts have many obstacles, such as the refusal of young people to participate and financial problems. The maintenance of date palm groves is physically demanding work, typically underpaid. Therefore, the younger generation is more inclined to pursue easier and higher-paying jobs, leaving the older generation to do the loading and unloading work. Young laborers want to move to cities for better education and professional opportunities, leading to a labor shortage in date palm groves.

Lack of adequate maintenance increases the risk of drought-exacerbated fires and the accumulation of flammable materials. Fires destroy palms and significantly reduce yields. Furthermore, insufficient palm maintenance, such as pruning and cleaning, can decrease productivity and fruit quality. This influences farmers' revenue and the price competitiveness of Moroccan dates on the world market. The following may have the potential to address these obstacles.

- Providing training and financial incentives for youth to demonstrate maintaining date palm orchards.
- Modernise agricultural techniques to make maintaining tasks extremely easy and boost efficiency. For example, mechanized techniques such as pruning and cleaning could minimize manual efforts and drudgery.
- Establishing risk management mechanisms for forest fires, such as firelines and awareness-raising on safe agriculture.

– **Lack of Investment in Modern Technologies**

Date palm cultivation faces a critical economic challenge because of insufficient modern technology investments at the international and national levels. The insufficient technological investment in traditional date-producing countries of the Middle East and North Africa region restricts agricultural efficiency. It reduces yields while making the industry less competitive in international markets. Date producers face challenges in optimizing costs and improving product quality for export market standards because they have not implemented advanced agricultural methods, especially in irrigation, soil management, and mechanized harvesting.

The problem in Morocco worsens because palm growers cannot fully use their natural resources or fight diseases and pests effectively because they lack modern technologies, resulting in significant economic losses. The absence of suitable postharvest technologies, including drying and packaging facilities, prevents date producers from maintaining product quality throughout storage and transportation, which lowers their market value.

The solution to these challenges requires actively promoting investments in modern agricultural technology. Implementing modern agricultural technologies would boost date palm cultivation productivity and profitability while making the sector more sustainable and resistant to present and future environmental and economic threats.

– **Marketing Channel Challenges**

Date market channels should be improved to maximize producers' income and make them more competitive in other markets. Although technologies play a crucial role in this process, other key aspects should also be considered, such as transport infrastructure, marketing strategies, and institutional support.

– *Transport and Storage Infrastructure*

Adequate infrastructure for transporting and storing dates is essential to preserve product quality and minimize postharvest losses. Improving rural roads and temperature-controlled storage facilities can significantly reduce losses and improve market access. Investing in modern storage and conditioning centers extends the shelf life of dates, ensures better stock management, and stabilizes supply.

Efforts are underway in Morocco to improve this infrastructure as part of the Green Morocco Plan. For example, developing complexes to recover date palm by-products in regions such as Drâa-Tafilalet aims to strengthen local capacities in date processing and storage (Sedra 2015).

– *Marketing and Branding Strategies*

Developing regional brands and quality certifications, such as Protected Geographical Indication (PGI), can help differentiate Moroccan dates in international markets. Branding initiatives increase the product's visibility and build consumer trust. Marketing innovations promoting dates' nutritional features and health attributes could induce demand.

In this regard, and as an example, in Morocco, the International Dates Fair of Erfoud and similar experiments are important as they help to further promote Moroccan dates by linking the producers to investors and consumers and sharing the best marketing mechanisms and branding potentials (NAP 2017).

– *Institutional Support and Trade Policies*

Local and national bodies support the date trade. Subsidies that promote exports and bilateral trade agreements open up new opportunities to sell products worldwide while freeing companies from many obstacles. Agricultural cooperatives are far from playing a purely passive role, as they link producers to increase their bargaining power on the market and provide mutual benefits in technological improvement and international market access.

The Green Morocco Plan and the Generation Green 2020–2030 strategy are fundamental initiatives to develop the date palm sector through institutional support, modern infrastructure, and technological investments (Naji 2022).

The integration of these elements enables date producers to enhance product quality and shelf life while expanding their market share and achieving higher profitability. The measures will boost Moroccan date competitiveness and maintain the long-term viability of this vital crop for national economic development.

## 5 Nutritional Analysis and Varieties

### 5.1 Nutritional Composition

The health advantages of dates stem from their abundant nutritional elements. Various world cultivation regions produce different date varieties, which present distinct nutritional profiles containing fundamental nutrients and bioactive compounds. Dates are a source of dietary fiber and contain high natural sugars, which contribute to their energy content, as noted by several studies (Harrak et al. 2005; Hegazi and Salem 1971; Benchelah and Maka 2008; Al-Farsi and Lee 2008; Nasir et al. 2015; Verma 2016; Hadi et al. 2023).

Dates contain B vitamins such as niacin (B3), pyridoxine (B6), and folate (B9), which aid in energy metabolism and nervous function. Dates supply vitamin K, which enables blood clotting and supports bone health, as well as small amounts of vitamin A and vitamin C, which promote vision health, skin integrity, and immunity (Amira et al. 2020; Al-Farsi and Lee 2008; Nasir et al. 2015; Verma 2016; Lecoq 1933; Booij et al. 1992; Bouaziz and Ould El Hadj 2010; Patron et al. 1954).

The potassium content in dates supports blood pressure regulation and muscle activity. Dates contain iron, which supports red blood cell production and oxygen transport, and magnesium, which supports bone development and muscle function to prevent anemia. The essential mineral content of dates includes copper, which enhances iron absorption, and manganese, which supports both bone development and metabolic processes (Benchelah and Maka 2008; Amira et al. 2020; Al-Farsi and Lee 2008; Nasir et al. 2015; Patron et al. 1954; Gujarathi et al. 2022; Irandegani et al. 2019; Husnah et al. 2021).

Antioxidant carotenoids and polyphenols present in dates protect against oxidative damage and help prevent cancer and age-related diseases (Al-Farsi and Lee 2008; Nasir et al. 2015; Biglari et al. 2008).

The diverse nutritional profiles of different date species establish their role as functional foods while demonstrating the necessity to encourage their dietary consumption for delivering essential nutritional advantages (Al-Farsi and Lee 2008; Benamara et al. 2017).

The nutritional abundance of dates, along with their species-dependent flavor profiles and textures, establishes them as essential food components for daily nutrition and cultural and economic operations (Toutain 1967; Chetto et al. 2005; Nasir et al. 2015; Harrak et al. 2003).

## 5.2 Classification of Varieties

– **Quality and Classification Criteria**

Date varieties receive market value and consumer approval based on essential criteria determining their quality classification. Several key factors, including size and texture, sugar content, color, and physical defects, determine date quality (Physicochemical criteria).

Size and Weight: Dates are classified based on size and weight. Larger and heavier varieties, such as Majhoul or Medjool, are generally considered of higher quality and command higher prices in the market (Harrak et al. 2003; Reynes et al. 1994; Acourene et al. 2001). The Majhoul and Medjool dates refer to the same variety of dates, spelled differently in French and English. Majhoul and Medjool dates originate from the Tafilalet region in Morocco and have spread to various locations worldwide, especially in the Middle East and North America.

Texture: Date varieties range from soft to firm. People select dates based on their purpose, either for eating raw or preparing recipes in the kitchen. Soft dates are preferred due to their fresh taste and rich flavor profile (Acourene et al. 2001).

Sugar Content: The Brix measurement, which represents the sugar content of dates, is a fundamental factor in determining their sweetness. High sugar content impacts date preservation and flavor, resulting in better preservation ability (Harrak et al. 2005; Reynes et al. 1994; Acourene et al. 2001; Ghnimi et al. 2018).

Color: Date color ranges from light yellow to dark brown based on the variety and stage of maturity (Figs. 3 and 4). Uniformly colored dates with no discoloration usually receive better quality assessments (Harrak et al. 2003; Reynes et al. 1994; Acourene et al. 2001).

Physical Imperfections: Dates with no physical flaws, such as cracks, spots, or molds, receive better classification. Dates that remain physically intact are essential for export or premium product sales (CODEX 1985).

Moisture Content: The amount of water in dates controls their texture and determines their storage ability. Dates with suitable moisture levels remain soft but do not become too wet or too dry (Harrak et al. 2005; Reynes et al. 1994; Acourene et al. 2001; Ghnimi et al. 2018; UNECE 2010).

These criteria serve producers and distributors by evaluating date quality while establishing commercial categories. Products available to consumers enable them to select items based on their desired quality and taste profile. Dates' nutritional profile also supports their consumption because of their nutritional benefits.

Quantification of the microbial load and date infestation rate is necessary to guarantee date safety and hygiene before market distribution. It is essential to characterize dates and verify that they meet hygiene and quality standards before commercialization (Hasnaoui et al. 2010).

**Fig. 3** Different stages of date maturity (Zarouit et al. 2024)

Establishing standardized criteria ensures that the global date market operates uniformly, making international trade possible for this valuable fruit. The standards for date production include international regulations such as CODEX STAN 143 (1985). In addition to the standard NM 08.1.135 (2014), Morocco has regulatory texts such as the Decree No. 2399-18 (2018) and the Decree No. 2-17-433 (2017).

– **Variability of Characteristics Among Varieties**

Different date varieties show significant variations in their characteristics, affecting their taste profile and market demand domestically and internationally. The genetic adaptations to regional microclimates and specific cultural techniques produce this diversity (Toutain 1967; Battesti 2013; Harrak and Boujnah 2012). The genetic diversity of date palms in Morocco includes about 450 varieties and cultivars, making up nearly half of its genetic heritage. The remaining half of the genetic material consists of "Khalts" or "Sayrs," which are date seed-derived palms that differ genetically from one another (INRA 2011).

The dates of different genotypes, varieties, cultivars, and Khalts have different nutritional, organoleptic, technological, and commercial qualities. This variability is linked to each genotype's unique physical, chemical, and sensory properties (Harrak and Boujnah 2012). Each variety exhibits unique characteristics in size, color, texture, and flavor profile (Fig. 4). The Majhoul variety surpasses other varieties, closely followed by the Boufeggous and Bourhare varieties. For example, Majhoul dates contain an average of 93% pulp and only 7% pit. Additionally, some varieties change color from yellow to reddish-brown. In contrast, others retain a golden or light brown hue even when ripe (Fig. 4; Table 1). The brown color is predominant in Moroccan dates. However, the Bouskri variety stands out with its greenish-brown hue that persists even when fully ripe (Table 1). Consumers often use these color variations to indicate freshness and quality (Harrak and Boujnah 2012). The golden or honey-like appearance is particularly prized by Moroccan consumers (INRA 2005).

Moroccan dates generally have a sugar content between 65 and 80%. They mainly contain sucrose (sugar) and two simple sugars (glucose and fructose), which allows for a distinction between sucrose dates and simple sugar dates. The Majhoul variety shows the highest concentration of simple sugars, reaching around 80%, while the Bouskri variety has the lowest concentration at around 27% but contains a significant proportion of sucrose, approximately 43%. Thus, Majhoul is described as naturally sweet with honey and caramel

**Fig. 4** Maturity stages of four Moroccan date varieties: *Boufagous*, *Majhoul*, *Bousthammi*, and *Khalt* (Zarouit et al. 2024)

**Table 1** Geographic origin, maturity period, some morphological characteristics, and market value of the main Moroccan date varieties harvested at the *Tamr* stage (Misbah et al. 2022)

| Variety | Photo | Geographic distribution | Maturity period | Color | Consistency | Market value |
|---|---|---|---|---|---|---|
| *Ahardane* |  | Drâa, Anti-Atlas, Saghro, Bani, Oriental | Early | Light brown | Semi-soft | Medium (valued for its early ripening) |
| *Aziza* |  | Figuig | In season | From greenish-yellow to brown | Semi-soft | Appreciated in its native region |
| *Boufeggous* |  | Ziz, Drâa, Tafilalet, Oriental, Anti-Atlas, Bani, Saghro, Ferkala, Ghéris, Guir, Todra, between Saghro and the High Atlas | In season | Brown | Soft | Good |
| *Bouittob* |  | Anti-Atlas, Bani | Moderately late | Light brown | Dry | Medium |
| *Bourhare* |  | Drâa, Bani, Saghro, Tafilalet | Moderately late | Golden yellow to brown | Semi-soft | Good |
| *Bouskri* |  | Drâa, Tafilalet, Oriental, Anti-Atlas, Bani, Saghro, Todra, between Saghro and the High Atlas | Moderately late | Greenish brown | Dry | Good (sweet taste and excellent shelf life) |
| *Bouslikhene* |  | Tafilalet, Saghro | Moderately early | Light brown | Semi-soft | Low |
| *Bousthammi* |  | Bani, Drâa, Tafilalet, Saghro, Anti-Atlas | Moderately late | Black | Soft | Low (but appreciated for local consumption, excellent taste) |
| *Bouzeggar* |  | Drâa, between Saghro and the High Atlas, Ferkala, Ghéris | Late | Reddish brown | Semi-soft | Low |
| *Iklane* |  | Anti-Atlas, Bani, Drâa, Saghro | Late | Reddish brown | Semi-soft | Low |
| *Jihel* |  | Drâa, Bani, Anti-Atlas, Tafilalet, Saghro, between Saghro and the High Atlas | Late | Marron clair | Dry | Good |

(continued)

**Table 1** (continued)

| Variety | Photo | Geographic distribution | Maturity period | Color | Consistency | Market value |
|---|---|---|---|---|---|---|
| *Majhoul* |  | Ziz, Drâa, Tafilalet, Oriental, Djbel Bani, Saghro | Late in Drâa, moderately late in the upper Ziz, and in season in its native region in Tafilalet | Dark brown | Semi-soft | Excellent |
| *Otoukdime* |  | Between Saghro and the High Atlas, Todra | Moderately late | Ligh brown | Dry | Medium |

Soft (> 30% Moisture), Semi-soft (between 20 and 30% moisture), and dry (< 20% moisture) (Lobo et al. 2014)

nuances, making it particularly prized for direct consumption. In contrast, the smaller and firmer Bouskri variety is appreciated for its sweet taste, making it ideal for culinary uses.

Dates are known for their low protein content, generally between 2 and 4% for the main Moroccan varieties. However, they are rich in minerals such as calcium, magnesium, potassium, phosphorus, and iron, making them nutritionally valuable.

Most of the main Moroccan date varieties, representing approximately 66.5%, are fiber-rich due to the fibrous pulp surrounding the pit. Additionally, these dates contain vitamins and phenolic compounds with antioxidant properties, varying amounts depending on the genotypes and their origin.

Regarding sensory attributes, a Date Sensory Committee from the National Institute of Agronomic Research identified several aromas in Moroccan dates. Through the nasal route, the primary aromas are fruity, floral, cereal, woody, and caramelized notes. The retronasal route detected aromas including honey, caramel, fruits, flowers, chocolate, cereals, and licorice (INRA 2011).

These substantial variations between varieties have significant implications for producers, processors, and consumers, as they dictate harvesting methods, post-treatment techniques such as drying or freezing, and marketing strategies (Harrak and Boujnah 2012).

## 6 Innovations and Technologies

### 6.1 Agricultural Improvements

– **Introduction to Advanced Irrigation Methods**

The introduction of advanced irrigation methods represents a crucial innovation in improving agricultural practices for date palm cultivation, especially in regions where water is a limited resource. New methods, such as drip irrigation, have revolutionized how water is managed as they express the importance of delivering water right to plant roots. These systems reduce evaporation and runoff and provide plants with precise amounts of water depending on each plant's needs (Hamdane 2023; Bourziza et al. 2017; Benmihoub et al. 2021). Advanced irrigation systems deliver multiple substantial benefits when compared to conventional irrigation practices. Water usage decreases by 50% through these systems, which becomes essential for regions that experience frequent and severe droughts. The methods deliver enhanced plant health and better fruit quality through their ability to manage growth conditions while minimizing water stress that impacts fruiting.

Drip irrigation, which simultaneously delivers water and fertilizer, is suitable for date palm areas and those with undulating terrain and poor soil. This approach promotes sustainable agriculture in various fields: it can conserve and protect the earth while ensuring that resources such as land and water are used most efficiently (Bourziza et al. 2017).

The adoption of advanced irrigation equipment, along with training expenses, leads to long-term advantages of higher yields, decreased operational costs, and environmental benefits that make the initial investment worthwhile. Date palm growers implementing these irrigation systems will achieve sustainable and profitable farming practices that match modern environmental conditions.

– **Genetic Selection for More Resilient Varieties**

The development of date palm varieties through genetic selection represents a significant agricultural innovation that targets disease and pest management and climate resilience. Scientists use this method to enhance palm resistance to biological and environmental stressors and to develop dates with improved agricultural and commercial characteristics (Ferry 2010; Saaidi et al. 1981; Hammadi and Ferchichi 2020).

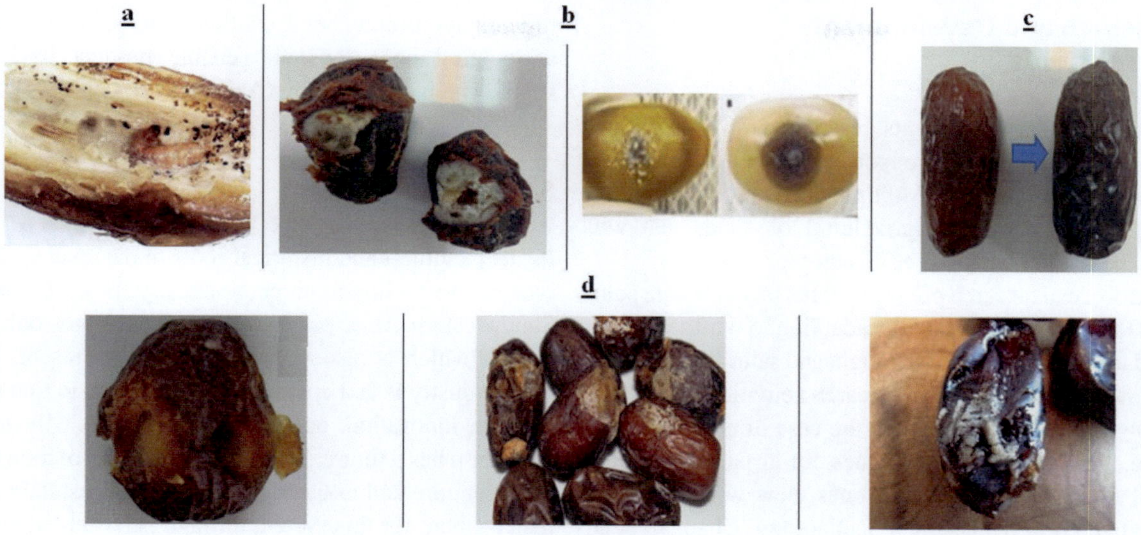


**Fig. 5** Types of imperfections likely to affect dates during storage **a** biological, **b** microbiological, **c** biochemical, **d** physical (Misbah et al. 2022)

Genetic selection programs use date palm genetic diversity to select and breed beneficial traits, including disease resistance, drought and salinity tolerance, and higher productivity. Research programs focus on selecting varieties that demonstrate resistance to Fusarium wilt, a fungal disease that causes severe damage to palms. The ability to withstand drought represents a vital selection goal because it protects crops from water scarcity in dry areas.

Modern genetic selection employs molecular marking and genomic editing techniques to achieve precise gene modification, which speeds up the development of new varieties for specific needs. The new methods boost traditional selection efficiency and help scientists understand how genes control preferred characteristics (Hazzouri et al. 2020; Ahmad et al. 2021).

Implementing these resilient varieties leads to better date palm cultivation sustainability because it reduces chemical usage for disease control and reduces production risks from climate variations. Genetic selection efforts create dual benefits for food security and sustainable ecological agricultural management, which becomes essential because of global climate change.

## 6.2 Post-harvest Technologies

Postharvest technologies need modern treatment and preservation innovations to extend date shelf life while reducing losses and maintaining fruit quality until consumer receipt (Sarraf et al. 2021; Mohammed et al. 2021). Modern treatment methods employ cooling and freezing chambers to reduce biological processes that lead to fruit spoilage (Fig. 5). The combination of temperature management and humidity control can be used in date preservation to create an environment ripe for extended maintenance of texture, color, and even taste safety.

Radiation and thermal treatments are important methods that postharvest technologies use against harmful insects. They also decrease microbial counts without using chemicals. These methods preserve dates' nutritional value, so the food safety of dates can be guaranteed (Misbah et al. 2022; Mohammed et al. 2021).

This technique allows gas level modification of the oxygen and carbon dioxide in the air, slowing down date ripening, thus lowering spoilage risks altogether. Therefore, this kind of packaging is beneficial.

Products can be tracked throughout the distribution chain, from harvest to customer, using traceability and tracking systems enabled by digital technology such as GPS and Radio Frequency Identification (RFID). This guarantees product freshness and quality. The implemented technologies enhance logistical efficiency and deliver precise information about date origin, treatment, and handling to consumers and distributors, thus fulfilling increasing consumer demands for food transparency (Boughzala and Ben Mahmoud 2022; Kennedy et al. 2020; Patelli and Mandrioli 2020; Mihi and Tarai 2020; Singha and Ranjit 2024).

Producers must make substantial investments to implement these postharvest technologies to meet international quality standards, boost global market competitiveness, and satisfy consumer demand for sustainable, high-quality dates (Misbah et al. 2022).

## 7 Research and Development Perspectives

The date palm sector presents researchers with multiple research and development opportunities, creating extensive investigation potential. Multiple current and upcoming projects seek solutions to agricultural obstacles that will enhance the market potential of this crop.

The Moroccan research program works to create date palm varieties that resist diseases while adapting to weather conditions and fighting against both drought and salinity challenges (Singha and Ranjit 2024). The research activities provide for preserving production stability in the case of environmental influence. New cultivation techniques, for instance, replacing chemical fertilizers with natural ones, new water systems, and integrated pest management technology (Channab et al. 2024; Sedra 2015), have helped researchers increase productivity and reduce the pressure on the environment caused by date palm cultivation.

Integrating renewable energy sources into date palm irrigation presents an exciting opportunity. Solar or wind energy can operate irrigation systems, enabling significant reductions in ecological impact and enhancements in irrigation sustainability.

Easy setup and performance make solar energy a good power option in date palm-growing arid areas. Solar panels enable pumps to irrigate the fields, which reduces operating costs significantly in the long term.

Wind power is a good alternative in areas with much wind. Combining wind energy with battery storage systems enables a power supply when wind is absent.

Research teams are now exploring hybrid energy systems, which unite solar power with wind power to optimize irrigation energy availability according to weather patterns. This method will deliver a stable and dependable power supply.

The method follows sustainable development principles and economically benefits the environment over time.

Studies focus on enhancing postharvest management approaches, which aim to increase date storage duration while maintaining nutritional and sensory characteristics (Misbah et al. 2022). Research explores alternative preservation techniques, including cooling chambers, active packaging, and modified atmosphere packaging methods. The research aims to create new date-derived products, including nutraceutical extracts, food industry ingredients, cosmetics, and natural health products (Ourradi 2022), to expand market possibilities for Moroccan dates.

Scientists are developing new uses for date cultivation by-products, including co-products, biomass waste, and processing residues (Khardi et al. 2024), for animal feed, eco-friendly construction materials, energy generation, and additional value-added products. Artificial Intelligence (AI) research is currently being studied for its potential to enhance agricultural operations and enable product tracking and supply chain optimization (Aziz et al. 2019).

## 8 Conclusions

Its deep cultural and historical roots in date palm cultivation continue to be important economically for arid regions like Southwestern Asia, particularly (although not only) North Africa, which is also very important. At present, the date palm industry is at a crucial point. Its ability to find strategic support, innovation, and upward expansion will determine whether it has a future. With the introduction of modern ways in agriculture and usage modes handled so astutely as water management for this system through persistent commitment towards development and research, there exists not only hope but also potential. Let everyone in the well-proven field throw their best efforts into the struggle for their survival so that we can achieve these environmental and economic changes for generations to come. Therefore, farmers, traders, researchers, and people from all walks of life must use advanced technology. Agribusiness and the Moroccan date palm sector can thus rely on these measures for sustainable growth, turning what seems to be an obstacle today into an opportunity tomorrow.

**Acknowledgements** We are grateful to Asmae MISBAH and Yousra ZAROUIT for permitting us to use photographs from their publications.

**Competing Interests** The authors declare that they have no competing interests in the work submitted for publication, directly or indirectly.

## References

Aarkoub, B.: Optimal valorization of water resources: economic impacts of irrigation water transactions in the tafilalet oases. In: Boulanouar, Kradi, C. (eds.) Proceedings of the International Symposium on the Sustainable Development of Oasis Systems, March 08–10, Erfoud, Morocco—B (2005). (in French)

Abouraïcha, E.F., Jaiti, F., El Hadrami, I.: Issues in Date Palm Cultivation in Morocco. Date Palm Biotechnologies. 45–54. IRD Edition (2010). (in French)

Acourene, S., Belguedj, M., Tama, M., Talebi, B.: Characterization and quality evaluation of dates and identification of rare date palm cultivars in the Ziban Region. Agron. Res. **8**, 19–23 (2001). (in French)

Ahmad, R., Anjum, M.A., Naz, S., Balal, R.M.: Applications of molecular markers in fruit crops for breeding programs—a review. Phyton-Int. J. Exp. Bot. **90**(1) (2021). https://doi.org/10.32604/phyton.2020.011680

Al-Farsi, M.A., Lee, C.Y.: Nutritional and functional properties of dates: a review. Crit. Rev. Food Sci. Nutr. **48**(10), 877–887 (2008). https://doi.org/10.1080/10408390701724264

Aziz, L., Kpoton, S.M., Bouamri, R.: Innovation platform for sustainable date palm production in the Tafilalet (Morocco). Altern. Rurales (7), December (2019). https://doi.org/10.60569/7-a2 (in French)

Battesti, V.: Humans and Plants, Agriculture. Gardens in the Desert. IRD Editions (2005). https://doi.org/10.4000/books.irdeditions.10168 (in French)

Battesti, V.: Agrobiodiversity of date palm (*Phoenix dactylifera* L.) in the Siwa Oasis (Egypt). J. Ethnoecol. **4** (2013). https://doi.org/10.4000/ethnoecologie.1538 (in French)

Benamara, S., Djouab, A., Boukhiar, A., Iguergaziz, N., Benamara, D.J.: Date Palm Fruit (*Phoenix dactylifera* L.): Ordinary Fruit or Health Food?—Literature Review. Phytotherapy. Lavoisier (2017). https://doi.org/10.1007/s10298-017-1113-4 (in French)

Ben Brahim, M.: Water Management in the Pre-Saharan Oases of Morocco. UNESCO (2021). (in French)

Benchelah, A.C., Maka, M.: Dates: nutritional interest. Phytotherapy **6**, 117–121 (2008). https://doi.org/10.1007/s10298-008-0296-0(inFrench)

Benmihoub, A., Akli, S., Benabid, N.: Challenges, practices, and constraints for agroecological valorization of land in the Sahara: the case of a Peri-oasis perimeter in the M'Zab Valley (Algeria). New Medit: Mediterranean J. Econ. Agric. Environ. (Jun) (2021). https://doi.org/10.30682/nm2102f (in French)

Ben Salah, M.: Marketing of dates on the world market: potential, constraints, and perspectives. Oasis agriculture and sustainable development. In: Proceedings of the 6th International Meeting on Aridoculture and Oasis Crops. Arid Regions Journal **46**(1/2020) (2018) (in French)

Biglari, F., AlKarkhi, A.F.M., Easa, A.M.: Antioxidant activity and phenolic content of various date palm (*Phoenix dactylifera*) fruits from Iran. Food Chem. **107**, 1636–1641 (2008). https://doi.org/10.1016/j.foodchem.2007.10.033

Bouaziz, S., Ould El Hadj, M.D.: Contribution to the study of the physico-chemical and biochemical characteristics of some traditional date vinegars obtained from some varieties of the Ouargla region. Ann. Sci. Technol. **2**(1) (2010) (in French)

Boughzala, Y., Ben Mahmoud, N.: Valorization of the date sector in Tunisia through the association of "modern" and "traditional" knowledge and techniques: difficulties, successes, and perspectives. ISTE OpenScience—Published by ISTE Ltd., London, UK (2022). (in French)

Bourziza, R., Hammani, A., Kuper, M., Bouaziz, A.: Performance of subsurface drip irrigation for young date palm irrigation. Moroccan J. Agric. Vet. Sci. **5**(1), 5–12 (2017). (in French)

Booij, G., I'Iombo, J.M., Risterucci, M., Coupe, D., Thomas, M., Ferr, Y.: Study of the chemical composition of dates at different maturity stages for the varietal characterization of various date palm cultivars (*Phoenix dactylifera* L.). Fruits **47**(6), 667–678 (1992) (in French)

Channab, B.D., El Idrissi, A., Ammar, A., Dardari, O., Marrane, S.E., El Gharrak, A., Akil, A., Essemlali, Y., Zahouily, M.: Recent advances in nano-fertilizers: synthesis, crop yield impact, and economic analysis. Nanoscale (2024). https://doi.org/10.1039/D3NR05012B

Chetto, A., Harrak, H., El Hachami, N.: Date Marketing in Morocco: Failures, Preferences, and Expectations. Edition: National Institute of Agronomic Research (2005). (in French)

CODEX Standard for Dates: CODEX STAN 143-1985 (1985)

Decree of the Minister of Agriculture, Maritime Fisheries, Rural Development, and Water and Forests No. 2-17-433 (2017, 25 October) concerning the quality and sanitary safety of marketed dates and date paste

Decree of the Minister of Agriculture, Maritime Fisheries, Rural Development, and Water and Forests No. 2399-18 (2018, 26 July) establishing the specifications for dates and date pastes marketed

Fanny, H.: Water Communication and Adaptation to Climate Change: The Case of the Sustainability of Khettaras in the Southeastern Moroccan Oases. Master's Thesis in Communication. University of Quebec in Montreal (2016). (in French)

Ferry, M.: The date palm sector crisis in Mediterranean countries. What research to address it? In: Ferry, M., Greiner, D. (eds.). The Date Palm in Oasis Agriculture of Mediterranean Countries. Zaragoza: CIHEAM. 129–156 (Options Méditerranéennes: Series A. Mediterranean Seminars; n. 28) (1993). (in French)

Ferry, M.: Status of Date Palm Genetic Improvement. Date Palm Biotechnologies. Edited by Frédérique Aberlenc-Bertossi. IRD Editions (2010). https://doi.org/10.4000/books.irdeditions.10787 (in French)

Ghnimi, S., Al-Shibli, M., Al-Yammahi, H.R., Al-Dhaheri, A., Al-Jaberi, F., Jobe, B., Kamal-Eldin, A.: Reducing Sugars, Organic Acids, Size, Color, and Texture of 21 Emirati Date Fruit Varieties (*Phoenix dactylifera* L.). Journal NFS, 12, August: 1–10 (2018). https://doi.org/10.1016/j.nfs.2018.04.002

Gros-Balthazard, M., Newton, C., Ivorra, S., Tengberg, M., Pintaud, J.C., Terral, J.F.: Origins and domestication of the date palm (*Phoenix dactylifera* L.). J. Ethnoecol. **4** (2013). https://doi.org/10.4000/ethnoecologie.1524 (in French)

Gujarathi, R., Arora, M., Kulkarni, M., Patwardhan, R., Kapgate, S.: Evaluating the role of date compounds in iron deficiency anemia in children. Int. J. Health Sci. **6**(S4), 12663–12675 (2022). https://doi.org/10.53730/ijhs.v6nS4.12239

Hadi, S.T., Shaban, D.M., Morgab, M.A.H., Jasim, S.M.: Chemical composition and effective compounds of dates and their use in a snack to provide energy to athletes. Funct. Foods Health Dis. **13**(8), 388–397 (2023). https://doi.org/10.31989/ffhd.v13i8.1147

Hakimi, F., Sebbar, A., Bouamri, R., Sidikou, A.A.H., El Janati, M., Bouaziz, A.: Effects of compost and compost tea on soil properties and nutrient uptake of the Moroccan date palm cultivar "Mejhoul" under organic cultivation. J. Ecol. Eng. **25**(7), 224–240 (2024). https://doi.org/10.12911/22998993/188334

Hamdane, A.: Conference Paper. The Future of Irrigation in Arid Areas. Tunisia, November (2023). (in French)

Hammadi, H., Ferchichi, A.: Molecular studies on date palm (*Phoenix dactylifera* L.): review. J. Oasis Agric. Sustain. Develop. **2**(1), 1–11 (2020). (in French)

Harrak, H., Chetto, A.: Valorization and Marketing of Dates in Morocco. INRA Edition (2001). (in French)

Harrak, H., Boujnah, M., Hamouda, A.: Physical and Morphological Characterizations of Main Moroccan Date Varieties. Al Avantia, 107 (2003). (in French)

Harrak, H., Hamouda, A., Boujnah, M., Gaboune, F.: Sugar content and technological and nutritional qualities of main Moroccan date varieties. In: International Symposium on the Sustainable Development of Oasis Systems, 8–10 March 2005, Erfoud, Morocco (2005). (in French)

Harrak, H., Boujnah, M.: Technological Valorization of Dates in Morocco. INRA Edition (2012). (in French)

Hasnaoui, A., El Houmaizi, M.A., Asehraou, A., Sindic, M., Deroanne, C., Hakkou, A.: Chemical composition and microbial quality of dates grown in the Figuig oasis of Morocco. Int. J. Agric. Biol. **12**(2) (2010)

Hazzouri, K.M., Flowers, J.M., Nelson, D., Lemansour, A., Masmoudi, K., Amiri, K.M.A.: Prospects for the study and improvement of abiotic stress tolerance in date palms in the post-genomics era. Front. Plant Sci. **11**, 293 (2020). https://doi.org/10.3389/fpls.2020.00293

Hegazi, S.M., Salem, S.A.: Chemical composition of Egyptian dried dates. J. Sci. Food Agric. Tome 22 (1971)

Husnah, R., As'ad, S., Sinrang, W., Usman, N., Ahmad, M., Hidayanty, H.: Addition of Sukkari dates (*Phoenix dactylifera* L) and Fe supplementation in increasing ferritin levels in young women with anemia. Ann. R.S.C.B. **25**(4), 17377–17389 (2021)

ICAMAS: International Centre for Advanced Mediterranean Agronomic Studies. Agricultural Development and Policies in the Mediterranean Region (1998). (in French)

INRA: National Institute of Agronomic Research: Study the Marketing and Processing of Dates and the Valorization of Date Palm Residues, Dates, and Their By-Products in Morocco. INRA Editions (2005). (in French)

INRA: National Institute of Agronomic Research: Atlas of Date Palm in Morocco. INRA Editions (2011). (in French)

Irandegani, F., Arbabisarjou, A., Ghaljaei, F., Navidian, A., Karajibani, M.: The effect of a date consumption-based nutritional program on iron deficiency anemia in primary school girls aged 8 to 10 years old in Zahedan (Iran). Pediatr. Health Med. Ther. **10**, 1–6 (2019). https://doi.org/10.2147/PHMT.S225816

Karray, J.: Water Balance of an Intercropping System (Olive Tree-Vegetable Crops) in Central Tunisia: Experimental Approach and Modeling Trial. PhD Thesis. National School of Agronomy of Montpellier (2006). (in French)

Kennedy, A., Stitzinger, J., Burke, T.: Food Traceability. In: Demirci, A., Feng, H., Krishnamurthy, K. (eds.) Food Safety Engineering. Food Engineering Series. Springer, Cham (2020). https://doi.org/10.1007/978-3-030-42660-6_10

Khardi, A., Nogot, A., Abdellaoui, M., Jaiti, F.: Valorization of date palm by-products to contribute to the sustainability of Moroccan Oases. Agric. J. **33**(3) (2024). https://doi.org/10.1051/cagri/2023027 (in French)

KIADPAI: Khalifa International Award for Date Palm and Agricultural Innovation:. Majhoul: The Core of Dates—Origins, Distribution, and International Market. Abdelouahhab Z. and Abdallah—Edition (2023)

Kradi, C.: Solidarity Agriculture in Fragile Ecosystems in Morocco. INRA Edition (2012). (in French)

Lecoq, R.: Nutritional value and vitamin content of Muscat Dates. Bull. Bot. Soc. France **80**(3), 338–348 (1933). https://doi.org/10.1080/00378941.1933.10833846(inFrench)

Lobo, M.G., Yahia, E.M., Kader, A.A.: Biology and postharvest physiology of date fruit. In: Dates: Postharvest Science, Processing Technology and Health Benefits. Wiley-Blackwell Edition, pp. 57–80 (2014)

MAPMDREF Document, prepared with EU funding. Available at CRP2 in Ifrane: Distribution of Date Plantations in Morocco. Source: Ecosystem Management Plan for the Date Sector (2020). (in French)

Mihi, A., Tarai, N.: Can Date Palm Plantations (DPP) and Oasis Formation Be Used to Combat Desertification and Sand Encroachment in Arid Areas? Options Méditerranéennes, A 124—Research and Innovation as Tools for Sustainable Agriculture, Food and Nutrition Security. MEDFORUM 2018. Bari, Italy, 18–20 Sep 2018 (2020)

Misbah, A., Essarioui, A., Noutfia, Y.: Post-harvest technologies for preserving date quality during storage. AFRIMED AJ Al Awamia **134**, 30–59 (2022). (in French)

Mohammed, M., Alqahtani, N., El-Shafie, H.: Development and evaluation of an ultrasonic humidifier to control humidity in a cold storage room for postharvest quality management of dates. Foods **10**, 949 (2021). https://doi.org/10.3390/foods10050949

Naji, A.: Morocco Develops the Date Palm Sector. Developing a Renewable Project (2022). (in French) from https://www.meer.com/en/71656-morocco-develops-dates-palm-sector

NAP: North Africa Post: Sweet Outlook for Morocco's Date Palm Cultivation (2017). (in French) from https://northafricapost.com/20447-sweet-outlook-moroccos-date-palm-cultivation.html

Nasir, M.U., Hussain, S., Jabbar, S., Rashid, F., Khalid, N., Mehmood, A.: A review on the nutritional content, functional properties, and medicinal potential of dates. Sci. Lett. **3**(1), 17–22 (2015)

NM Standard 08.1.135: Dates—Specifications (2014). (in French)

Ourradi, H.: Innovative Food, Cosmetic, and Energy Valorization of Moroccan Date Seeds (*Phoenix dactylifera* L.) in a Circular Economy Context. PhD Thesis, University of Sultan Moulay Slimane, Beni Mellal (2022). (in French)

Patelli, N., Mandrioli, M.: Blockchain technology and traceability in the agrifood industry. J. Food Sci. **85**, 3670–3678 (2020). https://doi.org/10.1111/1750-3841.15477

Patron, A., Patron, S., Swinzow, H.: The chemical composition of Moroccan dates considered from the perspective of their nutritional value. Fruits **9**(10) (1954). (in French)

Renevot, G., Bouaziz, A., Ruf, T., Raki, M.: Irrigation practices for date palm in the oasis systems of Tafilalet, Morocco. In: International Symposium: Sustainable Agriculture in the Mediterranean Region. Rabat, Morocco, May 14–16 (2009). (in French)

Reynes, M., Bouabidi, H., Piombo, G., Risterucci, A.M.: Characterization of major date varieties cultivated in the Djerid Region in Tunisia. Fruits **49**(4), 289–298 (1994). (in French)

Rosse, E., Sabir, M., Laouina, A.: Sustainable Management of Water and Soils in Morocco: Valorizing Traditional Mediterranean Techniques. IRD Editions (2007). (in French)

Ruf, T., Kleiche-Dray, M.: Irrigation water management in the Haouz of Marrakech: a century of confrontations between public, private, and community management models. EchoGéo (2018). https://doi.org/10.4000/echogeo.15258

Saaidi, M., Toutain, G., Bannerot, H., Louvet, J.: Selection of date palm (*Phoenix dactylifera* L.) for resistance to Bayoud Disease. Fruits **36**(4) (1981). (in French)

Sabri, A. (2017). Water needs during juvenile and adult phases, efficiency, and productivity of date palm (*Phoenix dactylifera* L.) under different water regimes: case of New Orchards in the extension zones of Errachidia. PhD Thesis in Agronomic Sciences. Hassan II Agronomic and Veterinary Institute, Rabat, Morocco. (in French)

Sarraf, M., Jemni, M., Kahramanoglu, I., Artés, F., Shahkoomahally, S., Namsi, A., Ihtisham, M., Brestic, M., Mohammadi, M., Rastogi, A.: Commercial techniques for preserving date palm (*Phoenix dactylifera*) fruit quality and safety: a review. Saudi J. Biol. Sci. **28**, 4408–4420 (2021). https://doi.org/10.1016/j.sjbs.2021.04.035

Sedra, M.: The Date Palm as a Basis for the Development of Oases in Morocco: Date Palm Techniques and Oasis Creation. INRA Editions (2003). (in French)

Sedra, M.: Date palm status and perspective in Morocco. In: Al-Khayri, J., Jain, S., Johnson, D. (eds.) Date Palm Genetic Resources and Utilization. Springer, Dordrecht 257–323 (2015). https://doi.org/10.1007/978-94-017-9694-1_8

Singha, S., Ranjit, S.: Crop improvement strategies and principles of selective breeding. In: Karmaoui, A. (ed.) Water-Soil-Plant-Animal Nexus in the Era of Climate Change. IGI Global, pp. 93–113 (2024). https://doi.org/10.4018/978-1-6684-9838-5.ch005

Spoerry, S.: The Return of Water to the Khettaras of Jorf, an Oasis in the Tafilalet, Southeast Morocco. Water Management Dynamics. Diploma in Tropical Agronomy. Institute of Warm Regions SUPAGRO. Montpellier (2007). (in French)

Sraïri, M.T., Bentahar, K.: How Do Oasis Agricultural Systems Use Labor and Water? Effects on the Incomes of Mixed Crop-Livestock Farms. Alternatives Rurales (2021). https://doi.org/10.60569/8-a4 (in French)

Tengberg, M.: Cultures and Uses of Date Palm in the Ancient Middle East, Nanterre. Cahier des thèmes transversaux ArScAn, IX, 237–242. hal-02264170 (2009) (in French)

Tengberg, M., Battesti, V., Newton, C.: The date palm: origin and cultivation in Egypt and the Middle East. J. Ethnoecol. **4**, 1216 (2013). https://doi.org/10.4000/ethnoecologie(inFrench)

Toutain, G.: Date Palm Cultivation and Production. Al Awamia (1967). (in French)

UNECE Standard DDP-08: Concerning the Marketing and Commercial Quality Control of Dates. UNITED NATIONS, New York and Geneva (2010)

Verma, P.: Effect of dates in lowering LDL level in geriatric cases. Health Sci. J. **10**(5), 8 (2016). https://doi.org/10.21767/1791-809X.1000100508

Zarouit, Y., Zekkouri, H., Ouhda, M., Aksasse, B.: Date fruit detection dataset for automatic harvesting. Data Brief **52**, 109876 (2024). https://doi.org/10.1016/j.dib.2023.109876

Zenchi, H., Abdoun, F.: Date palm: botany and ecology. Berber Encycl. (2015). https://doi.org/10.4000/encyclopedieberbere.3343(inFrench)

# Biovalorization of Algae: An Ecological Approach to Soil Fertilization and Biological Enrichment

Najwa Hassou, Redouane Ahnyne, Rachid Rahhal, Ahmed Errhif, and Said Gmouh

**Abstract**

Algae valorization is a promising ecological strategy to replace chemical fertilizers with sustainable, algae-derived alternatives. This process utilizes algae to convert sedimentary nutrients into biofertilizers, biostimulants, and biocontrol agents. Given the increasing global food demand and environmental degradation caused by synthetic fertilizers, the use of algae offers a nature-based solution for enhancing soil fertility and crop productivity while reducing ecological harm. Algae, particularly microalgae, thrive in diverse habitats and are efficient photosynthetic organisms capable of producing valuable biomolecules. These organisms can be cultivated in wastewater, reducing resource consumption while producing nutrient-rich biomass. Their extracts can stimulate plant growth, improve soil structure, and support beneficial microbial communities. In addition, algae help sequester carbon and contribute to biodiversity conservation. Various methods such as hydrothermal treatment and photobioreactor cultivation are used to harvest and process algae into usable agricultural products. Case studies have demonstrated successful implementations of algal fertilizers in improving soil health, especially in degraded and nutrient-poor soils. While promising, large-scale deployment faces technical and economic barriers, including high production costs, limited public awareness, and regulatory challenges. Nonetheless, algae-based soil amendments represent a vital component of a circular, green economy and a pathway toward climate-resilient agriculture. Their use supports environmental goals such as carbon sequestration, pollution reduction, and food system sustainability.

**Keywords**

Biovalorization · Microalgae · Biofertilizers · Soil fertility · Sustainable agriculture

**Abbreviations**

| | |
|---|---|
| °C | Degré Celsius |
| μm | Micromètre |
| C (dans biochimie) | Carbone |
| C/N ratio | Rapport Carbone/Azote |
| $CO_2$ | Dioxyde de carbone |
| DOM | Dissolved Organic Matter (Matière organique dissoute) |
| EC | Electrical Conductivity (Conductivité électrique) |
| EU | European Union (Union Européenne) |
| K | Potassium |
| mg/g | Milligrammes par gramme |

N. Hassou (✉) · R. Ahnyne · R. Rahhal
Ministry of Health and Social Protection, Higher Institute of Nursing Professions and Health Techniques in Rabat (ISPITS), Rabat, Morocco
e-mail: najwahassou@gmail.com

N. Hassou
Team Research of Virology, Oncology, and Biotechnologies, Laboratory of Virology, Oncology, Biosciences, Environment and New Energies, Faculty of Sciences and Techniques Mohammedia, University Hassan II of Casablanca, Casablanca, Morocco

N. Hassou · A. Errhif
Health, Environment and Biotechnology Laboratory, Faculty of Sciences Ain Chock, Hassan II University of Casablanca, Casablanca, Morocco

R. Ahnyne
Research Laboratory in Management of Organizations, Business Law, and Sustainable Development (LARMODAD), FSJES Souissi, Mohammed V University of Rabat, Rabat, Morocco

R. Rahhal
Laboratory of Ecology and Environment, Faculty of Sciences Ben M'Sick, Hassan II University of Casablanca, Casablanca, Morocco

S. Gmouh
Laboratory of Engineering and Materials (LIMAT), Faculty of Sciences Ben M'Sik, Hassan II University of Casablanca, Casablanca, Morocco



O. Jamal Eddine et al. (eds.), *Plant Pathology, Fungal Diversity, and Biotechnological Advances in Agriculture*, Sustainable Landscape Planning and Natural Resources Management, https://doi.org/10.1007/978-3-031-97666-7_9

| | |
|---|---|
| N | Azote |
| NPK | Azote-Phosphore-Potassium (Engrais triplet) |
| $O_2$ | Dioxygène |
| OC | Organic Carbon (Carbone organique) |
| P (dans engrais) | Phosphore |
| P | Phosphore |
| PBR | Photobioreactor (Photobioréacteur) |
| SDG(s) | Sustainable Development Goal(s) (Objectifs de développement durable) |
| $Se^0$ | Sélénium élémentaire |
| WHC | Water Holding Capacity (Capacité de rétention d'eau) |
| WTP(s) | Water Treatment Plant(s) (Station(s) de traitement des eaux) |

## 1 Introduction

In 2021, around 3.1 billion people worldwide lacked access to or could not afford a healthy diet. With global food demand expected to rise by 50% by 2050, agricultural output must double, significantly increasing the use of nitrogen and phosphorus fertilizers. However, excessive application of these chemicals has harmed soil health, disrupted natural processes, and polluted aquatic ecosystems. Their overuse has also been linked to rising antibiotic resistance and the failure to meet several 2030 Agenda targets. Therefore, identifying alternative solutions that are effective, cost-efficient, sustainable, and environmentally sound is imperative (Basnet et al. 2023; Mottaleb et al. 2023; Hemathilake and Gunathilake 2022; Humpenöder et al. 2022; da Cunha Dias et al. 2021).

Algae bio valorization offers a sustainable approach by using algal biotechnology to convert nutrient-rich sediments into soil fertilizers and biological enhancers. This method builds on the observation that nitrate-laden floodwaters trigger algal blooms, which create temporary ecosystems. Through hot water extraction, algae can be transformed into valuable agricultural inputs. Depending on biomass composition, the outputs may include fertilizers, biostimulants, or pest control agents such as natural antifungals or pesticides (Etemadifar et al. 2025; Vishwakarma et al. 2024; Giulianetti de Almeida et al. 2023; Yaashikaa et al. 2022).

Management excess nutrients leading to algal blooms is an underestimated, highly relevant challenge. The removal of algae from water bodies is costly, and the most successful available methods lead to a complete dry out of the pond, which can be restricted due to the possible environmental, social, and economic impacts. These methods are pursued in favor of sustainable approaches that do not modify the visible characteristics of ponds. A 3-step process allowing for the biovalorization of sedimentary nutrients is proposed: the separation of algae from water and their hydrothermal treatment are followed by nutrient transformation into solid, liquid and volatile fractions, which are soil fertilization, biostimulant and biocontrol agents, respectively (Hernández-Corroto et al. 2023).

## 2 Algae: An Overview

Algae are photosynthetic organisms found worldwide in different environments, such as lakes, rivers, oceans, and sewage. Algae can survive a wide range of temperatures, salinities, and pH levels, and can grow alone or with other organisms. Algae can be classified by their pigmentation, how they store material, and cell wall structure. The three main groups of algae are Rhodophyta (red), Phaeophyta (brown), and Chlorophyta (green). Algae can also fall into macroalgae (the large form, typically called seaweeds) and microalgae (the unicellular form that is < 200–500 $\mu$m in size) (Jardim et al. 2025; Anburaj et al. 2024; Ding et al. 2023; Shah et al. 2021).

Algae are photosynthetic organisms that create $O_2$ and organic matter from $CO_2$ and water, and as such, they are important primary producers. Their photosynthetic pigments increase light absorption and they can grow in very diverse environments including snow and soil. They can be harvested from natural sources or grown in tanks or photobioreactors to produce polysaccharides, lipids, pigments, proteins, and other valuable compounds. Algae are essential in the food industry and as feed for aquaculture (e.g., Chlorella sp. used as a food for omnivorous fish (Ogundele et al. 2023) and cosmetics (e.g., Chondrus crispus cosmetics). They have additional industrial applications to harvest in biorefineries, biodiesel, and use as a source of protein (e.g., Spirulina). Algae require light, nutrients, appropriate temperatures, and $CO_2$ to grow, and usually have a mutualistic association with bacteria or fungi that help to increase the alga and provide nutrients for the algae more efficiently. Industrial applications include algae harvesting, processing, and making of food products like pasta and biscuit powders (Raksasat et al. 2025; Lobus and Kulikovskiy 2023; Lobus 2022; Maberly and Gontero 2022; Renè 2025).

Algae and microalgae absorb carbon dioxide better in closed systems than in open systems. Open systems tend to be less expensive but lack environmental control, and are at a high risk of contamination. Closed systems allow for better control of growth and lower pollution potential. They are a better system for quality products because they have a controlled environment. Algae is cultivated and has no competition with food crops, and it has higher energy output. When we determine the financial memo of algae production and technology, this is a significant endeavor for high-value products and biomass/energy (Parray et al. 2024; Ahmad et al. 2022a, b, c; Godbole et al. 2021).

### 2.1 Types of Algae

Microalgae and macroalgae are kingdoms of autotropic eukaryotes contain phycobionts that photosynthesize and have a sustainable supply of organic feeds. Microalgae eukaryotic cells are 0.5–45 μm in diameter, while prokaryotic cyanobacteria are 0.2–13 μm in diameter. Macroalgae, or seaweeds, are multicellular, with greater than 45 μm diameter, usually including length of millimeters to meters, typically lacking a true root or shoot system, as well as a vascular, conducting system with xylem and phloem (Guadalupe Osorio-Reyes et al. 2023). The term 'thallus' is applied to the vegetative body of macroalgae, regardless of architectural type, as well as the phycobiont of fungi and lichens (Akolpoglu et al. 2025; Chen et al. 2024; Nayana et al. 2023; Pencik et al. 2023).

Microalgae are commonly found in terrestrial and aquatic ecosystems where they are the foremost producers of organic matter in marine ecosystems, and they are, with exceptions of some heterotrophic microorganisms, the only producers of organic matter in lakes, rivers, creeks, lagoons, and oceans. These organisms also provide an energy source to aquatic communities. Macroalgae, mainly consisting of green (Chlorophyta), red (Rhodophyta), and brown (Phaeophyceae) seaweeds, grow as attached species on the bottoms of shallow coastal waters, thereby altering the physical parameters of the habitat and providing habitat and refuge for organisms colonizing it (Fig. 1). The number of studies showing their utilization has also become overwhelming which show sequestering biopolymers from alcgallian cells that are required by various economic sectors and human organizations including having the biopolymer needed as part of a legal compliance or health policy (Cheng et al. 2024; Yang et al. 2024a, b; Lobus and Kulikovskiy 2023; Pozzer et al. 2022).

### 2.2 Ecological Importance of Algae

Ecological Role of Algae Apart from their use as biofertilizers, Microalgae have other environmental role as they are at the base of aquatic food chain, and are primarily responsible for most of the Earth oxygen production. Microalgae can be found in saltwater, freshwater (rivers, lakes, ponds) are damp habitat (soils). These plants could be filamentous, globular, elliptical, spiral or disc; they could be alone or in colonies and they are motile or almost stagnant. Also, with the fisthighest growth rate among the plants (20 times greater than most of terrestrial plants): microalgae require a few nutrients (heterotrophic or autotrophic), light (high or low irradiance), and a few chemical and physical parameters (turquoise water, low salinity, pH or temperature). Reuse wastewater for microalgae growth has significant environmental consideration given that it reduces cost for microalgae-based biotechnologies (Rai et al. 2025; Raza et al. 2024; Nur and Dewi 2024; Guadalupe Osorio-Reyes et al. 2023; Olabi et al. 2023).

## 3 Biovalorization: Concepts and Methods

Biovalorization refers to the process of transforming algae biomass into valuable bio-products with an increased economic value in mind. This can happen if bioconversion technologies are sourced efficiently in order to bioconvert the biowastes of algae into bio-products with a commercial need (Swaminaathan et al. 2024; Guadalupe Osorio-Reyes et al. 2023; Kolesovs and Semjonovs 2023).

Within the EU, bio valorization is viewed as being in the early growth stage and has a significant role of bringing to path a bio-based sustainable economies. Society, users, local communities perceive and know marine biovalorization as a social and environmental process with positive and significant monetary returns. The understanding and growth of biovalorization is occurring at different levels at either ends of the marine ecosystem, that is, regional seas eco domains and, Mediterranean waters and national/multinational government. Early adopters, viz. algae and macro product-based industries, contribute more than 80% of the business income generating 70–80% jobs, while early elitists either benthonic coasts exploitation/harvesting are driven by a niche market/money model in market opportunities (Alves et al. 2025; Singh et al. 2024a, b; Cheirsilp et al. 2023a, b; Harindintwali et al. 2022).

Marine biovalorization and its related socio-economic sectors hold significant potential for innovation, employment, and wealth generation, fostering the development of new marine-based industries and bioprocessing business models. In a perspective of decades biovalorization can offer insights and experiences not perceived at present in marine biodiversity. Spillover could take decades like the case of modern biotechnology. This can be efficiently achieved by collaborative, standardizing, and international approaches (Dominic and Baidurah 2025; Wibisono et al. 2025; Sharma et al.2024a, b, c; Vidhya et al. 2024).

### 3.1 Definition of Biovalorization

Biovalorization refers to the use of biotechnological methods to convert biological waste—especially from food production—into valuable resources. In agriculture, it focuses on

**Fig. 1** Diversity of seaweeds: red, brown, and green algae (copyright (c) 2019 Macrovector/ Shutterstock)

transforming mineral and organic waste into bio-stimulants that enrich soil and support chemical-free fertilization. Despite their potential, such wastes are often incinerated, landfilled, or composted, yielding limited economic benefits. Agronomic uses remain insufficiently explored (Yi et al.2024; Ufitikirezi et al. 2024; Guadalupe Osorio-Reyes et al. 2023; Kurmanbayeva et al.2023).

Crops like maize, alfalfa, and wheat contribute to carbon and nitrogen sequestration, prevent erosion, and enhance soil quality. Meanwhile, algae biomass—produced in large quantities and subsidized—still holds low market value. Research into converting this biomass into effective biostimulants is essential. These products, which boost plant growth and nutrient uptake, may contain soluble compounds, extracts, or microbial agents such as bacteria, fungi, or algae (Saravanan et al. 2023; Tawfik et al. 2022; Razzak et al. 2022).

### 3.2 Methods of Biovalorization

The use of biowastes can meet nutrient demands for microalgae cultivation, yielding enriched biomass and reducing fertilizer use. Algal biotreatments face challenges from nutrient limitations, complex biodegradation, and poor nutrient regulation. Understanding algae-bacteria interactions can improve nutrient availability. Upscaling lab technologies and managing nutrient biovalorization effects in complex environments is essential. Biowaste sources may also produce biofertilizers from freshwater or marine algae, enhancing soil fertility and crop production. Natural biostimulants from algae and humic substances serve as eco-friendly alternatives to synthetic regulators. Algal extracts improve seed germination, root growth, and nutrient uptake by affecting gene expression. Humic acid from biowaste enhances soil fertility and crop yield by stimulating microbial activity and nutrient release control. Formulations of marine algae and humics could offer benefits for soil improvement and require further optimization. Algae from effluent bioconversion can also address environmental issues while promoting sustainable agriculture (Loyal et al. 2024; Hull et al. 2024; Szparaga et al. 2023; Santoro et al. 2023).

## 4 Soil Fertilization Techniques

Algae are remarkable for soil fertilization due to their potential as biological agents for converting waste into crop nutrients and enhancing soil quality. They also contribute economically through bioremediation and reducing the carbon footprint from wastewater treatment. Furthermore, the biomass needed for these applications can be produced affordably from abundant waste nutrient sources, while cost-effective cultivation sites are available in various agricultural settings (Selvam et al. 2024; Eltarahony et al. 2024; Cheirsilp et al. 2023a, b; Zhang et al. 2022).

Soil fertilization with algal waste recovery is a growing area for urban development on arable land. Microalgae from agricultural waste increase biomass and yield of crops, including lettuce, peppers, and cucumbers. Algal waste harvesting from marine waters can be accomplished with affordable practices that are environmentally friendly; scientists have identified algae species that thrive in saline effluents; and there are basic production systems that allow one to efficiently harvest algae and use it for sustainable soil fertilization methods (Rápalo-Cruz et al. 2024; Khan et al. 2023; Huang et al. 2022; Sharma et al. 2021a, b).

The Carbon–Nitrogen ratio influences climate change and promotes soil health and carbon sequestration through bioremediation of aquaculture effluents by Spirulina, an inexpensive crop fertilizer. The rewards of algae and aquaculture waste go back to early agricultural. Alternative biovalorization avenues for aquaculture waste can provide algae biomass and broaden ecological fertilization to improve soil health and carbon sequestration to counter climate change (Osorio-Reyes et al. 2023; Dagnaisser et al. 2022; Gonçalves 2021; Ullmann and Grimm 2021).

Understanding soil characteristics is crucial for optimizing production. Soil fertility definitions aid in selecting fertilizers; organic options are slow-releasing and costly, whereas inorganic ones are faster and more effective (Zhang et al. 2024a, b, c; Guadalupe Osorio-Reyes et al. 2023).

Inorganic fertilizers are appreciated for their affordability, clear formulation, and lack of application restrictions. Countries like China, Japan, France, Italy, the Netherlands, Pakistan, and Thailand have sandy soils via water reclamation or desertification. They convert richer soils for construction and utilize sandy soils for agriculture, especially rice paddies during extended summer droughts (Eswaran et al. 2025; Wu et al. 2024; Dulias 2023).

### 4.1 Microbial Interactions

Maintaining beneficial microorganisms in agricultural soils is crucial for sustainable production, yet practices often harm soil microbiomes, fostering plant diseases. Various plant-associated microorganisms, like bacteria and fungi, positively interact with plants, enhancing nutrient uptake, immune responses, and producing bioactive metabolites. Thus, integrating beneficial microorganisms into soil-algal bacterial consortia could improve the efficacy of algal biofertilizers in promoting plant growth, especially in marginal or disease-prone soils. Culturable beneficial bacteria from green and cyanobacterial algal biofertilizers were isolated, and bioassays tested their effectiveness on diverse crops. Results indicated increased phosphate availability and reduced Fusarium resistance in wheat and tomato, yielding positive outcomes overall (Singh et al. 2025; Murugesan et al. 2025; De et al. 2024; Ramakrishnan et al. 2023; Akimbekov et al. 2022; Lee and Ryu 2021; Luis Fuentes et al. 2016).

In the last few years, some algal biomass composting studies have shown improved soil productivity in Sri Lanka (Alobwede et al. 2019). In addition to correcting soil deficiencies, algal fertilizers can turn rocky soils into fertile land. Alternative biofertilizers derived from algal biomass are considered an environmentally friendly agricultural soil supplement as they help overcome ecological and climatic issues, increase soil fertility, and yield. This chapter review emphasizes the diversity of algal practitioners and the extent of algal biomass utilization in various biotechnological applications. The use of microalgal biomass has been employed as a biofertilizer in sugar and vegetable crops, with some positive findings and worthwhile outcomes reported as the broad prospects for soil health and crop yield. Some limitations of algal biofertilizers have also been reported (Shukla et al. 2025; Sen et al. 2025; Tahir et al. 2024; Gonçalves et al. 2023; Yurkevich et al. 2022; Brindza et al. 2021).

## 5 Algae Cultivation Practices

Studies of algal culturing investigate the factors affecting algal growth. The study of cell structure provided further information relating to the diversity of shapes associated with different species, inspirational examples include filamentous Spirulina, spheroid Chlorella, and branched Volvox; the shape showed mobility and for example Chlorella remained stationary whereas Spirulina rotated. Chlorella produced significantly lower biomass (10–100 times lower than Paramecium) when comparing efficiency with respect to light and nutrients. Different culture conditions will influence nutrient limitations and the competition of nutrient availability. The limited phosphorus will have an impact on organic removal, while Chlorella will often bloom throughout the summer. In addition, nitrogen scarcity limited organic removal across many different green algae and cyanobacteria, which also hinders the water treatment processes (Sharma and Malaviya 2025; Jahan et al. 2025; Mahari et al. 2024; An et al. 2024; Liu and Yao 2023; Wan et al. 2022).

Photobioreactors are required to grow phototrophic microorganisms, and the more efficiently they can use light, the more productively they grow these microorganisms. A nested mesh-tube PBR was therefore developed by increasing the surface area and improving light distribution to improve the mass transfer of the PBR, and algal growth. The DWG drying method combines the advantages of lower operational costs of the PBR and fixes the issue of clogging in solid removal. Unfortunately, there are limited practical options and effective methods for concurrently purifying the algal biomass and concentrating the algal biomass, Ideally, coagulants should be employed that do not affect the final product, but even beyond biofuels there are various processes identified to extract more algal oils (Ma et al. 2024; Kossalbayev et al. 2024; Shareefdeen et al. 2023; Borella et al. 2022; Kwan et al. 2021; Chuka-ogwude et al. 2021).

### 5.1 Cultivation Systems

Fertilization with algal biostimulants can boost crop production by enhancing soil structure and resolving nutrient deficiencies. A trial in South Australia used the microalgal biostimulant Desmodesmus subspicatus on vineyard soil, improving soil aggregation and porosity, decreasing bulk density, and increasing water retention. While yield differences were not significant, pH, acidity, and potassium levels in juice improved, benefiting wine density and stability, indicating potential for better vineyard soil health (Mernissi and Arroussi 2025; Minaoui et al. 2024; Ramakrishnan et al. 2023; Kholssi et al. 2022; Sible et al. 2021; Alobwede et al. 2019).

Establishing a sustainable circular economy necessitates shifting from a linear fertilization approach to one focused on resource reuse. Microalgae show promise as biostimulants for enhancing soil structure, nutrient status, and fertility, but their effectiveness under varying carbon and nutrient limitations is not well understood (Mernissi and Arroussi 2025; Miranda et al. 2024; Sarsekeyeva et al. 2024).

### 5.2 Harvesting Techniques

There are various methods and technologies for harvesting microalgae depending on the microalgal characteristics and end-use application. Generally speaking, harvesting processes can be categorised as primary harvesting processes, which detach microalgal cell material from the surrounding environment; and secondary harvesting processes, which dewater the microalgal biomass into pastes or cakes (Roy et al. 2022; Sanislav et al. 2021; Titus et al. 2021).

Well-established low-cost primary harvesting methods include sedimentation, flocculation, and flotation methods. Sedimentation can be effectively used for larger scale processes, given the differences in weight, but requires large tanks and long retention times before re-suspension will occur. Flocculation, which can be used in tandem with sedimentation, neutralizes the charges in colloidal suspensions to make particles aggregate and settle faster than sedimentation alone, but comes with significant coagulant costs and market limitations, as well as fish uptake toxicity and impurities in sludge. Electrocoagulation uses electrochemical processes to create metal hydroxide on-site that can then form sedimentation with a low shear rate when it uses energy (Goria et al. 2025; Zhou et al. 2025; Rao et al. 2024; Lu et al. 2022).

Flotation is a harvesting technique that can attach air bubbles to microalgae producing buoyant aggregates. Flotation does not include settling tanks, is adaptable to a range of pH levels, and is mainly suited for media where nutrient concentrations are low. Drift-Flotation (D-Float) improves upon this process with the use of a larger tank and fast velocities of air flow to assist in the separation of bubbles. D-Float effectively reducing the microalgal concentration of the harvesting media, however, it is a more costly option for dewatering sludge sediments. The D-Cell is a modest size unit to incorporate concentration, air injection, float separation, and flocculation (Tyagi et al. 2025; Ye et al. 2024; Demir-Yilmaz et al. 2023; Nie et al. 2022).

## 6 Environmental Impact of Algae Utilization

The use of microalgae-based biofertilizers in agricultural practices has been shown to enhance soil quality, fertility levels, and crop productivity while being safe for human health and the environment (Guadalupe Osorio-Reyes et al. 2023). They significantly increase the population of beneficial microorganisms, such as bacteria, fungi, and actinomycetes, which leads to enhanced levels of available nutrients, bio-stimulation mechanisms, and improved soil structure, porosity, and aggregation. Furthermore, due to high nutrient levels and specific polysaccharides, they act as supports for the establishment of beneficial microorganisms in the soil

(Yang et al. 2025; Zhang et al. 2024a, b, c; Song et al. 2024; Gonçalves et al. 2023; Osorio-Reyes et al. 2023).

The minerals, polysaccharides, and auxins released into the soil increase the coaggregation of fertile particles, which increases soil clay aggregation, increased porosity and aeration, and retention of water and nutrients. Microalgae-based biofertilizers increase carbon, nitrogen, and phosphorus in soils for increased crop productivity, nutrient efficiency, and fertility. Additionally, assist beneficial soil microbiota that transfer nutrients from the soil to plant roots and help to reduce soil-borne diseases and nematodes (Yang et al. 2025; Song et al. 2024; Osorio-Reyes et al. 2023; Gonçalves et al. 2023).

Utilizing biofertilizers offers a sustainable source of nutrients that not only enriches soil fertility and crop yield but may also reduce soil toxicity over time. Maintaining healthy agricultural soils is a sensible objective, which requires evaluating risk-free nutrients and biomaterials, witnessed in our goal to achieve zero-NPK saturated soils. Moreover, we need an evaluation of denser soils with higher carbon levels for soil-aggregate-stabilizers with regard to crop replanting. We recognize the use of residue amendments may saturate soils with nutrients and result in pollution in water bodies, leading to the need for soil testing and monitoring technologies (Du et al. 2023; Haroun et al. 2023; Chaudhary et al. 2022; Nosheen et al. 2021).

## 7 Case Studies in Algae Biovalorization

The biovalorization process aims to foster a circular economy to produce valuable products like fine chemicals, bioplastics, fertilizers, animal feed, and biofuels. Case studies detail the use of algal biomass for organic fertilization and soil biological enrichment. Two types of fresh algal biomass with distinct characteristics were examined for soil amending protocols. Experimental results include the composition of algal biomass, and the biological and chemical analysis of both algal slurry and treated soil. The application of biostabilization processes is discussed as a means to enhance the commercial viability of fresh algal biomass in soil amendment (Nargotra et al. 2024; Haque et al. 2023; Ledda 2015).

The first two case studies focus on the valorization of the heterotrophic freshwater microalga (*Chlorella vulgaris*), cultivated on aquaculture waste effluents for algal slurries to be used as soil fertilizers (Wang and Zhang 2022). *Chlorella vulgaris* slurries had more agronomic value than chemical fertilizers in soil comparisons and soil-clay mineral transitions. These slurries produced soil biological activity without altering chemical properties and enhanced soil water holding capacity to mitigate the impacts of polluted runoff. The laboratory protocol allows for optimizing *C. vulgaris* slurries to be used for fertilization. The last two case studies presented another freshwater algal species, *C. vulgaris*, just as in the previous examples, and the photoautotrophic green microalga (*Scenedesmus obliquus*), also using aquaculture liquid wastes. After stabilization, phototrophic algal thickets were ground into smaller particles. This method improved germination rates for alkali-tolerant plants by over 50% across soils and helped to stop the soil water erosion by providing a degree of additional soil protection (Sánchez-Quintero et al. 2025; Chabili et al. 2024; Alvarenga et al. 2023; Alobwede et al. 2022; Chu et al. 2021).

Effective knowledge transfer for broader biowaste application necessitates behavioral changes in regulation, market dynamics, technology, and business models. Target groups require guidance on the benefits of biowaste and capacity building. Public acceptance is key, as health concerns regarding pathogens and contaminants persist. Risk perception varies due to personal knowledge and media influence. The ambition gap is exacerbated by underfunded sustainability communication. Highlighting societal impacts like costs, environmental benefits, job creation, safer food systems, and stronger public–private partnerships is essential (Ansar et al. 2025; Dehaene et al. 2024; Kacprzak et al. 2022; Kliachina 2021; Ledda 2015).

## 8 Challenges and Limitations

Sustainability constraints and regulations necessitate a shift towards high-quality, bio-original feed. Microalgae have ecological potential for sustainable aquaculture as a feed source and as bioproducts, while avoiding nutrient-related effluents. Although market share for algal feeds is small, farmers are hesitant to adopt them presently due to concerns over algal blooms. The transition to biovalorization is needed to minimize dependence on fossil resources, minimize the environmental impact of wastewater and enable a viable processing industry (Ambily et al. 2025; Guadalupe Osorio-Reyes et al. 2023; Yap et al. 2021).

Bio-production schedules can be adapted for bioproduct extraction workflows like protein, biofertilizers, or biofuels. The use of wastewater from aquaculture or farms helps overcome resource limitations. Many algae species, including Cyclotella and Spirulina, can grow in wastewater; therefore, the cost of growing algae in wastewater can be lowered while also improving wastewater treatment. The emphasis on soil health, risk mitigation, and useful co-products builds a circular economy which ultimately enhances both industry growth and sustainability (Ethiraj and Samuel 2024; Ogundele et al. 2023; Gururani et al. 2022; Saravanan et al. 2021).

A recent analysis identifies EU NUTS 3 regions like Ireland, Scotland, Cambridge, and Aquitaine as suitable for industrial algae biomass (Allen et al. 2018). They are vital for regional biorefineries focused on algae and biopolymers.

Actions include choosing technologies, designing biorefineries, and improving algae harvesting and biofuels to enhance efficiency and biopolymer reuse in waste management (Suchocki, 2024; Fertahi et al. 2023; de Oliveira and Bragotto 2022; Allen et al. 2018).

## 9 Conclusion

Ongoing research on microalgae-derived fertilizers shows significant benefits, enhancing crop productivity while benefiting the environment and human health. Long-term use can reduce dependence on chemical fertilizers, preventing soil degradation and improving carbon sequestration for climate change mitigation. Microalgae outperforms conventional fertilizers in boosting crop yields. Their adaptable nature allows for tailored properties for different crops, optimizing growth and acting as biopesticides to control pests and losses. This burgeoning agricultural field showcases diverse biopesticide products, indicating a promising industry future. Research also focuses on agents to alleviate abiotic stress, exploring mechanisms to reduce stress from biotic and abiotic factors in macroalgae. With rising food demand, biofertilizers and biostimulants present lucrative opportunities for companies, especially when using cyanobacteria, green algae, and diatoms. Increased application frequency benefits significantly, with optimal results from single or mixed species in solid and liquid forms, considering costs and density in product selection.

A significant part of the global population faces food insecurity due to socioeconomic inequalities. To tackle this issue, sustainable food production and responsible resource use are vital. Agriculture must undergo a rapid green transition to combat climate change and enhance human health. Unsustainable practices exclude farmers from supply chains. Algal biomass offers health benefits, potentially helping conditions like diabetes. A nutritious diet can prevent diseases and improve life quality. Safety evaluations for foods are necessary, with national authorities playing a key role in monitoring food consumption for public health protection.

## References


Ahmad, S., Iqbal, K., Kothari, R., Singh, H.M., Sari, A., Tyagi, V.V.: A critical overview of upstream cultivation and downstream processing of algae-based biofuels: opportunity, technological barriers and future perspective. J. Biotechnol. **351**, 74–98 (2022)

Ahmad, A., Banat, F., Alsafar, H., Hasan, S.W.: Algae biotechnology for industrial wastewater treatment, bioenergy production, and high-value bioproducts. Sci. Total Environ. **806**, 150585 (2022)

Ahmad, A., Hassan, S., Banat, F.: An overview of microalgae biomass as a sustainable aquaculture feed ingredient: food security and circular economy. Bioengineered (2022)

Akimbekov, N.S., Digel, I., Tastambek, K.T., Marat, A.K., Turaliyeva, M.A., Kaiyrmanova, G.K.: Biotechnology of microorganisms from coal environments: from environmental remediation to energy production. Biology **11**(9), 1306 (2022)

Akolpoglu, M.B., Baltaci, S.F., Bozuyuk, U., Karaz, S., Sitti, M.: Navigating microalgal biohybrids through confinements with magnetic guidance. Matter (2025)

Allen, J., Unlu, S., Demirel, Y., Black, P., Riekhof, W.: Integration of Biology, Ecology and Engineering for Sustainable Algal-Based Biofuel and Bioproduct Biorefinery (2018)

Alobwede, E., Leake, J.R., Pandhal, J.: Circular Economy Fertilization: Testing Micro and Macro Algal Species as Soil Improvers and Nutrient Sources for Crop Production in Greenhouse and Field Conditions (2019)

Alobwede, E., Cotton, A., Leake, J. R., Pandhal, J.: The fate and distribution of microalgal nitrogen when applied as an agricultural soil fertiliser and its effect on soil microbial communities. Phycology (2022)

Alvarenga, P., Martins, M., Ribeiro, H., Mota, M., Guerra, I., Cardoso, H., Silva, J.L.: Evaluation of the fertilizer potential of *Chlorella vulgaris* and *Scenedesmus obliquus* grown in agricultural drainage water from maize fields. Sci. Total Environ. **861**, 160670 (2023)

Alves, F.D.S., Condezo Castro, T.A., Gonçalves, L.M., Pedroza, M.M., Coutinho de Paula, E., Cardoso, M.: Characterization of sludge from a cellulose pulp mill for its potential biovalorization. Nord. Pulp Pap. Res. J. **40**(1), 203–218 (2025)

Ambily, B., Limna Mol, V.P., Sini, H., Nevin, K.G.: CRISPR-based microalgal genome editing and the potential for sustainable aquaculture: a comprehensive review. J. Appl. Phycol. (2025)

An, Q., Wang, H., Wang, X.: Fish survival subject to algal bloom: resource-based growth models with algal digestion delay and detritus-nutrient recycling delay. Ecol. Modell. (2024)

Anburaj, R., Kathiresan, K., Prasannakumar, C.: Exploring the depths: a comprehensive review of marine algae. Recent Trends In **176** (2024)

Ansar, A., Du, J., Javed, Q., Adnan, M., Javaid, I.: Biodegradable waste in compost production: a review of its economic potential. Nitrogen (2025)

Basnet, S., Wood, A., Röös, E., Jansson, T., Fetzer, I., Gordon, L.: Organic agriculture in a low-emission world: exploring combined measures to deliver a sustainable food system in Sweden. Sustain. Sci. **18**(1), 501–519 (2023)

Borella, L., Sforza, E., Bertucco, A.: An internally LED illuminated photobioreactor to increase energy conversion efficiency: Design and operation. Energy Conversion and Management (2022)

Brindza, J., Vozár, Ľ., Miko, M., Gažo, J., Kovár, P., Sedláčková, V.H., Grygorieva, O.: Unique effects of alginite as a bituminous rock on soil, water, plants and animal organisms–review. Agrobiodiversity for Improving Nutrition, Health and Life Quality **5**(1)

Chabili, A., Minaoui, F., Hakkoum, Z., Douma, M., Meddich, A., Loudiki, M.: Effects of extraction methods on the plant biostimulant activity of the soil microalga *Chlorella vulgaris*. J. Appl. Phycol. 1–14 (2024)

Chaudhary, P., Singh, S., Chaudhary, A., Sharma, A., Kumar, G.: Overview of biofertilizers in crop production and stress management for sustainable agriculture. Front. Plant Sci. **13**, 930340 (2022)

Cheirsilp, B., Billateh, A., Intasit, R., Upaichit, A., Boonsawang, P., Louhasakul, Y.: Fungal pretreatment and acid post-treatment for fractionation and biovalorization of palm biomass wastes into fungal oil, bioethanol, and lactic acid. Ind. Crops Prod. **196**, 116503 (2023)

Cheirsilp, B., Maneechote, W., Srinuanpan, S., Angelidaki, I.: Microalgae as tools for bio-circular-green economy: zero-waste approaches for sustainable production and biorefineries of microalgal biomass. Biores. Technol. **387**, 129620 (2023)

Chen, H., Sosa, A., Chen, F.: Growth and cell size of microalga *Auxenochlorella protothecoides* AS-1 under different trophic modes. Microorganisms (2024)

Cheng, S., Yoshikawa, K., Cross, J.S.: Influence of synthetic and natural microfibers on the growth, substance exchange, energy accumulation, and oxidative stress of field-collected microalgae compared with microplastic fragment. Sci. Total Environ. **908**, 167936 (2024)

Chu, Q., Lyu, T., Xue, L., Yang, L., Feng, Y., Sha, Z., et al.: Hydrothermal carbonization of microalgae for phosphorus recycling from wastewater to crop-soil systems as slow-release fertilizers. J. Clean. Prod. **283**, 124627 (2021)

Chuka-ogwude, D., Ogbonna, J. C., Moheimani, N.R.: Depth optimization of inclined thin layer photobioreactor for efficient microalgae cultivation in high turbidity digestate. Algal Res. (2021)

da Cunha Dias, T.A., Lora, E.E.S., Maya, D.M.Y., del Olmo, O.A.: Global potential assessment of available land for bioenergy projects in 2050 within food security limits. Land Use Policy (2021)

Dagnaisser, L.S., Dos Santos, M.G.B., Rita, A.V.S., Chaves Cardoso, J., De Carvalho, D.F., De Mendonça, H.V.: Microalgae as bio-fertilizer: a new strategy for advancing modern agriculture, wastewater bioremediation, and atmospheric carbon mitigation. Water Air Soil Pollut. **233**(11), 477 (2022)

Dehaene, M., Lopez, D., Fried, J., Tornaghi, C., Rayns, F., Cadieux, V., et al.: Soil Nexus: Urban Soil Care for Food Production, Community Greenspace, and Environmental Resilience (2024)

Demir-Yilmaz, I., Ftouhi, M.S., Balayssac, S., Guiraud, P., Coudret, C., Formosa-Dague, C.: Bubble functionalization in flotation process improve microalgae harvesting. Chem. Eng. J. **452**, 139349 (2023)

de Oliveira, A.P.F., Bragotto, A.P.A.: Microalgae-based products: food and public health. Future Foods (2022)

Ding, S., Xu, X., Zhu, F., Wang, R., Pan, G., Cui, Y., et al.: Distribution characteristics and physiological and ecological functions of soil algae in different habitats. Ecol. Environ. **32**(10), 1873 (2023)

Dominic, D., Baidurah, S.: A review of biological processing technologies for palm oil mill waste treatment and simultaneous bioenergy production at laboratory scale, pilot scale and industrial scale applications with technoeconomic analysis. Energy Convers. Manag. X 100914 (2025)

Du, T., Hu, Q., He, H., Mao, W., Yang, Z., Chen, H., et al.: Long-term organic fertilizer and biofertilizer application strengthens the associations between soil quality index, network complexity, and walnut yield. Eur. J. Soil Biol. **116**, 103492 (2023)

Dulias, R.: Record of aeolian processes in Pleistocene deposits in the foreland of the European sand belt. Environ. Socio-Econ. Stud. (2023)

Eltarahony, M., El-Deeb, N., Abu-Serie, M., El-Shall, H.: Biovalorization of whey waste as economic nutriment for mycogenic production of single cell oils with promising antibiofilm and anticancer potentiality. J. Biol. Eng. **18**(1), 62 (2024)

Eswaran, H., Vearasilp, T., Reich, P., Beinroth, F.: Sandy soils of Asia: a new frontier for agricultural development? A Holistic Approach for Sustainable Development of Problem Soils in the Tropics, Proceedings of the Management of Tropical Sandy Soils for Sustainable Agriculture, Khon Kaen, Thailand **27**, 22–30 (2025)

Etemadifar, Z., Dianatdar, F., Moridshahi, R.: Pharmaceutical waste valorization and bioremediation: challenges and innovations. In: Biotechnological Applications in Industrial Waste Valorization, pp. 43–88. Springer Nature, Singapore (2025)

Ethiraj, S., Samuel, M.S.: A comprehensive review of the challenges and opportunities in microalgae-based wastewater treatment for eliminating organic, inorganic, and emerging pollutants. Biocatalysis Agric. Biotechnol. (2024)

Fertahi, S., Elalami, D., Tayibi, S., Taarji, N., Lyamlouli, K., Bargaz, A., et al.: The current status and challenges of biomass biorefineries in Africa: a critical review and future perspectives for bioeconomy development. Sci. Total Environ. **870**, 162001 (2023)

Giulianetti de Almeida, M.P., Mockaitis, G., Weissbrodt, D.G.: Got whey? Sustainability endpoints for the dairy industry through resource biorecovery. Fermentation (2023)

Godbole, V., Pal, M.K., Gautam, P.: A critical perspective on the scope of interdisciplinary approaches used in fourth-generation biofuel production. Algal Res. (2021)

Gonçalves, A.L.: The use of microalgae and cyanobacteria in the improvement of agricultural practices: a review on their biofertilising, biostimulating and biopesticide roles. Appl. Sci. (2021)

Gonçalves, J., Freitas, J., Fernandes, I., Silva, P.: Microalgae as biofertilizers: a sustainable way to improve soil fertility and plant growth. Sustainability (2023)

Goria, K., Kothari, R., Singh, H.M., Pathania, D., Buddhi, D., Sharma, A., Tyagi, V.V.: Harvest of algal biomass with bioflocculants for biofuel with upstream and downstream processes: state-of-the-art towards sustainable alternative. Energy Ecol. Environ. 1–27 (2025)

Guadalupe Osorio-Reyes, J., Martin Valenzuela-Amaro, H., Juan Pablo Pizaña-Aranda, J., Ramírez-Gamboa, D., Ricardo Meléndez-Sánchez, E., López-Arellanes, M., Dolores Castañeda-Antonio, M., Coronado-Apodaca, K., Gomes Araújo, R., Eduardo Sosa-Hernández, J., Melchor-Martínez, E., Iqbal, H., Parra-Saldivar, R., Martínez-Ruiz, M.: Microalgae-Based Biotechnology as Alternative Biofertilizers for Soil Enhancement and Carbon Footprint Reduction: Advantages and Implications (2023)

Gururani, P., Bhatnagar, P., Kumar, V., Vlaskin, M.S., Grigorenko, A.V.: Algal consortiums: a novel and integrated approach for wastewater treatment. Water **14**(22), 3784 (2022)

Haque, F., Fan, C., Lee, Y.: From waste to value: addressing the relevance of waste recovery to agricultural sector in line with circular economy. J. Cleaner Prod. (2023)

Harindintwali, J.D., Wang, F., Yang, W., Zhou, J., Muhoza, B., Mugabowindekwe, M., Yu, X.: Harnessing the power of cellulolytic nitrogen-fixing bacteria for biovalorization of lignocellulosic biomass. Ind. Crops Prod. **186**, 115235 (2022)

Haroun, M., Xie, S., Awadelkareem, W., Wang, J., Qian, X.: Influence of biofertilizer on heavy metal bioremediation and enzyme activities in the soil to revealing the potential for sustainable soil restoration. Sci. Rep. (2023)

Hemathilake, D., Gunathilake, D.: Agricultural productivity and food supply to meet increased demands. Future Foods (2022)

Hernández-Corroto, E., Marina, M.L., García, M.C.: Application of high-voltage electrical discharges to the recovery of phenolic compounds from winery wastes. In: Processing of Food Products and Wastes with High Voltage Electrical Discharges, pp. 95–119. Academic Press (2023)

Huang, H., Zhong, S., Wen, S., Luo, C., Long, T.: Improving the efficiency of wastewater treatment and microalgae production for biofuels. Resour. Conserv. Recycl. **178**, 106094 (2022)

Hull, T., King, K., Kruger, J., Christensen, E., Chamas, A., Pienkos, P., Dong, T.: Nutrient Recovery from Algae Using Mild Oxidative Treatment and Ion Exchange (2024)

Humpenöder, F., Bodirsky, B.L., Weindl, I., Lotze-Campen, H., Linder, T., Popp, A.: Projected environmental benefits of replacing beef with microbial protein. Nature **605**(7908), 90–96 (2022)

Jahan, S., Pruvost, J., Cogne, G., Titica, M., Fallowfield, H.: Inactivation of an indicator virus during microalgae-based wastewater treatment. J. Appl. Phycol. 1–14 (2025)

Jardim, V.L., Grall, J., Barros-Barreto, M.B., Bizien, A., Benoit, T., Braga, J.C., et al.: A common terminology to unify research and conservation of coralline algae and the habitats they create. Aquat. Conserv. Mar. Freshwat. Ecosyst. **35**(3), e70121 (2025)

Kacprzak, M., Kupich, I., Jasinska, A., Fijalkowski, K.: Bio-based waste' substrates for degraded soil improvement—advantages and challenges in European context. Energies (2022)

Khan, S., Thaher, M., Abdulquadir, M., Faisal, M., Mehariya, S., Al-Najjar, M.A., et al.: Utilization of microalgae for urban wastewater treatment and valorization of treated wastewater and biomass for biofertilizer applications. Sustainability **15**(22), 16019 (2023)

Kholssi, R., Lougraimzi, H., Grina, F., Lorentz, J.F., Silva, I., Castaño-Sánchez, O., Marks, E.A.: Green agriculture: a review of the application of micro-and macroalgae and their impact on crop production on soil quality. J. Soil Sci. Plant Nutr. **22**(4), 4627–4641 (2022)

Kliachina, I.: Evaluation of the Efficiency of Using Biodegradable Waste as a Component of the Reclamation Mixture (2021)

Kolesovs, S., Semjonovs, P.: Microalgal conversion of whey and lactose containing substrates: current state and challenges. Biodegradation (2023)

Kossalbayev, B.D., Yilmaz, G., Sadvakasova, A.K., Zayadan, B.K., Belkozhayev, A.M., Kamshybayeva, G.K., et al.: Biotechnological production of hydrogen: design features of photobioreactors and improvement of conditions for cultivating cyanobacteria. Int. J. Hydrogen Energy **49**, 413–432 (2024)

Kurmanbayeva, M., Makhatov, Z., Kusmangazinov, A., Karabalayeva, D., Yerezhepova, N.: Protein, amino acid and carbohydrate content of fungal treated annual and perennial wheat straw. J. Ecol. Eng. **24**(12) (2023)

Kwan, P.P., Banerjee, S., Shariff, M., Md. Yusoff, F.: Influence of light on biomass and lipid production in microalgae cultivation. Aquacult. Res. **52**(4), 1337–1347 (2021)

Ledda, C.: A New Biorefinery Model For Livestock Farming: Microalgae Cultivation For Animal Slurries Valorization [PDF] (2015)

Lee, S.M., Ryu, C.M.: Algae as new kids in the beneficial plant microbiome. Front. Plant Sci. (2021)

Liu, Y., Yao, X.: Photochemical release of dissolved organic nitrogen from algal detritus and sediment particles in Lake Taihu, China. Water (2023)

Lobus, N.V.: Biogeochemical role of algae in aquatic ecosystems: basic research and applied biotechnology. J. Marine Sci. Eng. (2022)

Lobus, N.V., Kulikovskiy, M.S.: The co-evolution aspects of the biogeochemical role of phytoplankton in aquatic ecosystems: a review. Biology (2023)

Lu, Z., Beal, C.M., Johnson, Z.I.: Comparative performance and technoeconomic analyses of two microalgae harvesting systems evaluated at a commercially relevant scale. Algal Res. (2022)

Luis Fuentes, J., Garbayo, I., Cuaresma, M., Montero, Z., González-del-Valle, M., Vílchez, C.: Impact of Microalgae-Bacteria Interactions on the Production of Algal Biomass and Associated Compounds (2016)

Ma, S., Huang, Y., Zhu, X., Xia, A., Zhu, X., Liao, Q.: Growth-based dynamic light transmission modeling and optimization in microalgal photobioreactors for high efficiency $CO_2$ fixation. Renew. Sustain. Energy Rev. **197**, 114414 (2024)

Maberly, S.C., Gontero, B.: Blue Planet, Red and Green Photosynthesis: Productivity and Carbon Cycling in Aquatic Ecosystems (2022)

Mahari, W.A.W., Waiho, K., Fazhan, H., Azwar, E., Shu-Chien, A.C., Hersi, M.A., et al.: Emerging paradigms in sustainable shellfish aquaculture: microalgae and biofloc technologies for wastewater treatment. Aquaculture 740835 (2024)

Mernissi, N.E.L., Arroussi, H.E.L.: Biofertilizers, soil conditioners, and biostimulants from microalgae. Algal Bioreactors (2025)

Minaoui, F., Hakkoum, Z., Chabili, A., Douma, M., et al.: Biostimulant effect of green soil microalgae Chlorella vulgaris suspensions on germination and growth of wheat (Triticum aestivum var. Achtar) and soil fertility. Algal Res. (2024)

Miranda, A.M., Hernandez-Tenorio, F., Villalta, F., Vargas, G.J., Sáez, A.A.: Advances in the development of biofertilizers and biostimulants from microalgae. Biology **13**(3), 199 (2024)

Mottaleb, K.A., Kruseman, G., Frija, A., Sonder, K., Lopez-Ridaura, S.: Projecting wheat demand in China and India for 2030 and 2050: implications for food security. Front. Nutr. **9**, 1077443 (2023)

Murugesan, S., Priya, P., Sivamurugan, V.: Microbial interactions in soil algae. In: Soil Algae: Morphology, Ecology and Biotechnological Applications, pp. 1–35. Springer Nature, Singapore (2025)

Nargotra, P., Ortizo, R.G.G., Wang, J.X., Tsai, M.L., Dong, C.D., Sun, P.P., et al.: Enzymes in the bioconversion of food waste into valuable bioproducts: a circular economy perspective. Syst. Microbiol. Biomanufact. **4**(3), 850–868 (2024)

Nayana, K., Vidya, D., Soorya, K., Dineshan, A., Menon, A.S., Mambad, R., Arunkumar, K.: Effect of volume and surface area on growth and productivity of microalgae in culture system. BioEnergy Res. **16**(2), 1013–1025 (2023)

Nie, X., Zhang, H., Cheng, S., Mubashar, M., Xu, C., Li, Y., et al.: Study on the cell-collector-bubble interfacial interactions during microalgae harvesting using foam flotation. Sci. Total Environ. **806**, 150901 (2022)

Nosheen, S., Ajmal, I., Song, Y.: Microbes as biofertilizers, a potential approach for sustainable crop production. Sustainability (2021)

Nur, M.M.A., Dewi, R.N.: Opportunities and challenges of microalgae in biocement production and self-repair mechanisms. Biocatal. Agric. Biotechnol. (2024)

Ogundele, O.D., Adewumi, A.J., Oyegoke, D.A.: Phycoremediation: Algae as an effective agent for sustainable remediation and waste water treatment. Environ. Earth Sci. Res. J. **10**(1) (2023)

Olabi, A.G., Shehata, N., Sayed, E.T., Rodriguez, C., Anyanwu, R.C., Russell, C., Abdelkareem, M.A.: Role of microalgae in achieving sustainable development goals and circular economy. Sci. Total Environ. **854**, 158689 (2023)

Osorio-Reyes, J.G., Valenzuela-Amaro, H.M., Pizaña-Aranda, J.J.P., Ramírez-Gamboa, D., Meléndez-Sánchez, E.R., López-Arellanes, M.E., et al.: Microalgae-based biotechnology as alternative biofertilizers for soil enhancement and carbon footprint reduction: advantages and implications. Mar. Drugs **21**(2), 93 (2023)

Parray, J.A., Singh, N., Haghi, A.K.: Introduction to waste algal biomass-based energy production. In: Microalgae as a Sustainable Source of Green Energy Generation and Bioeconomy, pp. 1–10. Springer Nature, Cham, Switzerland (2024)

Pencik, O., Molnarova, K., Durdakova, M., Kolackova, M., Klofac, D., Kucsera, A., et al.: Not so dangerous? PET microplastics toxicity on freshwater microalgae and cyanobacteria. Environ. Pollut. **329**, 121628 (2023)

Pozzer, A.C., Gómez, P.A., Weiss, J.: Volatile organic compounds in aquatic ecosystems–detection, origin, significance and applications. Sci. Tot. Environ. (2022)

Rai, I., Achankunju, J., Gordon, R., Vinayak, V.: Terraforming Mars with Microalgae, Especially Diatoms. Diatom Cultivation for Biofuel, Food and High-Value Products, pp. 203–230 (2025)

Raksasat, R., Suparmaniam, U., Rawindran, H., Lim, J.W.: 8 Open and Closed. Green Technologies for Wastewater Treatment and Bioenergy Production, 198 (2025)

Raksasat, R., Suparmaniam, U., Rawindran, H., Lim, J.W.: Open and closed photobioreactors for algal wastewater treatment. In: Green Technologies for Wastewater Treatment and Bioenergy Production, pp. 198–219. CRC Press

Ramakrishnan, B., Maddela, N.R., Venkateswarlu, K., Megharaj, M.: Potential of microalgae and cyanobacteria to improve soil health and agricultural productivity: a critical view. Environ. Sci. Adv. **2**(4), 586–611 (2023)

Rao, N.R.H., Beyer, V., Thielemans, W., Muylaert, K.: Techno-economic comparison of flocculation combined with dissolved air flotation versus sedimentation for microalgae harvesting. Algal Res. (2024)

Rápalo-Cruz, A., Gómez-Serrano, C., González-López, C.V., Morillas-España, A., Jiménez-Becker, S.: Utilization of treated wastewater derived from microalgae production for the irrigation of horticultural crops. J. Appl. Phycol. **36**(3), 1259–1268 (2024)

Raza, N., Rizwan, M., Mujtaba, G.: Bioremediation of real textile wastewater with a microalgal-bacterial consortium: an eco-friendly strategy. Biomass Convers. Biorefinery (2024)

Razzak, S.A., Lucky, R.A., Hossain, M. M., deLasa, H.: Valorization of microalgae biomass to biofuel production: a review. Energy Nexus (2022)

Renè, N.H.: Tracing the Sources of Nitrogen in Submarine Groundwater Discharge & Primary Producers in the Great South Bay Estuary (2025)

Roy, S., Azad, A.W., Baidya, S., Alam, M.K., Khan, F.: Powering solutions for biomedical sensors and implants inside the human body: a comprehensive review on energy harvesting units, energy storage, and wireless power transfer techniques. IEEE Trans. Power Electron. **37**(10), 12237–12263 (2022)

Sánchez-Quintero, Á., Parsy, A., Adrien, A., Spitzer, L., Jiménez-Lamana, J., Fernandes, S.C., Beigbeder, J.B.: Effects of CO2 and liquid digestate concentrations on the growth performance and biomass composition of Tetradesmus obliquus and Chlorella vulgaris microalgal strains. Front. Bioeng. Biotechnol. **12**, 1459756 (2025)

Sanislav, T., Mois, G.D., Zeadally, S., Folea, S.C.: Energy Harvesting Techniques for Internet of Things (IoT). IEEE Access (2021)

Santoro, D.F., Puglisi, I., Sicilia, A., Baglieri, A., La Bella, E., Lo Piero, A.R.: Transcriptomic profile of lettuce seedlings (Lactuca sativa) response to microalgae extracts used as biostimulant agents. AoB Plants **15**(4), plad043 (2023)

Saravanan, A., Kumar, P.S., Varjani, S., et al.: A review on algal-bacterial symbiotic system for effective treatment of wastewater. Chemosphere (2021)

Saravanan, A., Kumar, P.S., Badawi, M., Mohanakrishna, G., Aminabhavi, T.M.: Valorization of micro-algae biomass for the development of green biorefinery: perspectives on techno-economic analysis and the way towards sustainability. Chem. Eng. J. **453**, 139754 (2023)

Sarsekeyeva, F.K., Sadvakasova, A.K., Sandybayeva, S.K., Kossalbayev, B.D., Huang, Z., Zayadan, B.K., et al.: Microalgae-and cyanobacteria-derived phytostimulants for mitigation of salt stress and improved agriculture. Algal Res. 103686 (2024)

Selvam, J., Mal, J., Singh, S., Yadav, A., Giri, B.S., Pandey, A., Sinha, R.: Bioprospecting marine microalgae as sustainable bio-factories for value-added compounds. Algal Res. **79**, 103444 (2024)

Sen, S.K., Jain, P., Singh, R.K., Singh, A.K.: Role of Soil Algae in the Maintenance of Soil Fertility and Soil Health. In: Soil Algae: Morphology, Ecology and Biotechnological Applications, pp. 293–304. Springer Nature, Singapore (2025)

Shah, S.T., Basit, A., Ullah, I., Mohamed, H.I.: Cyanobacteria and algae as biocontrol agents against fungal and bacterial plant pathogens. Plant Growth-Promoting Microbes for Sustainable Biotic and Abiotic Stress Management, pp. 1–23 (2021)

Shareefdeen, Z., Elkamel, A., Babar, Z.B.: Recent Developments on the Performance of Algal Bioreactors for $CO_2$ Removal: Focusing on the Light Intensity and Photoperiods. BioTech (2023)

Sharma, G.K., Khan, S.A., Shrivastava, M., Bhattacharyya, R., Sharma, A., Gupta, D.K., et al.: Circular economy fertilization: Phycoremediated algal biomass as biofertilizers for sustainable crop production. J. Environ. Manage. **287**, 112295 (2021)

Sharma, N., Jaiswal, K.K., Kumar, V., Vlaskin, M.S., Nanda, M., Rautela, I., et al.: Effect of catalyst and temperature on the quality and productivity of HTL bio-oil from microalgae: A review. Renewable Energy **174**, 810–822 (2021)

Sharma, A., Sharma, P., Deepak, B., Mona, S.: Harnessing algal and agri-waste residue biochar for sustainable soil amelioration: a review. In: Proceedings of the Indian National Science Academy, pp. 1–10 (2024)

Sharma, P., Bano, A., Singh, S.P., Varjani, S., Tong, Y.W.: Sustainable organic waste management and future directions for environmental protection and techno-economic perspectives. Curr. Pollut. Rep. **10**(3), 459–477 (2024)

Sharma, K., Rajan, S., Nayak, S.K.: Water pollution: primary sources and associated human health hazards with special emphasis on rural areas. In: Water Resources Management for Rural Development, pp. 3–14. Elsevier (2024)

Sharma, A., Malaviya, P.: Floating treatment wetlands: a futuristic approach for industrial wastewater treatment. Emerging Developments in Constructed Wetlands (2025)

Shukla, S., Upadhyay, D., Mishra, A., Jindal, T., Shukla, K.: The impact on soil ecology of the algal community. In: Soil Algae: Morphology, Ecology and Biotechnological Applications, pp. 193–219. Springer Nature, Singapore (2025)

Sible, C.N., Seebauer, J.R., Below, F.E.: Plant biostimulants: a categorical review, their implications for row crop production, and relation to soil health indicators. Agronomy (2021)

Singh, M., Sharma, V., Gupta, R.: Biovalorization of potato peel waste: an overview. Roots, Tubers, and Bulb Crop Wastes: Management by Biorefinery Approaches, pp. 19–41 (2024)

Singh, V.P., Kumar, A., Srivastava, A., Kumar, A.: Advancing environmental sustainability: a comprehensive review on alleviating carbon footprint and its application in microalgal fermentation and bioremediation. Environment, Development and Sustainability, pp 1–46 (2024)

Singh, V., Mishra, S., Singh, D.: Soil Algae as a Feedstock for Biofuel Production and Value-Added Products. In Soil Algae: Morphology, Ecology and Biotechnological Applications, pp. 389–414. Springer Nature, Singapore (2025)

Song, X., Liu, J., Feng, Y., Zhou, C., Li, X., Yan, X., et al.: Microalgae-based biofertilizers improve fertility and microbial community structures in the soil of potted tomato. Front. Plant Sci. **15**, 1461945 (2024)

Suchocki, T.: Energy utilization of rapeseed biomass in Europe: a review of current and innovative applications. Energies (19961073) (2024)

Swaminaathan, P., Saravanan, A., Thamarai, P.: Utilization of bioresources for high-value bioproducts production: Sustainability and perspectives in circular bioeconomy. Sustain. Energy Technol. Assess. **63**, 103672 (2024)

Szparaga, A., Kocira, S., Kapusta, I., Zaguła, G.: Solid–liquid extraction of bioactive compounds as a green alternative for developing novel biostimulant from Linum usitatissimum L. Chem. Biol. Technol. Agric. **10**(1), 108 (2023)

Tahir, F., Ashfaq, H., Khan, A.Z., Amin, M., Akbar, I., Malik, H.A., et al.: Emerging trends in algae farming on non-arable lands for resource reclamation, recycling, and mitigation of climate change-driven food security challenges. Rev. Environ. Sci. Bio/technol. **23**(3), 869–896 (2024)

Tawfik, A., Niaz, H., Qadeer, K., Qyyum, M.A., Liu, J.J., Lee, M.: Valorization of algal cells for biomass and bioenergy production from wastewater: sustainable strategies, challenges, and techno-economic limitations. Renew. Sustain. Energy Rev. **157**, 112024 (2022)

Titus, B.D., Brown, K., Helmisaari, H.S., Vanguelova, E., Stupak, I., Evans, A., et al.: Sustainable forest biomass: a review of current residue harvesting guidelines. Energy Sustain. Soc. **11**, 1–32 (2021)

Tyagi, S., Rao, N.R.H., Thielemans, W., Muylaert, K.: Comparing sedimentation and dissolved air flotation for microalgae harvesting: influence of ionic strength on floc size and separation efficiency. J. Appl. Phycol. **37**(1), 221–231 (2025)

Ufitikirezi, J.D.D.M., Filip, M., Ghorbani, M., Zoubek, T., Olšan, P., Bumbálek, R., et al.: Agricultural waste valorization: exploring environmentally friendly approaches to bioenergy conversion. Sustainability **16**(9), 3617 (2024)

Ullmann, J., Grimm, D.: Algae and their potential for a future bioeconomy, landless food production, and the socio-economic impact of an algae industry. Organic Agriculture. springer.com (2021)

Vidhya, C.V., Nandhini, S., Sheela, J.M., Reenaa, M.: Bioenergy-based sustainable bioeconomy—perspectives and challenges. Circular Bioeconomy Perspectives in Sustainable Bioenergy Production, pp. 5–36 (2024)

Vishwakarma, P., Pal, V., Nagar, J., Rai, R.: Biofertilizers and soil amendments from algae: production and agronomic benefits. In: Value Added Products From Bioalgae Based Biorefineries: Opportunities and Challenges, pp. 343–363. Springer Nature, Singapore (2024)

Wan, L., Cao, L., Song, C., Cao, X., Zhou, Y.: Regulation of the nutrient cycle pathway and the microbial loop structure by different types of dissolved organic matter decomposition in lakes. Environ. Sci. Technol. **57**(1), 297–309 (2022)

Wang, L., Zhang, B.: Culture de microalgues sur eaux usées agricoles pour le recyclage de l'énergie, de l'eau et des nutriments des engrais. Dans Gestion intégrée des eaux usées et valorisation à l'aide de cultures d'algues, pp. 235–264 (2022)

Wibisono, D.A.S., Saw, C.Y., Wu, T.Y., Chau, C.F.: Advancing industrial food byproduct management: strategies, technologies, and progress in waste reduction. Processes (2025)

Wu, Y., Li, X., Zeng, H., Zhong, X., Kuang, S.: Analysis of carbon sink benefits from comprehensive soil and water conservation in the loess hilly gently slope aeolian sand region. Water (2024)

Yaashikaa, P.R., Devi, M.K., Kumar, P.S.: Algal biofuels: technological perspective on cultivation, fuel extraction and engineering genetic pathway for enhancing productivity. Fuel (2022)

Yang, W., Gao, P., Ye, Z., Chen, F., Zhu, L.: Micro/nano-plastics and microalgae in aquatic environment: influence factor, interaction, and molecular mechanisms. Sci. Tot. Environ. (2024)

Yang, Y., Li, R., Zhao, J., Qiu, Y., Song, M., Yin, D., Chen, X.: Stable isotope tracer IAA-induced cultivation of microalgae with contaminated carbon sources in multiple medias: Carbon fixation and biomass conversion. Chem. Eng. J. **499**, 156287 (2024)

Yang, L., Song, X., Feng, Y., Qiu, X., Yan, X., Ruan, R., Cheng, P.: Improving soil structure and function for continuous cropping of sweet melon using microalgae-based biological fertilizer. Algal Res. **85**, 103868 (2025)

Yap, J.K., Sankaran, R., Chew, K.W., Munawaroh, H.S.H., Ho, S.H., Banu, J.R., Show, P.L.: Advancement of green technologies: a comprehensive review on the potential application of microalgae biomass. Chemosphere **281**, 130886 (2021)

Ye, J., Zhu, Z., Song, Z., Xu, H., Xu, T., Liu, H.: Application of microbubble air flotation to harvest Microcystis sp. from agriculture wastewater: the regulation and mechanisms. Biotechnol. Bioeng. **121**(12), 3742–3753 (2024)

Yi, S., Zhang, A.H., Huang, J., Yao, T., Feng, B., Zhou, X., et al.: Maximizing polysaccharides and phycoerythrin in porphyridium purpureum via the addition of exogenous compounds: a response-surface-methodology approach. Mar. Drugs **22**(3), 138 (2024)

Yurkevich, M., Suleymanov, R., Ikkonen, E., Dorogaya, E., Bakhmet, O.: Effect of brown algae (*Fucus vesiculosus* L.) on humus and chemical properties of soils of different type and postgermination growth of cucumber seedlings. Agronomy **12**(9), 1991 (2022)

Zhang, S., Qamar, S.A., Junaid, M., Munir, B., Badar, Q., Bilal, M.: Algal polysaccharides-based nanoparticles for targeted drug delivery applications. Starch-Stärke **74**(7–8), 2200014 (2022)

Zhang, J., Yong, H., Lv, N.: Balancing productivity and sustainability: insights into cultivated land use efficiency in arid region of Northwest China. J. Knowl. Econ. (2024)

Zhang, Q., Guan, Y., Zhang, Z., Dong, S., Yuan, T., Ruan, Z., Chen, M.: Sustainable microalgae cultivation: A comprehensive review of open and enclosed systems for biofuel and high value compound production. In: E3S Web of Conferences, vol. 577, p. 01008. EDP Sciences, e3s (2024)

Zhang, Z., Xu, M., Fan, Y., Zhang, L., Wang, H.: Using microalgae to reduce the use of conventional fertilizers in hydroponics and soil-based cultivation. Sci. Tot. Environ. (2024)

Zhou, J., Tong, L., Guo, H.: Advances in primary recovery for algae bioreactors: the centrifugation role. Algal Bioreactors (2025)

# Valorization of Food Waste for Soil Water Retention

Redouane Ahnyne, Rachid Rahhal, Najwa Hassou, Said Gmouh, and Abdellatif Chakor

**Abstract**

In the context of climate change, increasing water scarcity, and the growing accumulation of organic waste, the valorization of food waste into water-retaining soil amendments emerges as a promising innovation for sustainable agriculture. This chapter presents the results of an in-depth literature review and experimental studies on the transformation of food waste into materials capable of enhancing soil water retention capacity. Food waste extracts, processed through biotechnological methods (composting, fermentation, enzymatic treatment), have shown beneficial properties in terms of porosity, biodegradability, and nutrient contribution. These characteristics promote moisture retention in the root zone and improve crop productivity. The integration of such bio-based materials into agricultural practices also presents notable economic benefits, including reduced irrigation costs, the valorization of by-products, and the creation of new market opportunities. This chapter will address: • The physicochemical mechanisms of interaction between valorized waste and soil, • The observed water retention efficiency criteria, • The impacts on soil health and crop yield, • The scalability potential and the technical and economic barriers to be overcome.

**Keywords**

Valorization · Food waste · Water retention · Biotechnology · Soil · Sustainable agriculture · Circular economy

**Abbreviations**

| | |
|---|---|
| AWC | Available Water Capacity |
| $CH_4$ | Methane |
| $CO_2$ | Carbon Dioxide |
| COmpSA | Composting Systems Analysis |
| DB | Desert Biochar |
| EU | European Union |
| FC | Field Capacity |
| FW | Food Waste |
| FW-NSB | Food Waste–Nano Structured Biocomposite |
| GHG | Greenhouse Gases |
| HVCs | High-Value Compounds |
| HVMs | High-Value Molecules |
| LAB | Lactic Acid Bacteria |
| NA | New Agriculture |
| NAT | New Agriculture Techniques |
| NPK | Nitrogen, Phosphorus, Potassium |
| OFMSW | Organic Fraction of Municipal Solid Waste |

R. Ahnyne (✉) · R. Rahhal · N. Hassou
Higher Institute of Nursing Professions and Health Techniques (ISPITS of Rabat), Ministry of Health and Social Protection—Morocco, Rabat, Morocco
e-mail: redouane@ahnyne.com

R. Ahnyne · A. Chakor
Research Laboratory in Management of Organizations, Business Law, and Sustainable Development (LARMODAD), Faculty of Law, Economics and Social Sciences, Souissi, Mohammed V University of Rabat, Rabat, Morocco

R. Rahhal
Laboratory of Ecology and Environment, Faculty of Sciences Ben M'Sik, University Hassan II of Casablanca, Casablanca, Morocco

N. Hassou
Team Research of Virology, Oncology, and Biotechnologies, Laboratory of Virology, Oncology, Biosciences, Environment and New Energies, Faculty of Sciences and Techniques Mohammedia, University Hassan II of Casablanca, Casablanca, Morocco

S. Gmouh
Laboratory of Engineering and Materials (LIMAT) , Faculty of Sciences Ben M'Sik, Hassan II University of Casablanca, Casablanca, Morocco

A. Chakor
Faculty of Legal, Economic and Social Sciences, Souissi, Mohammed V University, Rabat, Morocco



O. Jamal Eddine et al. (eds.), *Plant Pathology, Fungal Diversity, and Biotechnological Advances in Agriculture*, Sustainable Landscape Planning and Natural Resources Management, https://doi.org/10.1007/978-3-031-97666-7_10

| | |
|---|---|
| PGPR | Plant Growth Promoting Rhizobacteria |
| RD | Recycled Digestate |
| SWRC | Soil Water Retention Characteristic |
| VOC | Volatile Organic Compounds |
| WEFW | Water-Energy-Food-Waste Nexus |

## 1 Introduction

The growing energy and environmental challenges are increasing the need for a rapid development of innovative methods in valorization of food waste into soil water retaining material. Data from different valorization processes suggest that food waste valorization for soil water retention has a potential as a "new" mode to manage food wastes that cannot be prevented or quantitatively reduced, and may represent a possible sustainable resolution to rapidly divert this problematic waste from landfilling. Even though some positive findings and contributions to practical use of the valorized products have been accomplished, overall worldwide, the advancements in a "real" use of food waste as a porous and highly degradable soil water retention material to support agronomical and soil hydrology use is still relatively new and there are many barriers that need to be addressed. Food waste is one of the major components of municipal solid waste due to the global amounts produced and land growing.

Food waste is a demand for human consumption, not including consumer peeling and other parts of foods removed for proper, inedible parts, such as bones and peels. It consists of food loss and/or food recovered intended for human consumption, resulting from processing of packing of food materials, and features of product with the product, manufacturer, process and also of shopping and consumer handling. A huge amount of food waste is generated worldwide, which comprises approximately 30% of the total food production or will be about 1.3 billion ton/year (Roy et al. 2023; Gustavsson et al. 2011). The management of food waste, as part of municipal solid waste, results in unacceptable environmental impacts and poses potential risks to human health. Due to their energy consumption, these wastes are predominantly landfilled or incinerated, contributing to soil degradation through possible groundwater contamination and the emission of methane, a potent greenhouse gas. Significant research efforts are currently underway to reduce the volume of waste, leading to solutions such as recycling and composting. However, the remaining residual waste still needs to be disposed of through landfilling or incineration, thus continuing to degrade the quality of soil, water, and air (Koshariya et al. 2024; Liu et al. 2023a, b; Roy et al. 2023; Singh et al. 2022).

## 2 Understanding Food Waste

Food waste includes discarded uneaten food at all stages of its lifecycle (e.g., preharvest and postharvest). It includes edible items, as well as inedible, and can take on many forms based on local regulations and consumer behavior and habits, as well as culture. In 2018, the amount of food waste produced worldwide was about 1.3 billion tonnes, generating significant environmental and economic hurdles due to waste disposal costs and lost resources (Roy et al. 2023). This has been improved through circular bioeconomy frameworks, especially the valorization of food waste, to convert that waste into valuable resources. Research suggests animal and catered food wastes are energy-dense wastes (higher in energy), whereas plant food waste is less energy-dense waste (lower in energy). Urbanization will perhaps perpetuate the problem of food waste, ultimately having implications on water availability due to decomposition. Although there are challenges to food waste, which would need to be discovered in a fast and cost-effective way while maintaining safety, soil, can provide the potential for agriculture to naturally supply water as opposed to more complex approaches such as piped systems (Waqas et al. 2023; Hajam et al. 2023; Khalida et al. 2022; Perman et al. 2022; Lu and Xu 2021).

### 2.1 Definition and Types of Food Waste

Food waste is a multi-faceted issue that intersects with different elements of everyday life: environmental, social, and economic. There is also an almost emerging field of calculating food waste which is problematic, related to both lack of visibility and uncertainty of waste. This section outlines definitions and categories of food waste (Sharma et al. 2024a, b, c; Di Vaio et al. 2023; Lozano Lazo et al. 2023; Laso et al. 2022).

Food waste is a complicated and multifaceted problem simultaneously involving complicated aspects of socio-economic and environmental nature (Fig. 1). However, today it is regarded as one of the most poignant challenges confronting contemporary global society. For avant-garde scientific and technological issues, food waste accounting remains in its infancy. This problem may be linked to inherently high complexity of accounting process, both lack of visibility and non-visibility of waste. Various food processing

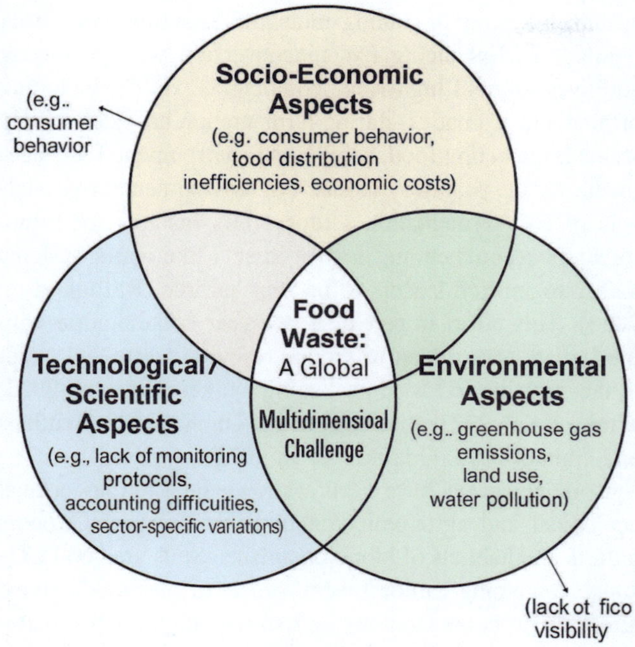


**Fig. 1** Multidimensional nature of food waste

sub-sector have different flow patterns, thus there is a need for sub-sector specific food waste monitoring protocols (Onyeaka et al. 2023; Roy et al. 2023; caldeira Carla et al. 2017; Faishal 2022; Brennan et al. 2021).

Fertile soil waste and food waste are closely interconnected. Fertile soil waste arises from the combustion processes to which food waste, particularly plant-based, is often subjected. The significant volume of food discarded throughout the marketing and consumption chain contributes to this issue, as much of this waste ultimately ends up being incinerated. Moreover, anthropological studies indicate that only a portion of post-harvest food actually reaches the market, with the remainder representing a hidden form of waste (Gascón Gutierrez 2018). Historically, the rise in commodity trading and market-oriented food systems has been accompanied by the emergence of new waste streams. In pre-industrial societies, however, unused food was largely reintegrated into the agro-pastoral cycle. These traditional systems, characterized by nearly complete physical and metabolic closure, eventually suffered from the accumulation of toxic substances in the soil, which gradually reduced fertility and crop yields (Le Pera et al. 2023; O'Connor et al. 2022; Anunobi 2022; Nanda and Berruti 2021).

### 2.2 Global Statistics and Impact

Annually, about 35% of food produced for human consumption is wasted globally. This waste, totaling 540 million tonnes at the retail and consumer stages, coexists with a significant number of starving individuals worldwide. Projections indicate that total food waste could rise by 66% by 2030, with an additional 93 million tonnes expected at the retail stage. From 2015 to 2018, over a billion tonnes of food worth 680 billion USD were wasted. Low and middle-income countries alone lose 345 billion USD annually due to food waste, with India wasting 33% of its food production, resulting in 240 billion USD loss. Agriculture also wastes about 32% of the water used yearly, underscoring the need to transform this waste into a valuable resource (Roy et al. 2023; Liu et al. 2023a, b; Singh et al. 2022; Rajeh et al. 2021).

## 3 Soil Water Retention: Importance and Challenges

Efficient water resources depend on understanding soil water retention characteristics, which reflect the moisture held after excess water drains away. This affects water availability for plants and irrigation efficiency. Food waste (FW) generates 1.3 billion tons annually, creating economic, environmental, and social issues for cities, including management costs, aesthetic problems, pest attraction, and food contamination risks. In Europe, 142 million individuals cannot afford regular quality meals. Wasted grains equate to calories for 780 million people, surpassing India's undernourished population. Traditional practices influence soil moisture, temperature control, water consumption, and labor productivity. However, challenges remain in sustainable food waste valorization concerning its effects on soil water retention and nutrient dynamics over time (Gatto and Chepeliev 2024; Durán-Sandoval et al. 2023; Singh et al. 2022; Skaf et al. 2021; Parfitt et al. 2021).

Several factors influence food waste on water retention, including type, quantity, size, and degradation. This study evaluates powder food waste (PP) and ground food waste (PG) in different sized watersheds. PP and PG take almost twice as long to degrade compared to other waste after 30 days. The moisture content of PC (279 mm) and PG (190 mm) in wet soils is crucial for evaporation. Waste-fed specimens show greater water retention than non-fed soils, with enhanced nutrient addition having a minimal effect (Eide 2025; Farooq et al. 2024; Mondal et al. 2023; Khatami et al. 2022; Tabbara 2021).

### 3.1 Role of Soil in Agriculture

Current approaches to sustainable land management emphasize the need to safeguard ecosystems by increasing agricultural productivity while preventing land degradation. For crops, soil's water retention is important, and assessing its organic carbon content is vital for soil's role in agriculture. The decline in biodiversity can be counterbalanced by increasing organic carbon, which will enhance soil and ecosystem services and fertility, thus benefiting crop growth in the Mediterranean region. Estimation of available phosphorus is important for environmental underuse of resources. Waste products such as biogas digestates serve as fertilizers and biostimulants, enhancing sustainable plant growth through positive rhizosphere interactions. Plant growth-promoting rhizobacteria (PGPR) and lactic acid bacteria (LAB) highlight the importance of following circular economy principles for growth strategies. Interactions can lead to effective soil sustainability approaches. Results may help formulate remediation protocols for biochar application in restoring soil quality in the degraded lands to support cultivation in semi-arid regions like the Mongolian Plateau (Koorneef et al. 2024; Tang et al. 2023; Lehmann et al. 2021; Sultana 2017).

### 3.2 Factors Affecting Soil Water Retention

Good soil water retention characteristics (SWRC) are essential for functions like crop growth and filtration. The elements impacting SWRC are soil type and structure, bulk density, organic matter, hydraulic pressure, temperature, contaminants, compaction, slope, aggregate state, and minerals. In dry regions, over 60 billion $m^3$ of freshwater is annually utilized for irrigation (Zhang et al. 2023; Rodrigues et al. 2023; Vereecken et al. 2022; Assouline 2021; He et al. 2021).

The retention of water in the root zone is one of the principle goals of managing water and soil. About 50% of rainfall evaporates before reaching ground, and the runoff takes place at 75 $mmh^{-1}$ or higher, with 40–50% of this runoff couting to the sea. Usually, 10–30% of water is consumed by crop plants, while the rest is lost due to evaporation or runoff. In arid areas, precipitation is greater than evaporation, and water-saving is particularly important. Limited rainwater poses a challenge for plant growth, necessitating strategies to stabilize crop production and retain soil moisture for farming (Assi et al. 2018; Stocker et al. 2023; Bhat et al. 2022; Babaeian et al. 2021; Amare and Desta 2021; Yu et al. 2021).

## 4 Biotechnological Innovations in Food Waste Valorization

Food waste (FW) from retail distribution poses significant challenges with high disposal costs and public scrutiny. The Italian mass-retail sector has explored converting unsold FW through fermentation into high-value molecules (HVMs), a sustainable process gaining industrial attention amid strict regulations. Enhancing FW bioconversion with packaging additives to yield high-value compounds (HVCs, including proline and aromatics) during fermentation has been widely studied, impacting food, health, and environmental management over the past two decades. An environmental risk analysis of FW fermentation setups offers insights for future research, recommending improvements like alarm systems and nitrogen-rich buffers to prevent hazards (Razouk et al. 2024). This effort is part of a two-year collaboration with the Italian government to reduce resource waste and adapt to the evolving EU food packaging market (Orr et al. 2023; Winkler et al. 2023; Riesenegger and Hübner 2022; Beullens and Ghiami 2022; Guarnieri et al. 2021).

Soils nowadays have been acknowledged to play critical ecological and agronomic functions as terrogenic process buffers and habitats of biotic organisms. Soils are good $CO_2$ sinks, containing carbon several orders of magnitude above atmospheric concentrations and storing high-quality material from composted waste. Low-quality compost, however, results in fecal contamination of sea soils, leading to colossal losses in marine ecosystems. Effective technologies are needed to enhance soil water retention and sustainability. Studies on biochar-amended sandy soil indicated disruptions in soil water retention despite the input of decomposed biomass, with a possibility of gaseous waste accumulation. In contrast, biotechnological advancement is promising in improving water retention, mechanical strength, slow water release, and bioabsorption of toxic PAHs from waste incineration (Zhang et al. 2024; Qi and Chmura 2023; Naorem et al. 2022; Sugimura et al. 2022; Lal et al. 2021; Terrer et al. 2021).

Urban areas and the food industry generate significant bio-waste, contributing to the waste issue. The EU alone produces 88 million tons of food waste annually, much of which is landfilled or burned. Although bio-waste can act as fertilizer, it often isn't used for that due to storage challenges. Nonetheless, bio-waste, rich in polysaccharides and proteins, can enhance soil moisture retention (Ch et al. 2025; Rasheed et al. 2025; Priya et al. 2025; Guo et al. 2024; Sudheer et al. 2023; Skrzypczak et al. 2022; Ur Rehman et al. 2022).

Composting is valued for waste recycling, especially biowaste from hygienic conversion. Key safety metrics, like aerobiological contamination and Salmonella survival, are examined. The understanding of Composting Systems Analysis (COmpSA) is enhanced through examples. Four studies are discussed: a review of COmpSA as a method and three focusing on $N_2O$ emissions, volatile organic compounds, and pollutants (Bremaghani 2024; Kumari et al. 2024a, b; Sharma et al. 2024a, b, c; Waqas et al. 2023 De Boni et al. 2022).

## 5 Economic Benefits of Valorizing Food Waste

The world's population is increasing at a very rapid pace, and this results in increased demand for food, thereby more waste product of food production and processing. This impacts the environment, which has a vital role in treating solid, liquid, and gaseous waste. Some treatment of waste may further deteriorate environmental quality. Hence, an integrated approach needs to be contemplated on the basis of environmental, social, and economic factors. Various technologies can valorize food wastage, for instance, by using it as a thickener during paper manufacturing. Hydrogelation, biocomposite formation, and adsorption of wastages from food can be of huge economic value. The present research is an endeavor to analyze some methods of valorizing food wastages on a commercial level, so that researchers and decision-makers can take well-informed decisions for efficient waste management and economic gain (Ch et al. 2025; Panda et al. 2024; Xie et al. 2024; Adi Sulianto et al. 2023; Zhu et al. 2022; Čechmánková et al. 2021).

Food loss is a critical issue in the food chain that occurs at various stages pre- and post-harvest. Over one-third of food produced globally—about 1.3 billion tonnes—is wasted annually and consists of edible and inedible parts. Edible loss is usually attributed to cultural practices, e.g., overbuying and perishability, while non-edible loss is attributed to preparation and processing. The type of waste varies by geographical area, culture, and income levels. Higher revenues, particularly from take-away and delivered food, have played a significant role in increased amounts of waste, with 57% in the UK and 42% in the U.S. being caused by this trend (Junaid and Gokce 2024; Singh et al. 2022; Barrera and Hertel 2021; Kummu et al. 2021).

## 6 Case Studies in Food Waste Valorization

Food waste is being valorized in order to avoid pollution. There is significant food waste in the form of households, agriculture, processing, and hospitality worldwide, leading to ecological disturbance as well as to the adverse effect on water and soil health (Roy et al. 2023). Biodegradation also leads to greenhouse gases (GHG), which are highly more active compared to $CO_2$. Ineffective landfilling of food waste also contributes to GHG emissions. Therefore, valorization processes must be established because such processes yield new biodegradable products by using various processing techniques (Sarker et al. 2024; Ansari et al. 2024; Okuthe 2024; Gupta et al. 2024; Râpă et al. 2024).

Food waste occurs at various stages from farm to fork, with about one-third attributed to major countries along the production, processing, marketing, consumption, and spoilage. Household, school, and restaurant leftovers and rotting food are obvious sources. Unharvested and unsellable crops are wastes, which means more than half is produced after the end stages of the food life cycle. Animal products provide the highest microbiological and anaerobic degradability, whereas fruits and vegetables are more challenged because they have lower anaerobic biodegradability in comparison to milk and meat. Spoilage occurs in the case of fruits, vegetables, meat, and milk. Livestock waste utilization in production is restricted by antibiotic and drug residues, and uncontrolled disposal of meat and milk can form carcinogens in water bodies. The wastes from fruits and vegetables contain heavy metals such as stainless steel, copper, aluminum, and zinc. Wild and native crops further help communities to control soil erosion and retain water since their erodible rates stand at 0.5–2.4 m in the riverbed of loamy soil (Bhatia et al. 2023; Capanoglu et al. 2022; Santagata et al. 2021).

In Italy, municipal solid waste management effectively separates biowastes. Furthermore, anti-food waste campaigns and innovative valorization practices gain support from food chain sector associations and stakeholders. Additionally, a mix of innovative valorization and processing methods is proposed for addressing various industrial and agri-food wastes alongside operations in Technopolis city (Priya and Ramesh Kumar 2025; Rodríguez-Espinosa et al. 2023; Slavova and Doneva 2022).

Compost application inefficiency arises from several factors, including insufficient stabilization, inadequate personnel training, and poor technology, which can result in incomplete sterilization and pathogen survival. Non-biodegradable waste may physically or chemically contaminate compost, risking foodborne pathogens in crops. Issues with microflora, specific weeds, polluted water, and suboptimal management practices also limit compost's effectiveness. Addressing these factors is vital for maximizing compost's benefits in plant protection (Manual 2023; Goodwyn et al. 2023; Lepesteur 2022; Zhan et al. 2022; Thakali and MacRae 2021; Black et al. 2021).

The use of bio-wastes is expected to positively impact soil ecological functionality by improving soil water potential and temperature buffering. Future agricultural strategies may leverage a spatially optimized distribution of composted agri-food wastes to boost soil water retention, benefiting both rainfed and irrigated crops and aiding soil restoration in water-scarce areas. These methods can predict the effects of compost on soil water potential and temperature across various conditions, showcasing their potential to enhance agricultural productivity amid climatic challenges. Integrating these practices could help close yield gaps by alleviating water stress, improving soil quality, and boosting water use efficiency (Mukhopadhyay 2025; Sanchez et al.

2024; Mustafa et al. 2022; Babla et al. 2022; Ozdemir et al. 2021; Anwar et al. 2021).

## 7 Impact on Soil Health and Crop Yield

Sustainability-driven approaches for valorizing surplus food and organic waste are increasingly gaining interest. Food waste can be repurposed through low- or no-tech solutions, such as organic fertilization, supplying essential nutrients. Transforming food waste into biofertilizers, biochar, and digestates has been shown to improve plant growth and soil health. Recycling techniques range from basic composting to more advanced methods like pyrolysis and anaerobic digestion, all offering environmental benefits by utilizing materials that would otherwise be discarded. Recent reviews have emphasized the potential of food waste as a biostimulant and soil improver. This article presents data from Bari biorefineries on by-product digestion and the effectiveness of resulting digestates in horticultural applications. While many studies have assessed the impact of digestate biofertilizers on crop yield and growth, fewer have focused on their effects on plant performance and soil quality under specific management practices. Variables such as soil adsorption capacity and application timing may influence competition with synthetic fertilizers. Research into reusing wasted bread as a soil amendment revealed significant enhancements in escarole development, including improved leaf quality and increased yields. The moisture content of the wasted bread affected the characteristics of the potting medium, thus influencing seedling growth. Ultimately, wasted bread proved to be an effective biochar, enhancing key traits of Mediterranean crops and reinforcing the contribution of food waste reuse to a circular bioeconomy (Pradhan et al. 2024; Kumari et al. 2024a, b; Upadhyay et al. 2023; Bian et al. 2022; Hashmi et al. 2021).

## 8 Barriers to Implementation

Barriers to implementation, from a waste management perspective, have been the subject of emerging interest in literature. The focus is on waste practices themselves and is relevant to common sustainability goals for waste management such as increased material recovery, resource conservation and successful disposal of waste that has no viable recovery options (Neville Menyuka et al. 2020). They analyzed waste practices in a comparative and contextual way, finding significant barriers confronted by organic waste beneficial use practices–namely composting and vermiculture (Bouchtaoui et al. 2024; Pereira and Fiore 2024; Rastegari et al. 2023; Chowdhury and Sarkar 2023).

Among the food waste recycling options, recycling through 'Feed' Fish, refers to vermiculture and composting with market or own consumption of seafood being a common application. In municipal contexts, the material is typically food waste but other feed materials are also observed. These are recyclables, first and foremost, but organic waste tends to dominate these flows and market conditions facilitate exploitation of food waste for this purpose. This includes relative abundance, low gate fees and a codigestion like substrate that can assist bioconversion of otherwise substandard feed materials (Li et al. 2024; Lelicińska-Serafin et al. 2023; Abrahamsson 2023; Ortiz-Sanchez et al. 2023; Misslin et al. 2022; Rut et al. 2021).

Two aspects of implementation barriers to food waste beneficial use are addressed as follows. On the one hand, barriers to best management practice creation for urban agricultural use of products are explored. This broad category includes deterring investment, standards, organizational capacity and supporting infrastructure. These are central areas of enabler effort typical in the development of novel waste management; they confront the systemic challenges of physiochemical limitations, feed market conditions and the dynamics of participant practices or experience. Additionally, there are several quality control considerations, notably regulatory parameters. On the other hand, the zoning restrictions of post-treatment facilities for high standard beneficial use products is increasingly enforced. Efforts toward recycling under the current regulatory climate are presented as well as the broader scope of barriers faced by less selective waste practices (Srinivasan and Yadav 2024; Costa et al. 2023; Kendall et al. 2022; Purwandani and Michaud 2021; Adhikari et al. 2021; Negash et al. 2021; Toxopeus and Polzin 2021).

## 9 Future Directions in Research

Food waste remains a global challenge despite numerous national and municipal waste management measures and regulations. Therefore, coordinated economic, ecologic and social actions must be taken towards the most beneficial outcome for the society. Innovative technologies, increasingly re-branded to seamlessly interlink the circular economy concept with the socio-economic vision 2030, focus on adopting a comprehensive approach, which unites ambitious technological innovation with the enhancement of corporate social responsibility (Razouk et al. 2024). In this context, circular economy approaches target keeping materials in use, namely, through re-designing existing processes making sure that the materials employed in these processes can be recycled or safely re-used, and they aim at sustaining the environment, simultaneously enabling the creation of local economic growth (Roy et al. 2023; Liu et al. 2023a, b; Faishal 2022; Zan et al. 2022; Cattaneo et al. 2021).

The food waste recovery hierarchy prioritizes actions aimed at preventing food waste and promoting sustainable management over disposal. Composting naturally converts part of the waste into manure, improving soil fertility and water retention. Modern chemical, biochemical, and microbiological methods also transform organic by-products into valuable materials and energy. Waste residues from these processes should be sorted and recycled according to the hierarchy, emphasizing the valorization of biological materials to conserve natural resources and support sustainable development. This approach can strengthen communities, reduce brain drain, and create high-quality jobs. Special attention is given to digestates, powerful organic fertilizers that enhance soil water retention, with proposals to optimize them through low-temperature drying and granulation, focusing on improving their NPK content (Mayerová et al. 2023; Cooke 2023; Häfner et al. 2022; Herrera et al. 2022; Karimi et al. 2022; Doyeni et al. 2021; Levin et al. 2021).

## 10 Conclusion

Annually, significant food waste is generated globally, wasting resources and harming the environment. Technological changes in production and consumption create complex waste streams, complicating management. This review addresses challenges and options for food waste valorization, emphasizing environmental and economic impacts, along with integrated strategies. Regional strategies are increasingly combining methods tailored to specific waste types and local characteristics.

Managing food and organic waste poses significant environmental and economic challenges. Reusing food waste is vital for cutting disposal costs and enhancing economic value through effective valorization processes. A proper valorization approach minimizes waste management costs while preserving resources. The economic impact of food waste closely ties to these methods. Developed nations have adopted strategies like source reduction, donation, and reuse in animal feed and organic amendments, while low- and middle-income countries use varied tactics. Food waste can act as a nutrient source and soil amendment for agriculture and urban uses. Studies show bioaugmentation boosts the value of urban waste digestates. While digestate type didn't affect vegetable production, lettuce grown in lignocellulosic-treated mediums exhibited favorable outcomes, benefiting urban soil mixtures and vegetable growth.

## References


Abrahamsson, S.: A defense of waste: the case of municipal food recycling in Sweden. Environ. Sociol. (2023)

Adhikari, M., Devkota, H.R., Cesuroglu, T.: Barriers to and facilitators of diabetes self-management practices in Rupandehi. BMC public health, Nepal-multiple stakeholders' perspective (2021)

Adi Sulianto, A., Adiyaksa, I.P., Wibisono, Y., Khan, E., Ivanov, A., Drannikov, A., et al.: From fruit waste to hydrogels for agricultural applications. Clean Technol. **6**(1), 1–17 (2023)

Amare, G., Desta, B.: Coloured plastic mulches: impact on soil properties and crop productivity. Chemical and Biological Technologies in Agriculture (2021)

Ansari, S.A., Kumar, T., Sawarkar, R., Gobade, M., Khan, D., Singh, L.: Valorization of food waste: a comprehensive review of individual technologies for producing bio-based products. J. Environ. Manage. **364**, 121439 (2024)

Anunobi, T.J.: Hazardous effects of plastic wastes on land biodiversity: a review. Zoologist (the) (2022). ajol.info

Anwar, H., Wang, X., Hussain, A., Rafay, M., Ahmad, M., Latif, M., et al.: Comparative effects of bio-wastes in combination with plant growth-promoting bacteria on growth and productivity of okra. Agronomy **11**(10), 2065 (2021)

Assi, A., Mohtar, R., Braudeau, E.: Soil Pedostructure-Based Method for Calculating the Soil-Water Holding Properties (2018)

Assouline, S.: What can we learn from the water retention characteristic of a soil regarding its hydrological and agricultural functions? Review and analysis of actual knowledge. Water Resources Research (2021)

Babaeian, E., Paheding, S., Siddique, N., Devabhaktuni, V.K., Tuller, M.: Estimation of root zone soil moisture from ground and remotely sensed soil information with multisensor data fusion and automated machine learning. Remote Sens. Environ. **260**, 112434 (2021)

Babla, M., Katwal, U., Yong, M.T., Jahandari, S., Rahme, M., Chen, Z.H., Tao, Z.: Value-added products as soil conditioners for sustainable agriculture. Resour. Conserv. Recycl. **178**, 106079 (2022)

Barrera, E.L., Hertel, T.: Global food waste across the income spectrum: implications for food prices, production and resource use. Food Policy (2021)

Beullens, P., Ghiami, Y.: Waste reduction in the supply chain of a deteriorating food item–Impact of supply structure on retailer performance. Eur. J. Oper. Res. (2022)

Bhat, S.A., Bashir, O., Haq, S.A.U., Amin, T., Rafiq, A., Ali, M., et al.: Phytoremediation of heavy metals in soil and water: an eco-friendly, sustainable and multidisciplinary approach. Chemosphere **303**, 134788 (2022)

Bhatia, L., Jha, H., Sarkar, T., Sarangi, P.K.: Food waste utilization for reducing carbon footprints towards sustainable and cleaner environment: a review. Int. J. Environ. Res. Public Health **20**(3), 2318 (2023)

Bian, R., Shi, W., Luo, J., Li, W., Wang, Y., Joseph, S., et al.: Copyrolysis of food waste and rice husk to biochar to create a sustainable resource for soil amendment: A pilot-scale case study in Jinhua, China. J. Clean. Prod. **347**, 131269 (2022)

Black, Z., Balta, I., Black, L., Naughton, P.J., Dooley, J.S., Corcionivoschi, N.: The fate of foodborne pathogens in manure treated soil. Front. Microbiol. **12**, 781357 (2021)

Bouchtaoui, E.M., Fahr, M., Smouni, A., Azim, K., Lahlali, R., Mokrini, F.: Harnessing compost and vermicompost for sustainable and effective management of plant-parasitic nematodes in agricultural systems: a critical review. Physiological and Molecular Plant Pathology, 102363 (2024)
Bremaghani, A.: Utilization of organic waste in compost fertilizer production: implications for sustainable agriculture and nutrient management. Law and Economics (2024)
Brennan, L., Langley, S., Verghese, K., Lockrey, S., Ryder, M., Francis, C., et al.: The role of packaging in fighting food waste: a systematised review of consumer perceptions of packaging. J. Clean. Prod. **281**, 125276 (2021)
Caldeira Carla, P.A.T.I.N.H.A., Sara, C.O.R.R.A.D.O., Serenella, S.A.L.A.: Food Waste Accounting—Methodologies, Challenges and Opportunities (2017)
Capanoglu, E., Nemli, E., Tomas-Barberan, F.: Novel approaches in the valorization of agricultural wastes and their applications. J. Agric. Food Chem. **70**(23), 6787–6804 (2022)
Cattaneo, A., Sánchez, M.V., Torero, M., Vos, R.: Reducing food loss and waste: five challenges for policy and research. Food Policy (2021)
Čechmánková, J., Skála, J., Sedlařík, V., Duřpeková, S., Drbohlav, J., Šalaková, A., Vácha, R.: The synergic effect of whey-based hydrogel amendment on soil water holding capacity and availability of nutrients for more efficient valorization of dairy by-products. Sustainability **13**(19), 10701 (2021)
Ch, A., Athulya, K.R., Rasheed, R.F., Kumar, A.C.: Biowaste to hydrogels: a sustainable solution for waste reduction with potential applications. ES Chem. Sustain. **2**, 1366 (2025)
Chowdhury, A., Sarkar, A.: Vermicomposting—the sustainable solid waste management. In: Waste Management and Resource Recycling in the Developing World, pp. 701–719. Elsevier (2023)
Cooke, J.: Study and prediction of the impact of anaerobic digestion process parameters on the composition of digestates, and their effect on the structural stability of soils (2023)
Costa, F., Frecassetti, S., Rossini, M., Portioli-Staudacher, A.: Industry 4.0 digital technologies enhancing sustainability: applications and barriers from the agricultural industry in an emerging economy. J. Clean. Prod. **408**, 137208 (2023)
De Boni, A., Ottomano Palmisano, G., De Angelis, M., Minervini, F.: Challenges for a sustainable food supply chain: a review on food losses and waste. Sustainability **14**(24), 16764 (2022)
Di Vaio, A., Hasan, S., Palladino, R., Hassan, R.: The transition towards circular economy and waste within accounting and accountability models: a systematic literature review and conceptual framework. Environ. Dev. Sustain. **25**(1), 734–810 (2023)
Doyeni, M.O., Stulpinaite, U., Baksinskaite, A., Suproniene, S., Tilvikiene, V.: The effectiveness of digestate use for fertilization in an agricultural cropping system. Plants **10**(8), 1734 (2021)
Durán-Sandoval, D., Durán-Romero, G., Uleri, F.: How much food loss and waste do countries with problems with food security generate?. Agriculture (2023)
Eide, L.H.: Alternative Protein Sources for Farmed Atlantic Salmon Evaluated Under Commercial-Scale Field Conditions (2025)
Faishal, A.: Laws and regulations regarding food waste management as a function of environmental protection in a developing nation. Int. J. Crim. Justice Sci. (2022)
Farooq, M., Islam, H.U., Khan, A., Khan, S., Mohammad, S., Waleed, A.A.: Remodeling in microbial fuel cell (MFC) design and parameters for the sustained production of electricity. Int. J. Environ. Agric. Biotechnol. **9**(1), 126–139 (2024)
Gascón Gutierrez, J.: Food Waste: A Political Ecology Approach (2018)
Gatto, A., Chepeliev, M.: Global food loss and waste estimates show increasing nutritional and environmental pressures. Nat. Food (2024)
Goodwyn, B., Millner, P., Punchihewage-Don, A.J., Schwarz, M., Hashem, F., Bowers, J., et al.: Survival and persistence of foodborne pathogenic and indicator bacteria in spatially separated integrated crop-livestock farms. Front. Sustain. Food Syst. **7**, 1287371 (2023)
Guarnieri, P., de Aguiar, R.C., Thomé, K.M., Watanabe, E.A.D.M.: The role of logistics in food waste reduction in wholesalers and small retailers of fruits and vegetables: a multiple case study. Logistics **5**(4), 77 (2021)
Guo, J., Zhang, Y., Fang, J., Ma, Z., Li, C., Yan, M., et al.: Reduction and reuse of forestry and agricultural bio-waste through innovative green utilization approaches: a review. Forests **15**(8), 1372 (2024)
Gupta, R.K., Ali, E.A., Abd El Gawad, F., Daood, V.M., Sabry, H., Karunanithi, S., Srivastav, P.P.: Valorization of fruits and vegetables waste byproducts for development of sustainable food packaging application. Waste Manag. Bull. (2024)
Gustavsson, J., Cederberg, C., Sonesson, U., Van Otterdijk, R., Meybeck, A.: Pertes et gaspillages alimentaires dans le monde: ampleur, causes et prévention, Rome, FAO (2011)
Häfner, F., Hartung, J., Möller, K.: Digestate composition affecting N fertiliser value and C mineralisation. Waste and Biomass Valorization (2022)
Hajam, Y.A., Kumar, R., Kumar, A.: Environmental waste management strategies and vermi transformation for sustainable development. Environ. Challenges (2023)
Hashmi, M.S., Akhtar, S., Ismail, T.: Biochemical characterization of vegetables wastes and development of functional bread: vegetables' wastes based functional bread. J. Microbiol. Biotechnol. Food Sci. **10**(4), 691–696 (2021)
He, M., Xiong, X., Wang, L., Hou, D., Bolan, N.S., Ok, Y.S., et al.: A critical review on performance indicators for evaluating soil biota and soil health of biochar-amended soils. J. Hazard. Mater. **414**, 125378 (2021)
Herrera, A., D'Imporzano, G., Zilio, M., Pigoli, A., Rizzi, B., Meers, E., et al.: Environmental performance in the production and use of recovered fertilizers from organic wastes treated by anaerobic digestion vs synthetic mineral fertilizers. ACS Sustain. Chem. Eng. **10**(2), 986–997 (2022)
Junaid, M.D., Gokce, A.F.: Global agricultural losses and their causes. Bull. Biol. Allied Sci. Res. **2024**(1), 66–66 (2024)
Karimi, B., Sadet-Bourgeteau, S., Cannavacciuolo, M., Chauvin, C., Flamin, C., Haumont, A., et al.: Impact of biogas digestates on soil microbiota in agriculture: a review. Environ. Chem. Lett. **20**(5), 3265–3288 (2022)
Kendall, H., Clark, B., Li, W., Jin, S., Jones, G. D., Chen, J., et al.: Precision agriculture technology adoption: A qualitative study of small-scale commercial "family farms" located in the North China Plain. Precision Agriculture, 1–33 (2022)
Khalida, A., Arumugam, V., Abdullah, L.C., Abd, L.: Dehydrated food waste for composting: an overview. Sci. Technol. (2022)
Khatami, K., Perez-Zabaleta, M., Cetecioglu, Z.: Pure cultures for synthetic culture development: next level municipal waste treatment for polyhydroxyalkanoates production. J. Environ. Manage. **305**, 114337 (2022)
Koorneef, G.J., Pulleman, M.M., Comans, R.N., Van Rijssel, S.Q., Barré, P., Baudin, F., De Goede, R.G.: Assessing soil functioning: What is the added value of soil organic carbon quality measurements alongside total organic carbon content? Soil Biol. Biochem. **196**, 109507 (2024)
Koshariya, A.K., Krishnan, M.S., Jaisankar, S., Loganathan, G.B., Sathish, T., Ağbulut, Ü., et al.: Waste to energy: an experimental study on hydrogen production from food waste gasification. Int. J. Hydrogen Energy **54**, 1–12 (2024)
Kumari, B., Prakash, O., Pal, P., Singh, P.K., Singh, M.P., Kumar, P., et al.: Advancement in sustainable management and valorization of solid waste through composting and vermitechnology. In: Recent Trends in Management and Utilization of Industrial Sludge, pp. 359–397. Springer Nature, Switzerland (2024a)

Kumari, R., Singh, A., Sharma, R., Malaviya, P.: Conversion of food waste into energy and value-added products: a review. Environ. Chem. Lett. **22**(4), 1759–1790 (2024)

Kummu, M., Heino, M., Taka, M., Varis, O., Viviroli, D.: Climate change risks pushing one-third of global food production outside the safe climatic space. One Earth (2021)

Lal, R., Monger, C., Nave, L., Smith, P.: The role of soil in regulation of climate. Philos. Trans. r. Soc. B **376**(1834), 20210084 (2021)

Laso, J., Ruiz-Salmón, I., Margallo, M., Villanueva-Rey, P., Poceiro, L., Quinteiro, P., et al.: Achieving sustainability of the seafood sector in the European Atlantic area by addressing eco-social challenges: the NEPTUNUS project. Sustainability **14**(5), 3054 (2022)

Le Pera, A., Sellaro, M., Sicilia, F., Ciccoli, R., Sceberras, B., Freda, C., et al.: Environmental and economic impacts of improper materials in the recycling of separated collected food waste through anaerobic digestion and composting. Sci. Total Environ. **880**, 163240 (2023)

Lehmann, A., Leifheit, E.F., Gerdawischke, M., Rillig, M.C.: Microplastics have shape-and polymer-dependent effects on soil aggregation and organic matter loss–an experimental and meta-analytical approach. Microplastics Nanoplastics **1**(1), 7 (2021)

Lelicińska-Serafin, K., Manczarski, P., Rolewicz-Kalińska, A.: An Insight into post-consumer food waste characteristics as the key to an organic recycling method selection in a circular economy. Energies **16**(4), 1735 (2023)

Lepesteur, M.: Human and livestock pathogens and their control during composting. Crit. Rev. Environ. Sci. Technol. **52**(10), 1639–1683 (2022)

Levin, K.S., Auerswald, K., Reents, H.J., Hülsbergen, K.J.: Effects of organic energy crop rotations and fertilisation with the liquid digestate phase on organic carbon in the topsoil. Agronomy (2021)

Li, Y., Qi, C., Tang, Y., Liu, B., Bian, B., Gao, L., Fan, L., Zhao, Z.: New model of nutrient utilization and salt regulation of anaerobic digestate from food waste. Desalination (2024)

Liu, K., Tan, Q., Yu, J., Wang, M.: A global perspective on e-waste recycling. Circular Economy (2023a)

Liu, Z., de Souza, T.S., Holland, B., Dunshea, F., Barrow, C., Suleria, H.A.: Valorization of food waste to produce value-added products based on its bioactive compounds. Processes **11**(3), 840 (2023)

Lozano Lazo, D.P., Bojanic Helbingen, C., Gasparatos, A.: Household waste generation, composition and determining factors in rapidly urbanizing developing cities: case study of Santa Cruz de la Sierra, Bolivia. J. Mater. Cycles Waste Manage. **25**(1), 565–581 (2023)

Lu, J., Xu, S.: Post-treatment of food waste digestate towards land application: a review. J. Clean. Prod. (2021)

Manual, G.: Best Practices and Strategies for Pathogen-Free Compost Production (2023). [HTML]

Mayerová, M., Šimon, T., Stehlík, M., Madaras, M., Koubová, M., Smatanová, M.: Long-term application of biogas digestate improves soil physical properties. Soil Tillage Res. **231**, 105715 (2023)

Misslin, R., Clivot, H., Levavasseur, F., Villerd, J., Soulié, J.C., Houot, S., Therond, O.: Integrated assessment and modeling of regional recycling of organic waste. J. Clean. Prod. **379**, 134725 (2022)

Mondal, R., Shanmughan, A., Murugeswari, A., Shanmugaraju, S.: Recent advances in fluorescence-based chemosensing of organoarsenic feed additives using luminescence MOFs, COFs, HOFs, and QDs. Chem. Commun. **59**(77), 11456–11468 (2023)

Mukhopadhyay, D.: Opportunities and challenges of organic waste recycling for sustainable agriculture in the Sub-Himalayan ecosystem of Eastern India. Perspectives on Waste Recycling in Indian Agriculture, 220 (2025)

Mustafa, A., Holatko, J., Hammerschmiedt, T., Kucerik, J., Skarpa, P., Kintl, A., et al.: Comparison of the responses of soil enzymes, microbial respiration and plant growth characteristics under the application of agricultural and food waste-derived biochars. Agronomy **12**(10), 2428 (2022)

Nanda, S., Berruti, F.: Municipal solid waste management and landfilling technologies: a review. Environmental Chemistry Letters (2021)

Naorem, A., Jayaraman, S., Dalal, R.C., Patra, A., Rao, C.S., Lal, R.: Soil inorganic carbon as a potential sink in carbon storage in dryland soils—a review. Agriculture **12**(8), 1256 (2022)

Negash, Y.T., Hassan, A.M., Tseng, M.L., Wu, K.J., Ali, M.H.: Sustainable construction and demolition waste management in Somaliland: regulatory barriers lead to technical and environmental barriers. J. Clean. Prod. **297**, 126717 (2021)

Neville Menyuka, N., Sibanda, M., Bob, U.: Perceptions of the Challenges and Opportunities of Utilising Organic Waste through Urban Agriculture in the Durban South Basin (2020)

O'Connor, J., Mickan, B.S., Siddique, K.H., Rinklebe, J., Kirkham, M.B., Bolan, N.S.: Physical, chemical, and microbial contaminants in food waste management for soil application: a review. Environ. Pollut. **300**, 118860 (2022)

Okuthe, G.: Valorizing fruit and vegetable waste: The untapped potential for entrepreneurship in Sub-Saharan Africa—a systematic review. Recycling (2024)

Onyeaka, H., Tamasiga, P., Nwauzoma, U.M., Miri, T., Juliet, U.C., Nwaiwu, O., Akinsemolu, A.A.: Using artificial intelligence to tackle food waste and enhance the circular economy: maximising resource efficiency and minimising environmental impact: a review. Sustainability **15**(13), 10482 (2023)

Orr, L., Goossens, Y., Heinrich, M., Brüggemann, N.: Jointly reducing food waste—the experiences of the German discussion forum for wholesale and retail. Sustainability (2023)

Ortiz-Sanchez, M., Solarte-Toro, J.C., Alzate, C.A.C.: Food waste valorization applying the biorefinery concept in the Colombian context: pre-feasibility analysis of the organic kitchen food waste processing. Biochem. Eng. J. **194**, 108864 (2023)

Ozdemir, S., Ozdemir, S., Ozer, H., Yetilmezsoy, K.: A techno-sustainable bio-waste management strategy for closing chickpea yield gap. Waste Management (2021)

Panda, J., Mishra, A.K., Mohanta, Y.K., Patowary, K., Rauta, P.R., Mishra, B.: Exploring biopolymer for food and pharmaceuticals application in the circular bioeconomy: an agro-food waste-to-wealth approach. Waste Biomass Valorization **15**(10), 5607–5637 (2024)

Parfitt, J., Croker, T., Brockhaus, A.: Global food loss and waste in primary production: a reassessment of its scale and significance. Sustainability (2021)

Pereira, V.R., Fiore, F.A.: Opportunities and barriers to Composting in a municipal context: a case study in São José Dos Campos, Brazil. Sustainability (2024)

Perman, E., Schnürer, A., Björn, A., Moestedt, J.: Serial anaerobic digestion improves protein degradation and biogas production from mixed food waste. Biomass and Bioenergy (2022)

Pradhan, S., Parthasarathy, P., Mackey, H.R., Al-Ansari, T., McKay, G.: Food waste biochar: a sustainable solution for agriculture application and soil–water remediation. Carbon Res. **3**(1), 41 (2024)

Priya, K., Ramesh Kumar, S.: Cultivating under-utilized vegetables using organic resources: unlocking nutritional security and sustainable agriculture. Int. J. Vegetable Sci. 1–23 (2025)

Priya, K., Rani, J., & Gwal, S. (2025). Transforming agricultural residues to value-added products: waste to wealth. Sustainable Management of Agro-Food Waste

Purwandani, J.A., Michaud, G.: What are the drivers and barriers for green business practice adoption for SMEs?. Environment Systems and Decisions (2021)

Qi, X., Chmura, G.L.: Invasive Spartina alterniflora marshes in China: a blue carbon sink at the expense of other ecosystem services. Frontiers in Ecology and the Environment (2023)

Rajeh, C., Saoud, I.P., Kharroubi, S., Naalbandian, S., Abiad, M.G.: Food loss and food waste recovery as animal feed: a systematic review. J. Mater. Cycles Waste Manage. **23**, 1–17 (2021)

Râpă, M., Darie-Niță, R.N., Coman, G.: Valorization of fruit and vegetable waste into sustainable and value-added materials. Waste (2024)

Rasheed, F., Kayani, W.K., Asghar, M.U., Farrukh, A., Gul, S., Khan, I., Nazli, N.: Unlocking the potential of biowaste: Sustainable strategies to produce valuable industrial products. J. Environ. Qual. (2025)

Rastegari, H., Nooripoor, M., Sharifzadeh, M., Petrescu, D.C.: Drivers and barriers in farmers' adoption of vermicomposting as keys for sustainable agricultural waste management. Int. J. Agric. Sustain. **21**(1), 2230826 (2023)

Razouk, A., Tiganescu, E., Julia von Glahn, A., Yaman Abdin, A., Jawad Nasim, M., Jacob, C.: The future in the litter bin—bioconversion of food waste as driver of a circular bioeconomy (2024)

Riesenegger, L., Hübner, A.: Reducing food waste at retail stores—an explorative study. Sustainability (2022)

Rodrigues, C.I.D., Brito, L.M., Nunes, L.J.R.: Soil carbon sequestration in the context of climate change mitigation: a review. Soil Systems (2023)

Rodríguez-Espinosa, T., Navarro-Pedreño, J., Gómez Lucas, I., Almendro Candel, M.B., Pérez Gimeno, A., Zorpas, A.A.: Soluble elements released from organic wastes to increase available nutrients for soil and crops. Appl. Sci. **13**(2), 1151 (2023)

Roy, P., Mohanty, A.K., Dick, P., Misra, M.: A review on the challenges and choices for food waste valorization: environmental and economic impacts. ACS Environmental Au (2023). acs.org

Rut, M., Davies, A.R., Ng, H.: Participating in food waste transitions: Exploring surplus food redistribution in Singapore through the ecologies of participation framework. J. Environ. Planning Policy Manag. **23**(1), 34–47 (2021)

Sanchez, E., Zabaleta, R., Navas, A.L., Torres-Sciancalepore, R., Fouga, G., Fabani, M.P., et al.: Assessment of pistachio shell-based biochar application in the sustainable amendment of soil and its performance in enhancing bell pepper (*Capsicum annuum* L.) growth. Sustainability **16**(11), 4429 (2024)

Santagata, R., Ripa, M., Genovese, A., Ulgiati, S.: Food waste recovery pathways: Challenges and opportunities for an emerging bio-based circular economy. A systematic review and an assessment. J. Clean. Prod. **286**, 125490 (2021)

Sarker, A., Ahmmed, R., Ahsan, S.M., Rana, J., Ghosh, M.K., Nandi, R.: A comprehensive review of food waste valorization for the sustainable management of global food waste. Sustain. Food Technol. **2**(1), 48–69 (2024)

Sharma, A., Soni, R., Soni, S.K.: From waste to wealth: exploring modern composting innovations and compost valorization. J. Mater. Cycles Waste Manage. **26**(1), 20–48 (2024)

Sharma, D., Nguyen, O., Palencia, F., Lecoutre, C., Garrabos, Y., Glockner, S., et al.: Supercritical water oxidation using hydrothermal flames at microscale as a potential solution for organic waste treatment in space applications–A practical demonstration and numerical study. Chem. Eng. J. **488**, 150856 (2024)

Sharma, K., Rajan, S., Nayak, S.K.: Water pollution: Primary sources and associated human health hazards with special emphasis on rural areas. In: Water Resources Management for Rural Development, pp. 3–14. Elsevier (2024)

Singh, A., Singhania, R.R., Soam, S., Chen, C.W., Haldar, D., Varjani, S., et al.: Production of bioethanol from food waste: status and perspectives. Biores. Technol. **360**, 127651 (2022)

Skaf, L., Franzese, P.P., Capone, R., Buonocore, E.: Unfolding hidden environmental impacts of food waste: an assessment for fifteen countries of the world. J. Clean. Prod. **310**, 127523 (2021)

Skrzypczak, D., Gil, F., Izydorczyk, G., Mikula, K., Gersz, A., Hoppe, V., et al.: Innovative bio-waste-based multilayer hydrogel fertilizers as a new solution for precision agriculture. J. Environ. Manage. **321**, 116002 (2022)

Slavova, G., Doneva, Y.: Application of the model of circular economy between the forestry and agricultural sector. Crisis management safety foresight inforest-based sector and SMS's operation in the global environment, 383–387 (2022)

Srinivasan, K., Yadav, V.K.: An empirical investigation of barriers to the adoption of smart technologies integrated urban agriculture systems. J. Decis. Syst. (2024)

Stocker, B.D., Tumber-Dávila, S.J., Konings, A.G., Anderson, M.C., Hain, C., Jackson, R.B.: Global patterns of water storage in the rooting zones of vegetation. Nat. Geosci. **16**(3), 250–256 (2023)

Sudheer, S., Bandyopadhyay, S., Bhat, R.: Sustainable polysaccharide and protein hydrogel-based packaging materials for food products: a review. Int. J. Biol. Macromol. **248**, 125845 (2023)

Sugimura, Y., Okada, T., Kuwae, T., Mito, Y., Naito, R., Nakagawa, Y.: New possibilities for climate change countermeasures in ports: organic carbon containment and creation of blue carbon ecosystems through beneficial utilization of dredged soil. Mar. Policy **141**, 105072 (2022)

Sultana, D.: Emergent Realities for Social Wellbeing: Environmental, Spatial and Social Pathways (2017)

Tabbara, M.: Assessment of Bioprocessed Post-Consumer Food Waste as Feed for Nile tilapia (*Oreochromis niloticus*) (2021)

Tang, B., Rocci, K.S., Lehmann, A., Rillig, M.C.: Nitrogen increases soil organic carbon accrual and alters its functionality. Glob. Change Biol. **29**(7), 1971–1983 (2023)

Terrer, C., Phillips, R.P., Hungate, B.A., Rosende, J., Pett-Ridge, J., Craig, M.E., et al.: A trade-off between plant and soil carbon storage under elevated $CO_2$. Nature **591**(7851), 599–603 (2021)

Thakali, A., MacRae, J.D.: A review of chemical and microbial contamination in food: What are the threats to a circular food system?. Environmental Research (2021)

Toxopeus, H., Polzin, F.: Reviewing financing barriers and strategies for urban nature-based solutions. J. Environ. Manag. (2021)

Upadhyay, S., Tiwari, R., Kumar, S., Gupta, S.M., Kumar, V., Rautela, I., et al.: Utilization of food waste for the development of composite bread. Sustainability **15**(17), 13079 (2023)

Ur Rehman, H.M.M., Khan, M., Rehman, M.M., Khan, S.A., Kim, W.Y.: High-performance humidity sensor for multipurpose applications by recycling of potato peel bio-waste. Sens. Actuators A **343**, 113662 (2022)

Vereecken, H., Amelung, W., Bauke, S.L., Bogena, H., Brüggemann, N., Montzka, C., et al.: Soil hydrology in the Earth system. Nat. Rev. Earth Environ. **3**(9), 573–587 (2022)

Waqas, M., Hashim, S., Humphries, U.W., Ahmad, S., Noor, R., Shoaib, M., et al.: Composting processes for agricultural waste management: a comprehensive review. Processes **11**(3), 731 (2023)

Winkler, T., Ostermeier, M., Hübner, A.: Proactive food waste prevention in grocery retail supply chains–an exploratory study. Int. J. Phys. Distrib. Logist. Manag. **53**(11), 125–156 (2023)

Xie, Y., Zhong, Y., Wu, J., Fang, S., Cai, L., Li, M., et al.: "From waste to wonder": comparative evaluation of Chinese cabbage waste and banana peel derived hydrogels on soil water retention performance. Gels **10**(12), 833 (2024)

Yu, J., Zhang, X., Xu, L., Dong, J., Zhangzhong, L.: A hybrid CNN-GRU model for predicting soil moisture in maize root zone. Agric. Water Manag. **245**, 106649 (2021)

Zan, F., Iqbal, A., Lu, X., Wu, X., Chen, G.: "Food waste-wastewater-energy/resource" nexus: integrating food waste management with wastewater treatment towards urban sustainability. Water Research (2022)

Zhan, J., Han, Y., Xu, S., Wang, X., Guo, X.: Succession and change of potential pathogens in the co-composting of rural sewage sludge and food waste. Waste Management (2022)

Zhang, J., Liu, J., Cheng, Y., Jiang, T., Sun, D.A., Saberian, M.: Water-retention behaviour and microscopic analysis of two biopolymer-improved sandy soils. Constr. Build. Mater. **403**, 133202 (2023)

Zhang, M., He, H., Zhang, L., Yu, G., Ren, X., et al.: Increased ecological land and atmospheric $CO_2$ dominate the growth of ecosystem carbon

sinks under the regulation of environmental conditions in national key …. Journal of …. (2024)
Zhu, J., Suhaimi, F., Lim, J.Y., Gao, Z., Swarup, S., Loh, C.S., et al.: A field study on using soybean waste-derived superabsorbent hydrogel to enhance growth of vegetables. Sci. Total Environ. **851**, 158141 (2022)

# Agricultural Palm Waste as a Sustainable Resource for Bio-Based Composite Materials in the Automotive Industry

Oussama Azmami, Adil El Arfaouy, Laila Sajid, Latifa Hajji, Omar Ait Sidi Ahmed, Fathallaah Bazi, and Said Gmouh

**Abstract**

This chapter aims to explore a novel approach to managing agricultural palm waste and increasing its value as a renewable resource by utilizing it to produce sustainable automotive materials. It focuses on the potential of palm fiber as a sustainable technical textile and reinforcement material in automotive composites, particularly for interior applications. By emphasizing its role in reducing environmental impact and promoting the circular economy, the chapter highlights how converting agricultural palm waste into high-value materials can drive sustainability while fostering economic opportunities across the agricultural, manufacturing, and automotive industries. The effect of palm fibers on the physical, mechanical, and thermal properties of conventional textile fibers such as wool and polyester, as well as polyester and epoxy resins, is the focus of this chapter. Palm fiber, especially from the washingtonia tree, can be efficiently processed into needle-punched nonwovens, which are then coated with waterborne polyurethane and waterborne acrylate finishes. These materials offer excellent visual appeal, mechanical strength, and flame retardancy, making them ideal for use as textile finishes in interior applications. Moreover, palm fibers demonstrate good adhesion to polymeric matrices, such as polypropylene, polyester and epoxy resins, used in automotive applications.

**Keywords**

Agricultural waste · Waste management · Palm fiber · Biocomposite · Automotive · Textile finishing

## 1 Introduction

Agricultural waste refers to the unwanted materials generated from farming activities, including crop residues, animal manure, spoiled produce, and discarded agrarian inputs like pesticides and plastics. It can be either biodegradable or non-biodegradable and arises during cultivation, harvesting, and processing (Khan 2024a; Chaudhary and Garg 2023). The rising global production of waste, particularly organic waste from agricultural activities, poses significant environmental issues, such as land use pressures, pollution, and the worsening of climate change (Khan 2024a; Chaudhary and Garg 2023). Additionally, the breakdown of organic matter in landfills releases methane, a potent greenhouse gas (Kiehbadroudinezhad et al. 2024). Therefore, adopting sustainable strategies for managing this waste is increasingly crucial. Natural fibers like palm fiber are agricultural waste derived primarily from the processing of palm trees, particularly during the extraction of palm oil or dates. It includes empty fruit bunches, palm fronds, trunks, and fibers left after oil extraction (Ooi et al. 2017). As a lignocellulosic biomass, palm fiber is rich in cellulose, hemicellulose, and lignin, making it a valuable raw material for compost, bioenergy, animal feed, paper production, and even biodegradable composites (Sajid et al. 2021a, b; Wang et al. 2021). However, when not properly managed, it can contribute to environmental pollution, including open burning and methane emissions from decomposition. With large-scale palm oil production in countries

O. Azmami · A. El Arfaouy · L. Hajji · S. Gmouh (✉)
Laboratory of Engineering and Materials (LIMAT), Faculty of Sciences Ben M'sick, Hassan II University of Casablanca, Casablanca, Morocco
e-mail: s.gmouh1@gmail.com

A. El Arfaouy · O. Ait Sidi Ahmed · F. Bazi
Laboratory of Analytical and Molecular Chemistry (LCAM), Faculty of Sciences Ben M'sick, Hassan II University of Casablanca, Casablanca, Morocco

L. Sajid
Laboratory of Expertise and Control (LEC), Higher School of Textile and Clothing Industries (ESITH), Casablanca, Morocco



O. Jamal Eddine et al. (eds.), *Plant Pathology, Fungal Diversity, and Biotechnological Advances in Agriculture*, Sustainable Landscape Planning and Natural Resources Management, https://doi.org/10.1007/978-3-031-97666-7_11

like Indonesia and Malaysia, palm fiber represents a waste management challenge and an opportunity for sustainable resource utilization (Naidu and Moorthy 2021). Agricultural waste valorization encompasses various approaches aimed at converting residues into valuable products. Standard methods include composting to produce natural fertilizers, anaerobic digestion to generate biogas and nutrient-rich digestate, and pyrolysis for biochar production, which enhances soil quality and sequesters carbon (Selvaraj et al. 2022; Zielińska and Bułkowska 2024). However, lignocellulosic biomass like palm fibers can be processed into biofuels or bioethanol, offering renewable energy alternatives (Kaniapan et al. 2021). Other strategies involve repurposing waste as animal feed or extracting valuable compounds like enzymes and antioxidants for use in pharmaceuticals, cosmetics, and food. Agricultural byproducts are also used to create bioplastics and biocomposites, particularly for applications in packaging, construction, and the automotive industry (Osorio et al. 2021). These approaches not only reduce environmental impact but also support sustainable development and resource efficiency. The demand for biocomposites in the automotive industry has been steadily increased due to the growing emphasis on sustainability, environmental regulations, and consumer preferences for eco-friendly products. Biocomposites, which combine natural fibers (such as hemp, flax, or palm fiber) with polymers, offer several benefits, including lightweight, improved fuel efficiency, and reduced carbon footprint compared to traditional materials like metal and plastic (Manu et al. 2022; Azmami et al. 2023). These materials are being used for interior components and exterior parts to reduce vehicle weight, enhance performance, and meet stringent environmental standards. Moreover, the automotive industry's interest in biocomposites is driven by a desire to incorporate renewable resources, reduce dependency on fossil fuels, and improve the recyclability and sustainability of the vehicles' life cycle. In this context, this chapter explores the potential of palm fibers as a sustainable textile and reinforcement material for composites used in automotive interiors and exterior parts. By integrating palm fibers into automotive textiles and biocomposites, the industry can move towards more sustainable manufacturing practices, aligning with the growing demand for greener, more efficient vehicles.

## 2 Technical Textile in the Automotive Industry

### 2.1 Overview of Textile Use in Vehicles

Textile materials are structurally developed materials derived from textile fibers. These fibers are mainly classified according to their origin. There are two major categories of fibers: natural and man-made. All natural fibers originate from either plant, animal, or mineral sources (see Fig. 1). Regarding man-made fibers, they are further divided into organic and inorganic fibers. Organic man-made fibers include those produced from regenerated cellulose such as viscose, modal, acetate, etc., or synthetic polymers such as acrylic, nylon, polyethylene, polypropylene, etc. Inorganic fibers include all fibers made from metals or ceramics. Textile materials are generally divided into two groups: conventional textiles and technical textiles. Conventional textile products are designed, developed, or used for everyday, decorative, or aesthetic applications. In contrast, technical textile products are mainly manufactured for their technical and performance properties rather than their decorative or aesthetic characteristics (Ismail 2016). The global textile market was valued at USD 993.6 billion in 2021 and is expected to grow at an annual growth rate of 4% from 2022 to 2030 (Global Textile Market Size and Share Report 2025). Increasing human needs have led researchers to develop technical textile products. The current estimate of the technical textiles market was approximately USD 180 billion in 2020 and is projected to reach USD 250 billion by 2026, with an average growth rate of 5% (Rasheed 2020). Technical textiles represent one of the fastest-growing segments in the textile industry. According to the technical textile trade fair "Techtextil," technical textile products are classified into twelve application-based categories: Mobiltech, Indutech, Medtech, Hometech, Clothtech, Agrotech, Buildtech, Sportech, Packtech, Geotech, Protech, and Oekotech (Rasheed 2020). Textile products used in transportation are referred to as Mobiltech. The market size of Mobiltech textiles was estimated at USD 28.9 billion in 2021 and is projected to grow at an annual rate of 3.2% from 2022 to 2030 (Automotive Textiles Market Size 2030). This field of application includes various transportation sectors: airplanes, ships, buses, cars, etc.

In the automotive industry, textiles are highly valued for their functionality and mechanical performance. They are used in three main parts of the vehicle: the interior, the

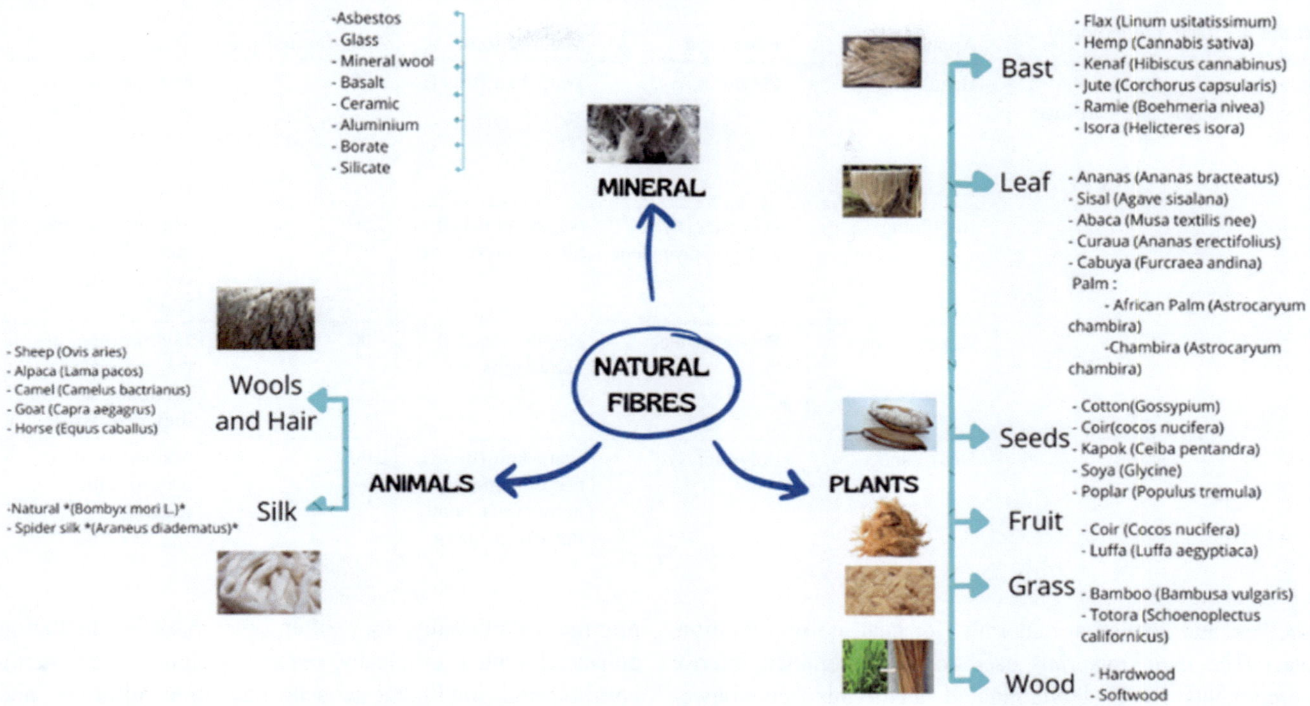


**Fig. 1** Classification of natural fibers

engine compartment, and the tires (Richaud et al. 2018). Their use has become essential to meet new consumer expectations, along with functional requirements such as filtration, protection, and acoustic and thermal insulation (Global Textile Market Size and Share Report 2025). The vehicle interior has evolved significantly over the past decade and has become one of the key factors in a consumer's vehicle purchase decision. As such, comfort, aesthetics (color and design harmony), and safety have become crucial elements in vehicle sales. Interior upholstery applications can be divided into five categories: seat and door coverings, rear parcel shelf, carpets, headliners, and safety components (seat belts and airbags) (Saricam and Okur 2018). Upholstery contributes to the interior's visual appeal and comfort through various aspects such as appearance, texture, mechanical, and physical properties. The surface fabric can be made using high-performance/technical fibers, and all types of fabrics (knitted, woven, nonwoven, braided, tufted, and leather) are used for its manufacture. Automotive seat upholstery traditionally consists of three layers: a surface fabric, a thin polyurethane foam layer, and a textile backing. The surface fabric, typically polyester, maybe a knit or woven. Cars built in the United States usually feature woven velvet upholstery, catering to consumer preferences. Velvet fabrics have long been used to reinforce the vehicle's luxurious look, despite their relatively expensive production and limited elasticity. Indeed, the European version of woven velvet is less common and is comparable to leather in terms of cost and application area (Pamuk and Çeken 2009). The polyurethane foam layer is typically 8 mm thick, offering softness and a cushioned feel. The textile backing is usually a knitted, nonwoven, or film substrate, providing dimensional stability and preventing excessive foam wear. For door panels, the textile covering used is similar to that for seat upholstery. It generally consists of a two-layer: surface fabric and a thin polyurethane foam layer. A textile backing is not necessarily used. As with seat coverings, the surface fabric is generally knit, woven, or nonwoven polyester. The covering can be applied to part or all of the door panel surfaces. However, polyurethane foam used in traditional laminates has faced environmental criticism (Sari 2021; Romero et al. 2021). Additionally, studies have shown its limited contribution to seating comfort. Fung and Parsons conducted a specific study on the relationship between seat cover materials and thermal comfort using human subjects (Fung and Parsons 1996). They identified two issues with foam padding: its inability to wick away perspiration and the tendency of thicker foam layers to wrap around the body, thus causing thermal discomfort (Fung and Parsons 1996). One alternative to increase seat comfort is to eliminate the foam layer. Recently, the textile industry has suggested replacing foam with a thick nonwoven material of equal weight. Spacer knits can also replace foam in vehicle interiors (Guo et al. 2021), but their low recovery after compression and high manufacturing cost limit their use (Guo et al. 2021). Apart from seats, other pre-assembled interior components include trunk linings, headliners, door panels, rear parcel

**Table 1** Basic properties of parts used for automotive interior trim applications (Mukhopadhyay and Partridge 1999)

| Application | Fiber used | Material form | Area covered ($m^2$) | Trends |
|---|---|---|---|---|
| Headlining | Polyester | Warp knit (tricot) | 2.00 | Moving towards needle punched nonwovens because of the price |
| Bootlining | Polyester, nylon and polypropylene | Needle punched or tufted nonwovens | 2.50 | Moving towards molded parts covered with textiles |
| Parcel shelves | Polypropylene | Needle punched nonwovens | 1.00 | Lightfastness of the material is becoming an important issue |
| Door panels | Polyester | Warp knit (tricot), needle punched nonwovens, some flat woven fabric | 2.00 | Mainly molded, backed with textiles |

shelves, and insulation materials (for heat, noise, vibration, etc.). The main materials used for pre-assembled interior components are coated/laminated needle-punched nonwovens and warp-knitted fabrics. A selection of preformed parts is presented in Table 1, including their approximate surface area, the fibers used, actual material forms, and market trends (Mukhopadhyay and Partridge 1999). There are three major industrial regions where nonwovens have developed into distinct industries: Japan, Western Europe, and the United States (Endlerová 2021). In Western Europe, about 50% of cars are equipped with nonwoven headliners (Mukhopadhyay and Partridge 1999). The Japanese are also very fond of these fabrics and are credited with improving the technique and appearance of nonwovens. Compared to established knit qualities, the new generation of nonwovens is cheaper, more resistant to abrasion and scratches, easier to process and handle, recyclable, and available in a wider range of colors and styles (Mukhopadhyay and Partridge 1999). Technical nonwovens refer to a broad range of fibrous assemblies essential for the functional performance of products used in various applications. The nonwoven market size is expected to increase from USD 40.5 billion in 2020 to USD 53.5 billion by 2025, with an annual growth rate of 5.7% between 2020 and 2025 (Nonwoven Fabrics Market 2025). This market growth is attributed to rising demand for nonwoven products and solutions in hygiene, medical, construction, and filtration. Nonwovens are widely used in the automotive sector for various components such as floor carpeting (43%), insulation (17%), trunk linings (13%), hood liners (10%), headliners (6%), seats (6%), packaging trays (3%), door panels (1%), and other areas (1%) (Palak and Karagüzel Kayaoğlu 2019). The growing development of needle-punched nonwovens in the automotive industry is due to their advantages, such as short processing times, lightweight, good abrasion resistance, and competitive pricing. Additionally, they offer other benefits, including enhanced vehicle efficiency, reduced weight, lower energy consumption, intelligent acoustic insulation, filtration, and improved protection. The choice of nonwovens in automotive applications depends on various characteristics such as abrasion resistance, stain resistance, heat and UV resistance, aesthetics and texture, formability, recyclability, low weight, durability, flame resistance, and tear strength (Arenas 2016). Nonwovens must maintain their practical and aesthetic durability throughout the vehicle's lifetime. However, one of the main causes of deterioration in automotive textile materials is solar radiation, especially ultraviolet rays, and climatic and physical abrasion, particularly in carpets (Srivastava et al. 2016). Recycling industrial textiles reduces the need for raw materials and, consequently, the environmental impact. A prerequisite for effective recycling is a collection system that efficiently retrieves textiles with environmental considerations in mind. Recycling is a key condition for a circular economy, meaning that materials can be recycled, reintegrated, and reused. In recent years, there has been a rising push to increase the volume of recovered textile materials. Rising waste management costs and the limited availability of landfill space have likely contributed to this growing interest in material or energy recovery from waste (Voncina 2016). An increasing variety of products are now made using recycled materials as the primary raw resource. One such resource includes textile waste from cars. In automobiles, fibers are used in various textile forms such as circular knits, warp knits, woven and nonwoven structures. Fibers are also used as components in multilayer composite structures. Therefore, reusing this waste is highly desirable. Djordjevic and Avramovic report that the automotive industry generates large amounts of textile waste. This waste has potential uses as thermal and acoustic insulation in lightweight structures and could lead to savings of over €100 million across Europe (Danihelová et al. 2019).

## 2.2 Textile Processing

### 2.2.1 Woven Fabrics

Woven fabrics are textiles made by interlacing two sets of yarns, warp (vertical threads) and weft (horizontal threads), at right angles. This interlacing process creates a durable, stable, and strong fabric (Mishra 2023). In the automotive industry, woven fabrics are commonly used due to their strength, versatility, and excellent dimensional stability. The most common materials used for woven automotive fabrics are nylon, polyester (PET), and polyamide (PA), which offer resistance to abrasion, fading, and wear (Messiry et al. 2022). These fabrics are often used for upholstery, seat covers, headliners, door panels, and seat belts. Additionally, woven fabrics provide a structured finish and can be easily dyed or coated to meet specific functional and aesthetic requirements. Their strength and resilience make them particularly suitable for automotive applications that demand performance and durability (Hassabo et al. 2022).

### 2.2.2 Knitted Fabrics

Knitted fabrics are textiles created by interlocking loops of yarn, which gives them a stretchable, flexible structure compared to woven fabrics. The interlocking loop construction allows knitted fabrics to be more elastic, comfortable, and resilient, making them ideal for applications where flexibility and movement are essential (Mamdouh et al. 2022). In the automotive industry, knitted fabrics are commonly used for seating surfaces, floor mats, car covers, and upholstery due to their softness and comfort. They are typically made from polyester, spandex, and nylon, which offer durability, resistance to wear, and excellent moisture-wicking properties (Liu et al. 2021). Knitted fabrics can also be easily shaped and are often used with other materials or coatings to provide additional strength, thermal insulation, or waterproofing. Their flexibility suits them particularly for automotive applications where comfort and stretch are needed, such as in seat covers and headrests.

### 2.2.3 Nonwoven Fabrics

Nonwoven fabrics are textiles made by bonding fibers together through mechanical, chemical, or thermal techniques, without weaving or knitting. These fabrics offer unique properties such as lightweight, cost-effective, and versatile performance, making them ideal for a wide range of automotive applications (Mukhopadhyay and Partridge 1999). In the automotive industry, polypropylene and polyester (PET) are commonly used to create nonwoven fabrics, which provide excellent insulation, soundproofing, and vibration damping. Nonwoven fabrics are widely used in components such as headliners, carpets, trunk liners, and filter materials (Arenas 2016). They can also be treated to enhance some properties such as flame resistance, moisture absorption, and durability. These fabrics are often used in areas that require lightweight materials that still provide performance benefits, like sound insulation and under-the-hood applications.

### 2.2.4 Faux Leather (Synthetic Leather)

Faux leather (Synthetic leather) is a man-made material designed to mimic the look and feel of natural leather while offering greater durability, lower cost, and easier maintenance (Zhu et al. 2021). It is typically made from polyvinyl chloride (PVC) or polyurethane (PU), with a textured surface that resembles genuine leather (Zhu et al. 2021). In the automotive industry, faux leather is commonly used for seat coverings, door panels, steering wheel covers, and dashboard trims due to its aesthetic appeal, soft texture, and resilience. Faux leather offers significant advantages over genuine leather, such as being more resistant to wear and tear, waterproof, and easy to clean, while being a more sustainable option in many cases (Hardcastle 2020). It can be produced in various colors, patterns, and finishes, allowing for versatile design options. Additionally, PU faux leather is often considered more environmentally friendly than PVC, as it is more biodegradable and free from harmful chemicals like phthalates (Hardcastle 2020). Faux leather provides an ideal balance of luxury appearance and functional performance in automotive interiors.

### 2.2.5 Laminated Fabrics

Laminated fabrics consist of multiple layers of fabric or material bonded together with an adhesive, resin, or polymer film (Fung 2002). This lamination process enhances the fabric's performance by combining the beneficial properties of different layers (Fung 2002). In the automotive industry, laminated fabrics often provide improved durability, waterproofing, thermal insulation, and soundproofing (Maity et al. 2032). These fabrics can be created by combining a woven or nonwoven base fabric with a polyurethane (PU) or polyvinyl chloride (PVC) coating, or by using other films to add specific properties (Sikdar et al. 2022). Typical automotive applications for laminated fabrics include seat covers, floor mats, trunk liners, underbody protection, and headliners (Azmami et al. 2021a). The laminated layers provide added strength and protection, ensuring the materials can withstand abrasion, moisture, and extreme temperatures. In addition, laminated fabrics offer flexibility, allowing them to conform to the shape of automotive parts while maintaining their enhanced properties.

### 2.2.6 Carbon Fiber Fabrics

Carbon fiber fabrics are high-performance textiles made from thin strands of carbon that are woven together to create a lightweight yet robust material (Zhang et al. 2024). The unique properties of carbon fiber, such as its high strength-to-weight ratio, stiffness, and resistance to fatigue and corrosion, make it a preferred choice in the automotive industry, particularly for performance and luxury vehicles (Noreikaitė et al. 2024). These fabrics are commonly used to manufacture structural components like bumpers, spoilers, roof panels, and reinforced body parts, where reducing weight while maintaining strength is critical for performance, fuel efficiency, and handling. Carbon fiber is also used in the interior design of high-end cars for dashboards, door panels, and trim elements, offering a sleek, modern appearance while extremely durable (Noreikaitė et al. 2024). While carbon fiber fabrics are more expensive than traditional materials, their ability to improve vehicle performance and aesthetics makes them highly desirable in premium automotive applications. The fabric is often combined with resins to form carbon fiber reinforced polymers (CFRP), providing even more strength and versatility for automotive parts.

### 2.2.7 Acoustic Fabrics

Acoustic fabrics are specially designed textiles used in the automotive industry to reduce sound transmission and control noise, vibration, and harshness within a vehicle (Noreikaitė et al. 2024). These fabrics are engineered to absorb, dampen, or block unwanted noise, providing a more comfortable and quiet driving experience (Noreikaitė et al. 2024). Acoustic fabrics are typically made from nonwoven materials, often incorporating fibers like polyester, polypropylene, or polyamide, along with specialized coatings or foam layers to enhance sound absorption. In automotive applications, acoustic fabrics are commonly used in headliners, door panels, floor mats, trunk liners, and engine compartments to reduce road noise, engine noise, and vibration (Madushika and Lanarolle 2022).

## 2.3 Textile Finishing

Coating is a crucial technique in the automotive industry for enhancing the functionality of technical textiles, providing added value through improved properties. Polymer coating involves the application of one or more layers of polymer onto a textile substrate's surface. Various coating techniques are employed depending on the desired characteristics of the final product. Despite the differences among these methods, they all share a common goal: achieving a well-coated fabric with specific properties. The coating process typically follows three main steps: dosage, transfer, and fixation (Kačarević et al. 2022). The dosage step controls the amount of polymer applied to the fabric. During the transfer phase, the coating material is deposited onto the textile surface. Finally, the fixation step ensures that the transferred material adheres properly to the fabric (Kačarević et al. 2022). The choice of an appropriate coating technique for textile finishing is primarily influenced by the intended application and performance specifications of the final product. Table 2 provides an overview of the most commonly used coating methods in the automotive industry. Certain polymer coating techniques and materials are particularly favored within this sector due to their durability, flexibility, and ability to meet aesthetic and functional demands. Table 3 outlines the most commonly used polymers in these popular coating methods for automotive applications. Polyvinyl Chloride (PVC) is a cost-effective option, providing durable and waterproof surfaces, making it ideal for seat coverings and floor mats. Polyurethane (PU) is favored in luxury interiors for its soft, leather-like feel, often used in upholstery and dashboard trims (KJ A 2024). For high-performance applications requiring flexibility and abrasion resistance, Thermoplastic Polyurethane (TPU) is commonly employed in seat covers and steering wheels. Polyethylene Terephthalate (PET) and Polyamide (PA) are strong, durable materials typically used for technical textiles such as headliners, seat belts, and under-hood components (Bouasria 2020; Akovali 2012). Finally, Silicone coatings are selected for their excellent heat resistance, making them perfect for textiles in engine compartments and other high-temperature environments.

## 2.4 Integration of Agricultural Waste into Automotive Textiles

Regarding the integration of agricultural waste into automotive textiles, natural fibers obtained from agricultural wastes are typically short, generally ranging from a few millimeters to 20–30 mm in length. For lignocellulosic fibers such as flax, jute, and palm fibers, the limited fiber length, high rigidity, and large diameter often make conventional textile manufacturing methods such as spinning and weaving unsuitable. In this context, nonwoven technology emerges as the most practical and effective method for processing these short fibers into structural textile components. Nonwoven processes allow the direct conversion of short fibers into fabric structures without the need for yarn formation, making them particularly suitable for incorporating agricultural waste into automotive textile applications. Recently, nonwovens have become one of the most common bio-based textile products in the automotive industry and have been used as an essential sound absorption and thermal insulation material. (Nick et al. 2002)., examined the sound absorption properties of composite nonwoven materials composed of Cotton/PET, Lyocell/PET, Flax/PP, and Hemp/

**Table 2** Most commonly used coating methods in the automotive industry (Bouasria 2020; Akovali 2012)

| Method | Description | Advantages | Limitations | Applications |
|---|---|---|---|---|
| Knife-over-roll | A blade spreads polymer across fabric supported by a roller; thickness is controlled by the gap | Precise thickness control, robust coatings | Not ideal for textured surfaces and limited to flat surfaces | Seat backing, sound insulation layers, trunk liners |
| Knife-over-air | Similar to knife-over-roll, but fabric is suspended in air, allowing more flexibility | Ideal for flexible or thicker coatings | Requires tight tension control | Dashboard skins, coated seat fabrics, technical underlayers |
| Dip-coating | Textile is immersed in polymer solution, then withdrawn and dried | Uniform coverage, good for porous or complex materials | Less precision, material runoff | Seat belt coatings, corrosion-resistant layers on technical textiles |
| Roller coating | Polymer is transferred using rollers for a uniform finish | Consistent, thin coating; high-speed | Less effective on textured or bulky fabrics | Surface coatings for interior trims, back-coating headliners |
| Transfer coating | Polymer is applied to release paper and transferred to fabric with heat and pressure | High-quality, smooth surface, good for complex shapes | Expensive, multi-step | Artificial leather for seats, armrests, and door panels |
| Pad-dry-cure | Fabric is dipped in polymer solution, squeezed, dried, and cured | Simple, economical | Less control over the thickness | Flame retardant treatments, water/oil repellency for seat covers and mats |
| Hot melt coating | Molten polymer was applied and solidified by cooling | Fast, durable, solvent-free | Requires heat-resistant fabric | Adhesive bonding layers, acoustic barrier coatings, under-hood insulation |
| Powder coating | Electrostatic powder applied and melted onto fabric during curing | Solvent-free, robust, long-lasting | Requires heat curing and electrostatic setup | Engine compartment fabrics, thermal shields, abrasion-resistant layers on trunk fabrics |
| Spray coating | Polymer is atomized and sprayed on the surface for localized or 3D application | Ideal for irregular shapes and surface treatment | Overspray waste, variable thickness | Surface touch-ups, innovative coatings (e.g., UV-reflective), dashboards |
| Screen coating | A mesh screen applies a patterned polymer layer | Custom patterns, decorative or functional | Limited to flat surfaces | Decorative elements, anti-slip patterns, branding elements |
| Foam coating | Lightweight foam is applied to the fabric for softness and bulk | Breathable, light, adds volume | Less control on uniformity | Seat cushioning layers, sound-dampening panels, soft-touch interior linings |
| Gravure coating | Engraved roller applies precise, thin layers of polymer | Very precise, fast production | High setup cost, best for continuous fabric | Coatings for filtration components, thermal barrier fabrics, airbag fabrics |
| Embossing | Involves pressing a textile between heated engraved rollers to create a raised or recessed pattern on the surface. The heat and pressure fix the pattern, especially when applied to thermoplastic or coated fabrics | Enhanced aesthetic appeal, cost-effective, no need for additional coatings, improved tactile feel, increased durability and design flexibility | Fabric stiffness, limited effect on certain fabrics, reduced strength, limited texture depth and pattern fade over time | Seat covers and upholstery, dashboard and door panels, floor mats, headliners, trim and accent panels, sun visors and bags and storage compartments |

PP blends, each with a 50% weight ratio between the natural and synthetic fiber components. Compared to the standard material, the Lyocell/PET composite with 0.9 dtex Lyocell fibers exhibited significantly higher sound absorption in the frequency range above 1200 Hz. This improved performance is attributed to the fine diameter of the Lyocell fibers, which enhances the material's ability to dissipate sound under real conditions. (Gu et al. 2017), investigated the physical properties, including sound absorption, of Kenaf-based nonwoven composites developed for automotive pillar trim applications. Seven types of Kenaf nonwovens blended with polypropylene and low-melt PET fibers were produced using a needle punching process. The study found that the sound absorption coefficient was strongly correlated with the weight and thickness of the nonwoven materials (correlation coefficients exceeding 0.83). Additionally, smaller pore sizes, resulting from greater needle depth and higher ratios of low-melt PET, also contributed to enhanced sound absorption performance. (Thirumurugan et al. 2021),

**Table 3** Most commonly used polymer coating in the automotive industry (Akovali 2012; Patil et al. 2017)

| Polymer | Coating methods | Advantages | Limitations | Applications |
|---|---|---|---|---|
| Polyvinyl chloride (PVC) | – Knife-over-roll<br>– Pad-dry-cure<br>– Hot melt coating | – High durability<br>– Water and chemical resistance<br>-Flexible and adaptable | – Can be stiff at lower temperatures<br>– Less eco-friendly | – Seat coverings (faux leather)<br>– Floor mats<br>– Door panels |
| (Polyurethane (PU) | – Roller coating<br>– Transfer coating<br>– Foam coating | – Soft, elastic, wear-resistant<br>– Aesthetic and durable | – Higher cost compared to PVC<br>– May require special curing | – Upholstery (seats, door trims)<br>– Dashboard coverings<br>– Steering wheel covers |
| Thermoplastic polyurethane (TPU) | – Spray coating<br>– Dip coating<br>– Roller coating | – Elasticity and high wear resistance<br>– Good for soft-touch finishes | – Sensitive to heat; needs careful processing | – Performance seat covers<br>– Steering wheel grip<br>– Protective trims |
| Polyethylene terephthalate (PET) | – Gravure coating<br>– Roller coating<br>– Dip coating | – Durable, dimensionally stable<br>– Often used in blends for strength | – Lower flexibility compared to TPU or PU | – Interior headliners<br>– Trunk linings<br>– Technical textile applications |
| Polyamide (PA) | – Pad-dry-cure<br>– Hot melt coating<br>– Spray coating | – High strength<br>– Excellent abrasion resistance<br>– Tough and durable | – Less flexible than TPU or PU<br>– Higher cost for premium grades | – Seat belts<br>– Airbag fabrics<br>– Under-hood textiles |
| Silicone-based coatings | – Spray coating<br>– Hot melt coating | – High-temperature resistance<br>– Non-stick, flexible, durable | – Expensive<br>– Can be too soft for some applications | – Heat-resistant automotive parts<br>– Engine compartment fabrics |

investigated the thermal conductivity and thermal resistance of needle-punched nonwoven fabrics made from jute, polypropylene, and recycled polyester fibers. The fibers were blended in various proportions and processed into cross-laid, needle-punched nonwovens. Experimental results revealed that nonwovens composed entirely of recycled polyester exhibited the highest thermal resistance compared to those made from jute or polypropylene. Nonwovens containing 25%–30% jute fibers demonstrated suitable thermal insulation properties, making them viable for applications such as dashboard insulation, carpets, seat padding, flooring, wheel housing, engine hood insulation, and boot liners. (Islam and Mominul Alam 2018), studied the acoustic properties of needle-punched nonwoven composites made from bamboo, banana, and hemp fibers, blended with polyester (PET) and polypropylene (PP), aiming to reduce sound transmission within automotive interiors. The fiber composition was maintained at 35% natural fibers, 35% PET, and 30% PP. Among the tested blends, the bamboo/PET/PP nonwoven exhibited superior performance, enhanced stiffness, higher tensile strength, lower elongation, reduced thermal conductivity and air permeability, improved sound absorption coefficient, and the highest sound transmission loss. These characteristics indicate its suitability for use in vehicle interiors for effective noise control. Therefore, natural nonwovens represent a sustainable alternative to traditional synthetic materials such as glass fibers, foams, and other man-made fibers, owing to their eco-friendly, reusable, and biodegradable nature. Their effectiveness in reducing interior noise enhances the overall comfort and perceived quality of the vehicle for passengers, while also contributing to the manufacturer's reputation for environmentally conscious and high-performance design.

### 2.4.1 Palm Waste as a Technical Textile in Automotive

In this study, we introduce a new agricultural waste from *Washingtonia* palm tree, into the textile industry to produce environmentally friendly needle-punched nonwovens for automotive interiors. The selection of this fiber is based on its abundance as an agricultural waste in Morocco and the fact that it has not yet been utilized in textile materials. While its mechanical properties are similar to those of certain palm fiber species (see Table 4), its current applications are primarily limited to reinforcing plaster and producing construction panels (Benzannache et al. 2021; Ferrández-García et al. 2018). First, Palm fibers were extracted using a mixed extraction process that combines two treatments (alkali and bleaching) in a single step using caustic soda (NaOH) and hydrogen peroxide (H2O2) with equal concentrations 0.70%. The treatment was carried out in a thermostatically controlled bath with a volume of 40

**Table 4** Mechanical properties of *Washingtonia* palm fiber and fibers from other palms

| Fiber | Diameter (μm) | Tensile strength (MPa) | Tensile modulus (GPa) | Elongation (%) | References |
|---|---|---|---|---|---|
| Washingtonia palm | 100–300 | 70–180 | 1.3–2.2 | 13.2–26.4 | Present work |
| Windmill palm | 202–418 | 82.0–113.7 | 0.82–1.25 | 39.5–47.5 | Zhai et al. (2012) |
| Oil palm | 200–300 | 150.9 | 2.9 | 30 | Anuar et al. (2019) |
| Sugar palm | 100–200 | 133.2–156.8 | 3.9–6.8 | – | Atiqah et al. (2018) |
| Date palm | 575 ± 60 | 187 ± 37 | 6.40 ± 1.17 | 3.31 ± 0.73 | Bezazi et al. (2020) |
| Palmyra palm | 300–320 | 276 ± 5 | 8.99 ± 0.12 | 3.08 | Shanmugam and Thiruchitrambalam (2014) |
| Doum palm | 353.37 ± 31.49 | 155.88 ± 12.03 | 6.29 ± 1.25 | 6.23 ± 2.92 | Fatma et al. (2019) |

L at a temperature of 96 ± 2 °C for 1 h. The extracted fibers were characterized in terms of tensile strength. The results obtained indicate a significant improvement in the tensile strength and tensile modulus of the extracted fibers compared to the raw palm fibers (260 MPa instead of 180 MPa and 4.8 GPa instead of 2.2 GPa). This can be explained by the action of the alkaline/bleaching treatment, which does not cause total fibrillation of the bundle fibers but rather eliminates hemicellulose and lignin. For this reason, the applied treatment increases the crystallinity of the fibers and therefore improves its mechanical properties (Hamidon et al. 2019; Santos et al. 2018). The mechanical properties obtained are better than those of some palm fiber species used as reinforcements in composite materials (see Table 4). Then, two nonwoven composites, Palm/Wool (P/W) and Palm/Polyester (P/PES), were produced by blending 75 wt% of palm fibers with 25 wt% of wool or polyester fibers. The polyester fiber used comes from the recycling of industrial waste. The recycled polyester fiber is characterized by its high abrasion and ultraviolet resistance, high stiffness, and low cost. However, recycled wool fibers are chosen for their good thermal comfort and resilience. Then, nonwovens were made using a DI-LOOM OD-II 10,069/2012 manufacturing unit (Fig. 2), consisting of five parts (Sajid et al. 2021c):

- Opener: The processing of chopped fibers begins with the bale opening operation. The selected opener depends on the bale characteristics, such as size and density. Depending on the requirements, the required degree of opening is achieved through several machine elements involving different opening levels.
- Feeder: This involves transporting the raw material to grinding cylinders and then to the carding machine.
- Carding: Carding is one of the basic techniques in the textile industry. Fed with raw fibers, the carding machine produces a web from which impurities (seeds, short fibers, dust, etc.) are removed. Depending on its design, it produces webs with parallel or randomly distributed fibers. The carding machine consists of three parts: The first part introduces the fibers via a "feed cylinder," the second part performs the actual carding operation, and the third part recovers the web using a "combing drum."
- Cross lapper: This involves folding a web into a web, comprising two belts, each following a closed path and associated with rollers for their respective guidance and drives.
- Needling: This involves physically interlocking the fibers constituting the webs. This part is similar to that used by paper mill felt suppliers to produce wet-wraps for paper machines. The needle-loom is equipped with a head into which a multitude of specially designed metal needles or barbs are inserted.

Afterwards, waterborne polyurethane (WPU) and waterborne polyacrylate (WPA) are used as cross-linking and finishing agents; these polymers are water-based and less toxic because organic solvents are not used. Nonwovens were coated with different percentages of polyurethane and polyacrylate (0, 5, 10, 15 and 20% by weight), dried at 105 ± 2 °C and hot calendered between two rollers at a pressure of about 20 ± 0.5 KN and a temperature of 140 ± 2 °C. These finishes improve the feel, visual appearance, tensile strength, abrasion resistance and climate of the developed nonwovens.

Figure 3 shows photographs of uncoated and coated nonwovens with 20% WPU and 20% WPA. All samples produced are flexible, and the surface of the coated samples is quite smooth and has a good visual appearance. Figures 3 indicate that polyurethane-coated nonwovens have an exciting feel and softness compared to those coated with polyacrylate. This is explained by the strong cohesion between the fibers used (Palm, Wool and Polyester) and the polyurethane. In addition, the flexibility of polyacrylate-coated nonwovens is greater than that of polyurethane-coated nonwovens.

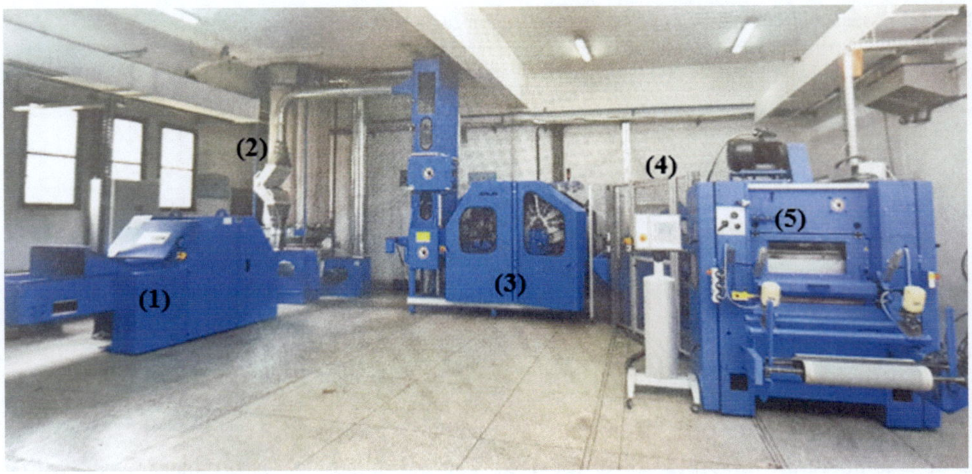


**Fig. 2** DI-LOOM OD-II 10,069/2012: (1) Opener; (2) Feeder; (3) Carding unit; (4) Cross lapper; (5) Needling unit

**Fig. 3** Visual appearance of (**a**, **b**) uncoated nonwovens, **c** 20% WPU-coated Palm/Wool **d** 20% WPU-coated Palm/Polyester, **e** 20% WPA-coated Palm/Wool and **f** 20% WPA-coated Palm/Polyester

Nonwovens are also characterized in terms of physical properties and the results are showed in Tables 5 and 6. The thickness of Palm/Wool nonwovens varies between 1.82 mm and 2.24 mm, and their surface density varies between 500.98 g/m$^2$ and 684.45 g/m$^2$. However, the thickness of Palm/Polyester nonwovens is between 1.16 mm and 1.45 mm, and their surface density is between 368.24 g/m$^2$ and 503.96 g/m$^2$. We notice that the thickness and surface density of Palm/Wool nonwovens are higher than those of Palm/Polyester. This is due to the low compactness of Palm/Wool nonwoven compared to Palm/Polyester. It is clearly seen that increasing the amount of polymers (WPU or WPA) increases the density of the nonwovens. However, as the proportion of polymers increases, the thickness of the sample decreases. The observed behavior is consistent with the literature and can be explained by the thermobonding of the applied polymer, which is enhanced by the calendering temperature and the roller clearance (Çinçik and Günaydin 2017; Pourdeyhimi 2019). Therefore, the pores of the nonwoven were closed and the thickness was reduced (Pourdeyhimi 2019; Muthukumar et al. 2019).

**Table 5** Physical characteristics of WPU-coated nonwovens

| Sample | Thickness (mm) | Surface density (g/m$^2$) | Porosity (%) | Water absorption capacity (%) |
|---|---|---|---|---|
| Palm/Wool 0 | 2.24 ± 0.02 | 500.98 ± 3.48 | 77.45 ± 0.26 | 768.65 ± 4.73 |
| Palm/Wool 5 WPU | 1.95 ± 0.01 | 578.68 ± 1.49 | 70.11 ± 0.14 | 752.11 ± 4.72 |
| Palm/Wool 10 WPU | 1.90 ± 0.01 | 608.23 ± 1.33 | 67.76 ± 0.16 | 743.30 ± 4.70 |
| Palm/Wool 15 WPU | 1.87 ± 0.01 | 643.56 ± 1.39 | 65.34 ± 0.12 | 735.54 ± 4.68 |
| Palm/Wool 20 WPU | 1.82 ± 0.01 | 684.45 ± 1.18 | 62.13 ± 0.15 | 729.75 ± 4.66 |
| Palm/Polyester 0 | 1.45 ± 0.01 | 368.24 ± 1.91 | 74.35 ± 0.25 | 484.07 ± 4.05 |
| Palm/Polyester 5 WPU | 1.34 ± 0.01 | 426.84 ± 1.23 | 67.92 ± 0.15 | 448.06 ± 4.00 |
| Palm/Polyester 10 WPU | 1.28 ± 0.01 | 446.51 ± 1.07 | 64.87 ± 0.15 | 430.79 ± 3.98 |
| Palm/Polyester 15 WPU | 1.22 ± 0.01 | 475.00 ± 1.12 | 60.79 ± 0.16 | 414.72 ± 3.97 |
| Palm/Polyester 20 WPU | 1.16 ± 0.01 | 503.96 ± 1.17 | 56.13 ± 0.15 | 404.07 ± 3.95 |

**Table 6** Physical characteristics of WPA-coated nonwovens

| Sample | Thickness (mm) | Surface density (g/m$^2$) | Porosity (%) | Water absorption capacity (%) |
|---|---|---|---|---|
| Palm/Wool 0 | 2.24 ± 0.02 | 500.98 ± 3.48 | 77.45 ± 0.26 | 768.65 ± 4.73 |
| Palm/Wool 5 WPA | 1.98 ± 0.01 | 570.40 ± 1.52 | 70.99 ± 0.16 | 462.14 ± 4.22 |
| Palm/Wool 10 WPA | 1.94 ± 0.01 | 597.35 ± 1.35 | 68.99 ± 0.14 | 453.80 ± 4.22 |
| Palm/Wool 15 WPA | 1.92 ± 0.01 | 635.03 ± 1.34 | 66.69 ± 0.14 | 445.40 ± 4.20 |
| Palm/Wool 20 WPA | 1.88 ± 0.01 | 675.68 ± 1.22 | 63.81 ± 0.12 | 438.76 ± 4.20 |
| Palm/Polyester 0 | 1.45 ± 0.01 | 368.24 ± 1.91 | 74.35 ± 0.25 | 484.07 ± 4.05 |
| Palm/Polyester 5 WPA | 1.37 ± 0.01 | 412.88 ± 1.22 | 69.65 ± 0.13 | 271.76 ± 3.80 |
| Palm/Polyester 10 WPA | 1.31 ± 0.01 | 431.90 ± 1.07 | 66.80 ± 0.13 | 259.88 ± 3.80 |
| Palm/Polyester 15 WPA | 1.27 ± 0.01 | 459.46 ± 1.11 | 63.57 ± 0.12 | 234.60 ± 3.79 |
| Palm/Polyester 20 WPA | 1.24 ± 0.01 | 487.48 ± 1.15 | 60.41 ± 0.12 | 222.80 ± 3.78 |

According to the literature, the car floor mat is the heaviest and thickest textile material than other textiles used in automotive interiors. Their weight and thickness must be greater than 1100 g/m$^2$ and 4 mm (Atakan et al. 2020; Gurudatt et al. 2005). Indeed, the obtained values of weight and thickness indicate that the developed nonwovens can be considered as automotive upholstery fabrics. In addition, they can be used as decorative covering after lamination on the surface of the main carpet (Atakan et al. 2020; Gurudatt et al. 2005). The density of the generated nonwovens is also a key consideration. The obtained values remain lower than those of the materials used in the market, leading to the reduction of the weight of the vehicles. The porosity of nonwovens is a critical parameter, its high values improve both the thermal and acoustic resistivity of the materials. However, the reduction of pores contributes to the improvement of aesthetic, mechanical and durability of nonwovens. Coating the developed nonwovens with WPU significantly reduces the porosity of nonwovens compared to WPA. The porosity of Palm/Wool nonwoven was decreased from 77.45% to 62.13% and 63.81% after coating with 20% WPU and WPA respectively. Similarly, the porosity of Palm/Polyester nonwoven was reduced from 74.35% to 56.13% and 60.41% after coating with 20% PU and PA respectively. The porosity values obtained are comparable to nonwovens developed for the automotive industry. Some automotive upholstery fabrics require water resistance, especially for seat coverings, carpets, etc. In our study, using natural fibers and water-based coating polymers resulted in highly water-sensitive materials. The water absorption capacity of Palm/Wool nonwovens is 1.6 times higher than that of Palm/Polyester nonwovens. In addition, polyurethane-coated nonwovens tend to absorb more water than polyacrylate-coated nonwovens. Given the high sensitivity of wool fiber to water, only a Palm/Polyester nonwoven was treated in our previous study to overcome this limitation.

The effect of CPTS treatment on the water sensitivity of Palm/Polyester nonwoven was evaluated by determining the water absorption capacity of these materials. Figure 4 indicates that using CPTS significantly reduces the water absorption capacity of the nonwoven products. The best water resistance results were found for Palm/Polyester (P/PES) treated with CPTS followed by 5% WPA (60.32%). This is due to the hydrophobic nature of the aliphatic chain of CPTS

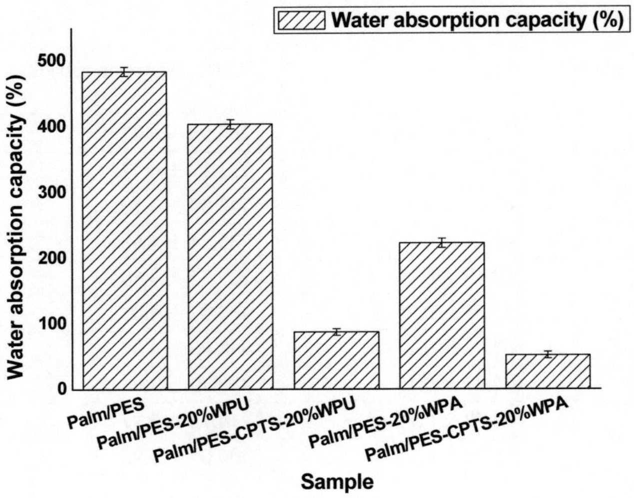


**Fig. 4** Water absorption capacity of uncoated and CPTS, 20% WPU and 20% WPA coated Palm/Polyester (P/PES) nonwoven

present on the surface of the coating, which contributes to low surface energy (Yin and Wang 2013; Azmami et al. 2021b). In addition, the acrylate group remains less sensitive to moisture compared to the polyurethane and hydroxyl groups of the nonwoven, which protects the nonwoven material against water. These results reveal that CPTS and WPA coated nonwovens can be selected as water-repellent textile materials for automotive applications.

The mechanical properties, especially the tensile strength and abrasion resistance of the developed nonwovens, are essential for evaluating the quality and mechanical performance of materials for automotive interiors. In general, the polymers used improve the tensile strength and abrasion resistance of the developed nonwovens. The tensile strength and elongation of WPU-coated and WPA-coated nonwovens are shown in Fig. 5. As observed, coating the nonwovens with WPU and WPA significantly improves the mechanical performance of the nonwovens. Figure 5 indicates that the tensile strength of the nonwovens increases as the polymer proportion rises. The best results are obtained with WPU. This polymer increased the tensile strength of the Palm/Polyester nonwoven to 136% and the Palm/Wool nonwoven to 410%. In contrast, WPA increased the tensile strength of Palm/Polyester and Palm/Wool nonwovens to 105% and 369%. The elongation at break of WPU-coated nonwovens ranges from 22.8% to 14.7% for Palm/Wool nonwovens and from 27% to 31.2% for Palm/Polyester nonwovens. However, the elongation at break of WPA-coated nonwovens ranges from 22.8% to 30.1% for Palm/Wool nonwovens and from 27% to 44.1% for Palm/Polyester nonwovens. The elongation results of WPA-coated nonwovens have the same trend as those of WPU-coated nonwovens. The high elongation rates with polyacrylate confirm their high elastic power compared to PU. Therefore, WPA-coated nonwovens are more flexible than WPU-coated ones. The mechanical properties of coated nonwovens with WPU and WPA content above 15% remain comparable to those of automotive carpets made with polyester fibers or a mixture of natural and synthetic fibers for automotive interior applications (Gurudatt et al. 2005; Thilagavathi et al. 2010; Kamath et al. 2005).

Abrasion resistance is a key factor in assessing the durability of nonwoven materials (Zhang et al. 2020). The abrasion resistance of nonwovens before and after coating was conducted using a Taber GESTER GT-C14B abrasion tester according to ISO 5470-1 standard. It was assessed based on fiber loss after a specified number of abrasion cycles. The minimum required fiber loss rate varies among car manufacturers. According to one of the most quality-conscious car manufacturers, fiber loss could be acceptable when it was less than 1 g at 1500 cycles (equivalent to a 10% mass loss) (Atakan et al. 2020). Figure 6 shows photographs of uncoated and coated nonwovens with 20% WPU and 20% WPA nonwovens after Taber abrasion. The uncoated Palm/Wool and Palm/Polyester samples were rapidly abraded and destroyed after 300 and 160 cycles, respectively, with mass losses of 14.23% and 12.86%, respectively. The resistance of Palm/Wool compared to Palm/Polyester was due to its high surface area and thickness. However, as more fibers per unit area contributes to more contact between fibers, it would be more difficult to abrade (Palak and Karagüzel Kayaoğlu 2019; Çinçik and Günaydin 2017). Coating the developed nonwovens with WPU and WPA results in a decrease in mass loss and demonstrates a significant increase in abrasion resistance. After coating with WPU, the Palm/Wool nonwoven shows an increase in abrasion resistance from 300 cycles to 25,000 cycles, while the Palm/Polyester nonwoven shows an increase in abrasion resistance from 160 to 5500 cycles. It is also noted that WPA is slightly less effective in abrasion resistance compared to WPU (22,000 cycles for Palm/Wool and 4500 cycles for Palm/Wool). However, the mass loss values obtained with WPA and WPU concentrations higher than 10% are competitive for automotive interiors (Palak and Karagüzel Kayaoğlu 2019; Atakan et al. 2020). Its properties are also equivalent to nonwoven composites made from a blend of banana, jute, and bamboo fibers with polypropylene for noise reduction in automotive interiors (Thilagavathi et al. 2010). Moreover, these nonwovens have higher mechanical properties than those provided by a blend of cotton, linen, and kenaf fibers with poly(tetramethylene adipate-coterephthalate) and polypropylene still for automotive applications (Kamath et al. 2005). Therefore, the developed nonwoven composites have various applications in automotive interiors, as a decorative surface layer for carpets, door trims, trunk liners, and cargo floor liners (Gurudatt et al. 2005; Thilagavathi et al. 2010; Kamath et al. 2005; Kim and Kim 2019).

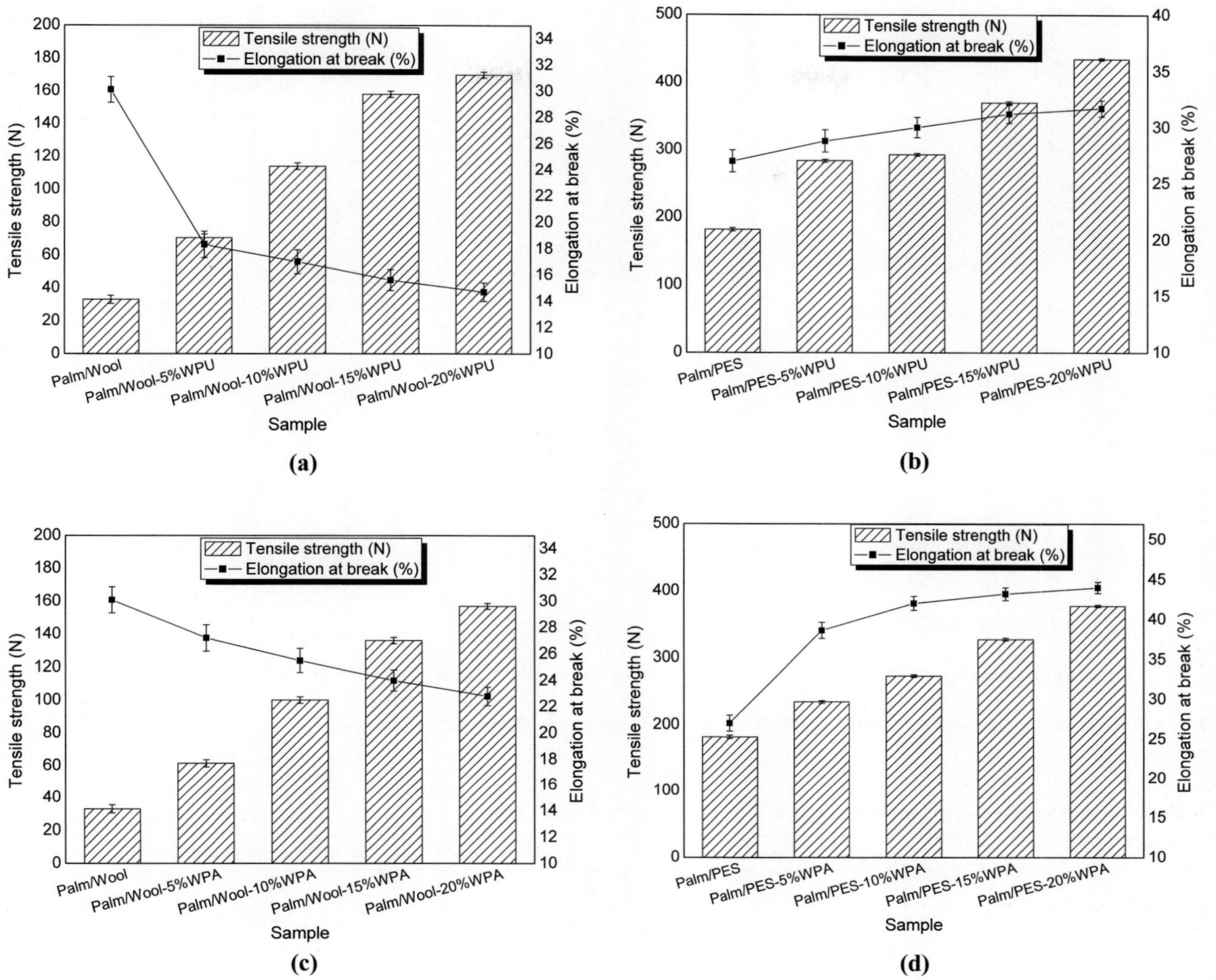


**Fig. 5** Tensile strength and elongation of uncoated, (**a**, **b**) WPU-coated and (**c**, **d**) WPA-coated nonwovens

The fire behavior of the developed nonwovens, as well as the effect of WPU and WPA coating on flammability, were evaluated according to ISO 6940, and the results are shown in Table 7. As observed, the Palm/Wool nonwoven material demonstrated high resistance to direct flame. Indeed, the Palm/Wool nonwoven is not burned during and after the flame removal (see Fig. 7a). Moreover, the final residue persists at a weight of 100%. This result is explained by the flame retardant character of the wool fiber (Allafi 2020; Guna et al. 2021). The latter does not ignite and is self-extinguishing.

This means that it extinguishes as soon as the flame goes out, leaving behind a soft ash residue that dissipates (Guna et al. 2021). The result obtained suggests that the palm fiber also helps in achieving the flame retardancy of nonwoven textiles. Nevertheless, as shown in Fig. 7d, the Palm/Polyester nonwoven tends to burn for 104 s after the flame has been removed with a hard and molten central residue. The final residue after the test is 47 wt% and the burned surface area equals 88.60 $cm^2$. This observation shows that the polyester fiber is susceptible to flame and promotes flame propagation throughout the nonwoven sample. The WPU coating of the nonwovens promotes rapid combustion of the materials. Palm/Wool nonwovens treated with WPU concentrations less than or equal to 10% did not burn during and after flame removal. Indeed, their flame-exposed surfaces were charred with final residues greater than 97wt%. At WPU concentrations equal to 20%, Palm/Wool nonwovens begin to burn for 84 s after flame removal. The final residue after the test can be up to 56% and the burned surface reaches 68 $cm^2$. The same behavior is observed in the case of the Palm/Polyester non-woven fabric with a final residue after the test that can get up to 36% and the burnt surface reaches 95 $cm^2$ (see Table 7). Regarding WPA-coated nonwovens, the results obtained remain poor compared to those of WPU-coated nonwovens. To address

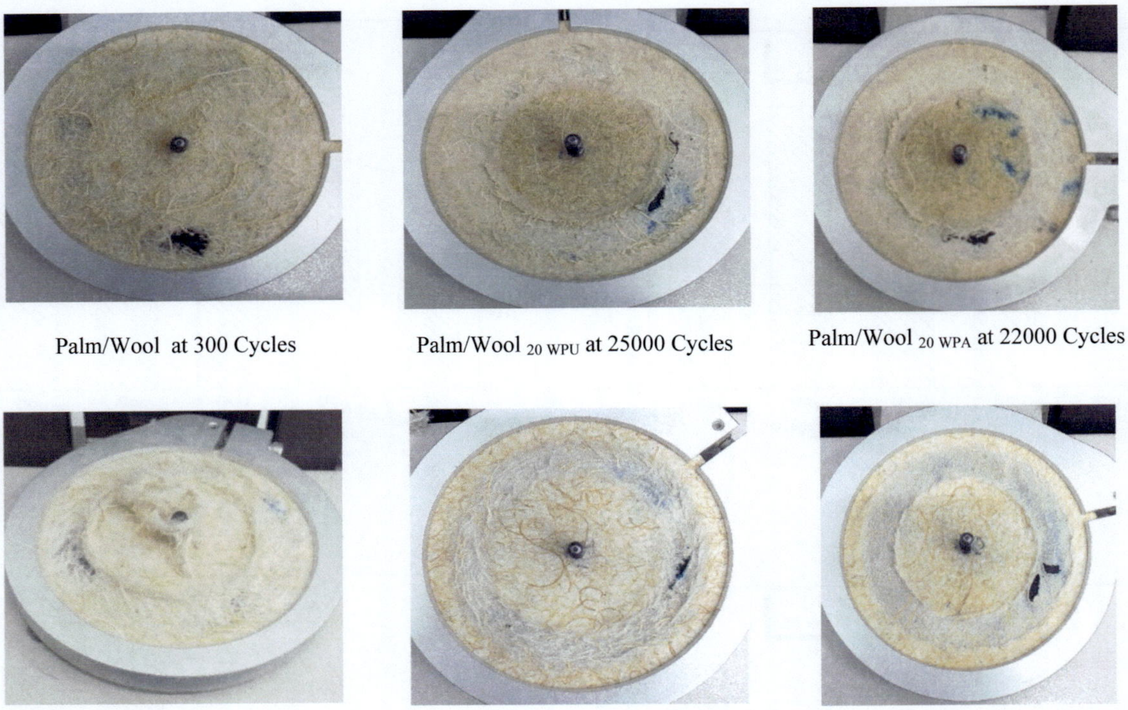

Palm/Wool at 300 Cycles | Palm/Wool $_{20\ WPU}$ at 25000 Cycles | Palm/Wool $_{20\ WPA}$ at 22000 Cycles

Palm/Polyester $_{20\ PA}$ at 160 Cycles | Palm/Polyester $_{20\ WPU}$ at 5500 Cycles | Palm/Polyester $_{20\ WPA}$ at 4500 Cycles

**Fig. 6** Images of uncoated and 20% WPU and 20% WPA coated nonwovens after Taber Abrasion test

**Table 7** Results of vertical flammability test of WPU and WPA coated nonwovens

| Sample | Time after flame removal (s) | Burned area ($cm^2$) | Residue (wt %) |
|---|---|---|---|
| Palm/Wool | 0 | 0 | 100 |
| Palm/Wool 5 WPU | 5 ± 1 | Carbonized area 7.5 ± 0.5 | 99 ± 0.2 |
| Palm/Wool 10 WPU | 8 ± 1 | Carbonized area 12 ± 0.5 | 97 ± 0.2 |
| Palm/Wool 20 WPU | 84 ± 2 | 68.00 ± 1 | 56 ± 0.5 |
| Palm/Wool 5 WPA | 32 ± 1 | 102 ± 0.5 | 50 ± 1 |
| Palm/Wool 10 WPA | 43 ± 1 | 119 ± 1 | 49 ± 1 |
| Palm/Wool 20 WPA | 114 ± 2 | 126 ± 1.5 | 46 ± 1 |
| Palm/Polyester | 104 ± 1 | 88.60 ± 0.5 | 47 ± 0.5 |
| Palm/Polyester 5 WPU | 106 ± 1 | 91.14 ± 0.5 | 42 ± 0.5 |
| Palm/Polyester 10 WPU | 114 ± 2 | 92.66 ± 1 | 40 ± 0.5 |
| Palm/Polyester 20 WPU | 133 ± 2 | 94.75 ± 1 | 36 ± 1 |
| Palm/Polyester 5 WPA | 122 ± 2 | 132 ± 1 | 34.1 ± 1 |
| Palm/Polyester 10 WPA | 132 ± 2 | 136 ± 1 | 31.5 ± 1 |
| Palm/Polyester 20 WPA | 137 ± 2 | 140 ± 1.5 | 25.8 ± 1 |

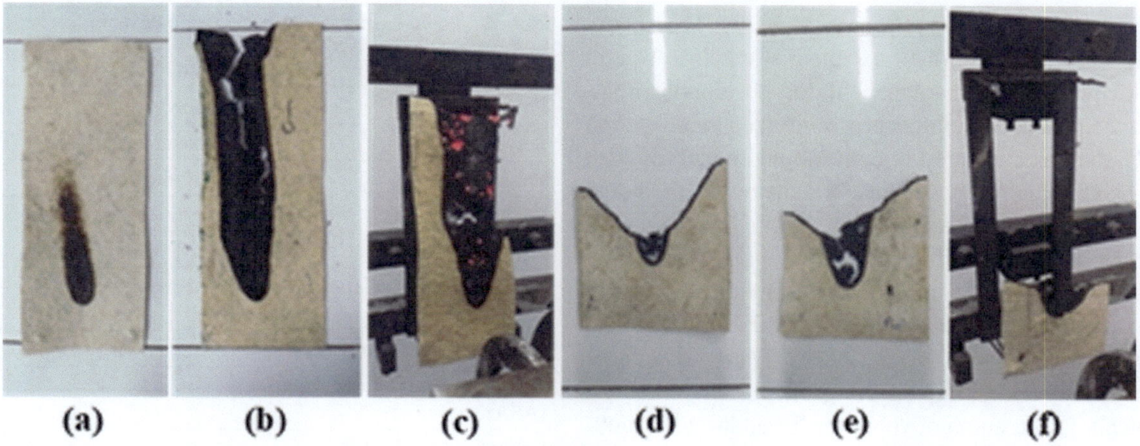


**Fig. 7** Images of samples after flammability test: **a** uncoated Palm/Wool, **b** 20% WPU-coated Palm/Wool **c** 20% WPA-coated Palm/Wool, **d** uncoated Palm/Polyester, **e** 20% WPU-coated Palm/Polyester and **f** 20% WPA-coated Palm/Polyester

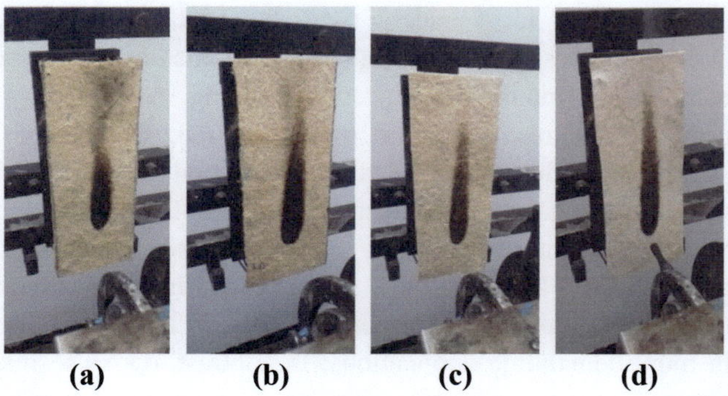


**Fig. 8** Images of CPTS-treated samples after flammability test: **a** Palm/Wool CPTS + 20%WPU, **b** Palm/Wool CPTS + 20%WPA, **c** Palm/Polyester CPTS + 20%WPU and **d** Palm/ Polyester CPTS + 20%WPA

this issue, treating nonwovens with the CPTS precursor improves their flammability. Thus, nonwovens treated with CPTS and then coated with WPU do not burn during and after flame extinction (see Fig. 8), and the final residue persists at 100% mass. The same behavior is observed with WPA-coated nonwovens (see Fig. 8), suggesting their application in fields where fire resistance is required, such as automotive.

## 3 Composite Materials in the Automotive Industry

### 3.1 Overview of Automotive Composites

Composite materials refer to constructions made up of two or more separate materials, each with its own set of attributes, that work together to generate a synergy that allows for better performance than individual components (Bhong et al. 2023). Because of their unique properties of lightness, strength, and durability, these materials have gained substantial attraction in various applications, including aerospace, automotive, marine, and construction (Bhong et al. 2023). According to Shekar and Ramachandra, composite materials are made up of a matrix (usually polymer, metal, or ceramic) and a reinforcement (such as glass or carbon fibers), which provide the composite with enhanced mechanical strength and stiffness (Shekar and Ramachandra 2018). This combination meets specific needs in increasingly demanding applications. In particular, the automotive industry leverages this technology to optimize vehicle performance, safety, and environmental impact (GuoH and Liu 2024). A composite material comprises two major components: the matrix and reinforcement. The matrix, often a polymer or metal, maintains component cohesion and dictates the composite's overall qualities. Reinforcement, on the other hand, plays a vital function in enhancing mechanical attributes like strength and stiffness. These reinforcements might be fibers, particles, or whiskers and are commonly constructed of glass fibers, carbon fibers, or Kevlar. The latter, for example, is a lightweight but incredibly robust material employed in the automobile sector to manufacture high-performance parts. The global market for composite materials is rapidly growing and is projected to reach approximately USD 159.35 billion by 2034 (Precedence Research 2024). This trend highlights their increasing popularity across a wide range of industries. In the automotive sector, demand is rising primarily due to the advantages of vehicle weight reduction, which enhances fuel efficiency and lowers $CO_2$ emissions. Composite materials are used extensively in modern car manufacture. Their applications range from ultra-lightweight body panels to structural components (such as side panels and beams), interior elements (dashboards and seats), and mechanical items like suspensions and drive shafts (Elmarakbi 2013). This adaptability is mainly owing to their excellent strength-to-weight ratio, which is a significant benefit for developing high-performance vehicles. Composites reinforced with carbon and glass fiber, as compared to conventional materials such as steel or aluminum, may reduce weight by 40–70% and 25–35%, respectively (Sarfraz et al. 2021). This decrease has a direct influence on fuel consumption: a 10% weight reduction results in a 5% increase in energy efficiency. For example, using carbon fiber can result in a 30% improvement compared to steel (Sarfraz et al. 2021). These advantages are particularly strategic given the growing environmental requirements in the automotive sector. In addition to lightness and energy efficiency, composites offer additional benefits: their resistance to corrosion increases vehicle durability, even in extreme conditions (salty roads, humidity, thermal variations). Their superior damping properties reduce vibrations and noise inside the cabin, providing better comfort for passengers (Ratna and Chakraborty 2023). Nowadays, these materials are found in various parts, including doors, roofs, rear panels, seats, windows, luggage compartments, cooling fans, and even in the brakes of luxury or racing vehicles. Driveshafts particularly benefit from carbon fiber composites, which allow for better compactness, elimination of joints, and resistance to high rotational speeds. Despite sometimes higher production costs than traditional materials, the performance, durability, and environmental compliance benefits fully justify their growing adoption (Aznaw 2025). Composites thus appear as a key technological solution for the sustainable mobility of the future (Aznaw 2025; Skosana et al. 2025). Composite materials are categorized based on four key criteria: the type of matrix, the form of reinforcement, the size of the reinforcement, and the field of application. This classification system enables engineers to select materials that closely align with the specific demands of various industries, including automotive, aerospace, and construction.

## 3.2 Polymer Matrices Used in Automotive

The selection of the matrix represents a key element in the design of composite materials, as it directly influences their mechanical and thermal performance as well as their durability. In the automotive sector, polymer matrices play a fundamental role in producing composites that are lightweight, high-performance, and suitable for mass production requirements. There are two main families of matrices: thermosetting resins, such as unsaturated polyesters and epoxies, and thermoplastic resins, such as polypropylene and polyamides. The most commonly used polymers in this field are presented in Table 8. These matrices are selected based on various criteria, including mechanical performance, chemical resistance, cost, weight, and compatibility with manufacturing processes. Polyester resins are the most widely used due to their low cost, ease of processing, and adaptability to methods such as compression molding or injection molding. Among them, orthophthalic resins are the most common. Other types, such as isophthalic resins (which have good chemical and moisture resistance) or chlorinated/bisphenol resins (which have better thermal stability), may be used in specific applications. However, their shrinkage during curing and toxic emissions in case of fire sometimes limit their use in more demanding applications. In contrast, epoxy resins, although more expensive, offer better mechanical performance, higher heat resistance, low shrinkage, and good adhesion to fibers. These materials are ideal for structural elements that require high fatigue resistance, dimensional stability, and durability (Gonçalves et al. 2022; Sakib and Iqba 2021). Polyimides, less commonly used but classified as high-performance resins, are resistant to very high temperatures and have good chemical resistance, while emitting little smoke in case of fire (Kondo et al. 2022). They are therefore intended for specific high-temperature applications. Polyamides (particularly PA6, PA6.6, and PA11) are also used for their good mechanical properties, but their sensitivity to moisture can affect their long-term performance (Kondo et al. 2022). On the thermoplastic resin side, materials like polypropylene are attractive due to their recyclability, low cost, and chemical inertness. However, the low adhesion between these polymers and fibers requires coupling agents. Recent development of composites using natural fibers helps reduce environmental impact while maintaining good performance for non-structural parts.

**Table 8** Polymer matrices used in automotive

| Thermoplastics | Thermosets |
|---|---|
| Polypropylene (PP) | Epoxy (EP) |
| Polyethylene (PE) | Unsaturated polyester (UPR) |
| Polyamide (PA) | Phenolic |
| Polycarbonate (PC) | Polyurethane (thermoset type) |
| Polystyrene (PS) | Vinyl ester |
| Acrylonitrile–butadiene–styrene (ABS) | Silicone |
| Polyether-ether-ketone (PEEK) | |
| Acrylonitrile styrene acrylate (ASA) | |
| Polyvinylchloride (PVC) | |

## 3.3 Reinforcement Used in Automotive

Composite reinforcement plays a vital role in improving the mechanical properties of materials, especially in the automotive sector. Although reinforcements can appear as particles, whiskers, or flakes, fibers are the most widely used form due to their superior strength-to-weight ratio. Notably, glass, carbon, and aramid fibers are widely favored, bringing distinct benefits to automotive component design.

### 3.3.1 Glass Fibers

Glass fibers are the most commonly used reinforcement due to their excellent cost-to-performance ratio. They are well-suited for reinforcing non-structural components such as body panels, passenger compartments, and other non-load-bearing parts (Lazzari 2025). These fibers exhibit excellent tensile strength and impact resistance. Furthermore, their resistance to heat and moisture makes them ideal for withstanding the extreme operating conditions vehicles frequently encounter. A key aspect of using fiber reinforcements lies in the interface with the polymer matrix. Raw or untreated fibers generally exhibit poor adhesion, which limits their mechanical effectiveness (Deng et al. 2022). To overcome this, surface treatments such as organosilane coatings for glass fibers are commonly applied. These treatments enhance chemical bonding at the fiber-matrix interface, leading to composites with improved strength, durability, and stability (Deng et al. 2022).

### 3.3.2 Carbon Fibers

Carbon fibers are renowned for their exceptional lightweight properties and mechanical strength. They are up to 75% lighter than steel and 40% lighter than aluminum (Ahmad et al. 2020). This makes them an ideal material for manufacturing lightweight structural parts, including body panels, engine components, and high-performance vehicle structures. However, due to their significantly higher cost than other structural materials, carbon fibers are often reserved for premium applications or components requiring exceptional performance characteristics (Ahmad et al. 2020).

### 3.3.3 Aramid Fibers

These fibers are characterized by their superior impact and tear resistance, making them particularly suitable for applications that require high levels of safety and protection (Wang and Gao 2021). In the automotive industry, aramid fibers are employed in safety-related components, such as impact protection panels and braking systems. Their unique combination of mechanical strength and energy absorption capacity ensures adequate protection during collisions, and enhancing overall vehicle performance (Wang and Gao 2021).

## 3.4 Composite Manufacturing Processes

### 3.4.1 Hand Lay-Up

The hand lay-up technique is one of the oldest and most fundamental composite production techniques. It entails manually inserting layers of reinforcing material (such as fiberglass or carbon fiber) into a mold, then manually applying resin to saturate the fibers (Sadashiva et al. 2021). This approach is commonly used for prototype development and low-volume production, particularly for oversized, simple-shaped items (Friedrich and Almajid 2013). In the automobile industry, the hand lay-up method is used in composite leaf springs, where unidirectional glass fiber/epoxy compositions allow for adequate strain storage and release along the spring's axis. Despite being labor-intensive, it is nevertheless useful for specialized parts and research and development.

### 3.4.2 Resin Transfer Molding (RTM)

RTM is a closed-mold method that involves placing dry reinforcements in a mold cavity and injecting resin under pressure. This technology provides for exact control of fiber content and part thickness, making it suited for medium-volume production (Schwartz 2010). RTM is frequently used in automobile structural components due to its automation possibilities and high part repeatability. However, issues such as air entrapment and resin front collision have been documented during multi-gate injection, which could result in voids in the final product (Shen et al. 2024).

### 3.4.3 Compression Molding

Compression molding is a high-volume manufacturing process perfect for making composite parts such as hoods, bumpers, roofs, and panels. Pre-impregnated materials, such as sheet molding compounds or glass mat thermoplastics, are crushed into heated molds to make the part (Farahmand 2025). It offers short cycle times, consistent quality, and good surface finishes, making it a popular choice for automotive body and structural parts (Farahmand 2025).

### 3.4.4 Vacuum Infusion Process (VIP)

The vacuum infusion process (VIP) entails inserting dry fibers in a mold and applying vacuum pressure to pull resin into the mold. This approach ensures even resin dispersion and high fiber content while keeping tooling costs low (Poorzeinolabedin 2017). VIP is often used to make big, high-performance carbon fiber reinforced polymer) parts for the automotive and aerospace industries. It offers precise control over fiber alignment and thickness while minimizing environmental effects and waste (Poorzeinolabedin 2017).

**Table 10** Natural fiber composites for automotive applications (Naik and Kumar 2021)

| Natural fiber | Components in automotive |
|---|---|
| Flax | Door panel cover, covered, instrument panels, covered inserts, arm rests, seat back panels |
| Hemp | Door panels, seat back panels, dashboard |
| Sisal/Flax | Door linings, interior panels |
| Jute | Door panels, Dashboard |
| Abaca | Body panel, floor panel |
| Banana | Wrapping paper |
| Coir | Seat covers, Seat mattresses, mats |
| Cotton | Insulation material, soundproofing, trunk panel |
| Wood | Foamed instrument panels, seat back panels, fiber in the seatback cushions, covered inserts and components, inserts, |
| Wool | Upholstery, rear seat covers |

### 3.4.5 Sheet & Bulk Molding Compound (SMC and BMC)

Sheet molding compound (SMC) and bulk molding compound (BMC) are composite materials that come in either sheet or dough-like shapes. These materials comprise chopped fibers, resins, and additives, making them ideal for compression molding into intricate shapes (Isham 2017).

SMCs are commonly utilized in the automotive sector for body panels, bumpers, engine covers, battery enclosures, and other components (Khan 2024b). BMCs are appropriate for under-the-hood parts, structural reinforcements, and electrical components due to their high strength and design flexibility (Khan 2024b).

### 3.4.6 Filament Winding

Filament winding makes cylindrical or tubular elements, such as driving shafts and pressure tanks. Continuous fibers are coiled onto a revolving mandrel at specified angles before being impregnated with resin (Asyraf et al. 2023). This approach provides parts with high strength-to-weight ratios, particularly in directions parallel to winding routes, and is best suited for axisymmetric structures (Asyraf et al. 2023). Table 9 summarizes the composite manufacturing processes along with their specific advantages, limitations, and application areas in the automotive industry.

## 3.5 Integration of Agricultural Waste into Automotive Composite

Natural fibers generated from agricultural waste, such as sisal, jute, kenaf, banana, and others (see Fig. 1), are increasingly being mixed into composite materials due to environmental and economic demands in the automobile industry. These fibers are frequently utilized to make non-structural interior components like dashboards, door panels, headliners, trunk linings, and seat backs. Table 10 outlines the natural fibers most commonly employed in the automotive industry for fabricating a range of structural and interior components. Natural fiber composites offer several advantages over glass fiber composites, including lower weight, reduced density, cost-effectiveness, environmental sustainability, renewability, and non-toxicity. These attributes have led leading automotive manufacturers, such as Mercedes-Benz, Audi, BMW, Volkswagen, Toyota, and Ford, to incorporate natural fiber composites into various vehicle components. In 2004, Mercedes-Benz initiated commercial production of spare wheel pan covers for the A-Class, incorporating abaca natural fiber reinforcements as a partial replacement for traditional glass fibers (Kalia and Avérous 2011). The development process spanned 18 months and resulted in a product that achieved notable reductions in both cost and weight, with minimal impact on overall quality. Furthermore, life cycle analysis revealed that using abaca fibers led to an approximate 10% reduction in primary energy consumption (Kalia and Avérous 2011). In addition, Mercedes-Benz also employed coconut fiber–rubber latex composites for the seats of the A-Class model and used flax–sisal fiber mat–reinforced epoxy composites for the door panels of the E-Class model, demonstrating a broader commitment to sustainable material innovations across its vehicle lineup (Shinoj et al. 2011). Audi utilizes flax–sisal mat-reinforced polyurethane composites, in combination with other materials, for the production of door trim panels, reflecting the company's commitment to incorporating sustainable materials in automotive design (Ghosh et al. 2024). Ford incorporates kenaf fibers into the production of its Mondeo model. The Mondeo's door panels are manufactured using kenaf-reinforced polypropylene (PP) composites, while flax fibers are used in the creation of floor trays (Anuar and Zuraida 2011). Similarly, a blend of kenaf and flax fibers has been used in the package trays and door panel inserts of the Opel Vectra, highlighting a broader industry trend toward adopting natural fiber composites in automotive interior components (Shanmugam et al. 2021). (Sreenivas et al. 2020), explored the use of Kenaf fibers in the fabrication of composite materials, considering factors such as stacking sequence (layer by layer), fiber-to-matrix volume ratio, fiber orientation and chemical surface modification to enhance fiber-matrix adhesion. Their findings

**Table 9** Composite processing methods and automotive applications

| Method | Application | Advantages | Limitations |
|---|---|---|---|
| Hand lay-up | Leaf springs, prototypes, simple body panels | Low cost, design flexibility, suitable for large parts | Manual process, low repeatability, operator-dependent quality |
| RTM | Structural parts (doors, body shells, reinforcements) | Good thickness control, ideal for complex shapes | Risk of voids/air traps, costly tooling |
| Compression molding | Hoods, roofs, bumpers, side panels | Fast cycle times, good surface finish, compatible with SMC/GMT | High tooling cost, some material waste |
| Vacuum infusion process | Large CFRP parts (body panels, hoods, structural covers) | High fiber volume, low tooling cost, good dimensional accuracy | Long cycle times, requires vacuum materials (films, bags,...) |
| Filament winding | Drive shafts, tanks, cylindrical components | Excellent axial strength, continuous structure | Limited to symmetrical shapes, mandrel required |
| SMC / BMC | Bumpers, enclosures, engine parts, interior panels | Fast molding, good cost-performance ratio | Limited shape complexity, less recyclable |

indicate that kenaf-based composites have potential applications in automotive components, including dashboards, door panels, and underfloor parts. Volkswagen has utilized cellulose fibers in the manufacturing of seatbacks, door panels, boot-lid finish panels, and boot liners across several models, including the Passat Variant, Golf, A4, and Bora (Rahman et al. 2023). The BMW Group has extensively integrated natural fiber–reinforced polymer composites into its automotive manufacturing. In 2004 alone, the company utilized approximately 10,000 tonnes of natural fibers (Verma and Senal 2019). Each BMW 7 Series vehicle incorporates around 24 kg of renewable raw materials, including flax and sisal in the interior door lining panels. Additionally, cotton is employed for soundproofing, wool is used in the upholstery, and wood fibers are incorporated into the seat back structures. These eco-efficient composites achieve sustainability goals while adhering to increasingly stringent standards, such as those imposed by the European Union and Japan, which require the recyclability of vehicle components.

### 3.5.1 Palm Waste in Automotive Composites

Palm trees, which belong to the Arecaceae family, are extensively spread in arid and semi-arid locations, and they constitute a renewable source of biomass with significant material uses. Date palm, sugar palm, palmyra palm, doum palm, and washingtonia palm are all neglected in Morocco, despite their ability to withstand challenging conditions and yield considerable fibrous leftovers. The washingtonia palm petiole, in particular, has sparked interest because of its fibrous nature and potential use in natural fiber-reinforced composites. Palm fibers are generally lignocellulosic, consisting primarily of cellulose (38–55%), hemicellulose (12–28%), and lignin (15–32%) (Shanmugam and Thiruchitrambalam 2014; Fatma et al. 2019). Palm fibers have high dimensional stability, moderate stiffness, and low density (0.9–1.2 g/cm$^3$), making them ideal for lightweight composites. The mechanical characteristics of palm fibers vary greatly depending on the species, treatment, and extraction process. Table 4 shows the typical mechanical and physical properties of fibers from several palm species. The results highlight the significant variability in mechanical properties among different palm species. Notably, palmyra and date palms demonstrate superior tensile strength and stiffness, indicating their suitability for load-bearing applications. In contrast, windmill palm exhibits higher elongation at break, making them more appropriate for applications requiring enhanced flexibility. Palm fibers are considered well-suited for automotive applications due to their ability to be processed into both continuous and short forms (Hamouda and Aly 2020). This adaptability facilitates the manufacturing of complex components with isotropic properties, ultimately improving production efficiency and supporting the advancement of the automotive industry (Hamouda and Aly 2020). (Shinoj et al. 2011)., report that oil palm fiber serves as an effective reinforcement within polyurethane matrices for the production of various components, including flexible foam seating, rigid foam insulation panels, automotive suspension bushings, carpet underlays, and hard plastic parts for electronic devices. Additionally, polyurethane is utilized to fabricate molded architectural elements such as door frames, columns, and window headers (Shinoj et al. 2011). (Arumugaprabu et al. 2021), examined the sustainable use of natural resources and agricultural waste, highlighting the potential of date palm fiber for industrial and automotive applications. Owing to its excellent vibration-absorbing properties, date palm fiber is considered suitable for manufacturing automotive components such as springs and shock absorbers. (Mastura et al. 2017), identified sugar palm fiber as the most effective natural fiber for hybridization with glass fiber-reinforced polymer composites in developing an automotive anti-roll bar. The selection process employed a novel material selection methodology that integrates the Analytic Hierarchy Process (AHP) with Quality Function Deployment for Environment (QFDE), emphasizing environmental impact as a key criterion (Mastura et al. 2017). On the other hand, Weyerhaeuser, a leading pulp and paper company, developed a thermoplastic composite known as THRIVE™, composed of engineered palm cellulose blended with polypropylene (Khatri et al. 2023). This composite was found to be both cost-effective and readily available. Compared to short, abrasive glass fibers, THRIVE™ offered advantages such as reduced wear on manufacturing equipment and lower energy consumption during production. The composite fibers met the automotive industry's requirements for hardness, durability, and thermal resistance, while also being approximately 10% lighter than comparable glass fiber composites, an attribute with significant implications for improving vehicle performance (Khatri et al. 2023). (Muthalagu et al. 2021), investigated the performance of epoxy composites reinforced with Kevlar and date palm fibers, comparing them to traditional glass fiber materials. Using hand layup techniques and finite element analysis, they assessed tensile strength and ductility across various fiber compositions. The results demonstrated that a 50% reinforcement with Kevlar and date palm fibers significantly enhanced ductility, and the composite exhibited comparable or superior mechanical properties to glass fiber. These findings support using palm fiber-reinforced epoxy composites as a safer and more sustainable alternative for automotive bumper beam applications (Muthalagu et al. 2021).

## 4 Conclusion

Agricultural waste is increasingly used in the composite materials industry, driven by growing awareness of the environmental drawbacks of synthetic fibers. In the automotive sector, they are widely adopted for manufacturing interior components of both passenger and commercial vehicles. Biocomposites reinforced with natural fibers show great promise for producing sustainable, lightweight, cost-effective, recyclable parts. In particular, palm fibers have recently gained attention for their contribution to industrial sustainability and environmentally responsible waste management. Palm fibers extracted from various palm species like date palm, oil palm, washingtonia palm and sugar palm, have demonstrated excellent reinforcing capabilities when combined with different matrix materials, enhancing the performance of automotive composites. Their use also supports industrial sustainability and responsible waste management. Chemical modifications can improve their mechanical and physical properties, making them even more suitable for demanding applications. These biocomposites are particularly promising for various automotive components, including door panels, trunk liners, seat backs, spare tire linings, friction pads, and headliners. Moreover, developing new hybrid biocomposites that combine date palm fibers with other natural fibers could further broaden their applicability within the automotive industry.

Nevertheless, the broader use of palm fibers in the automotive industry faces several challenges. One key issue is the variability in fiber characteristics, which stems from differences in palm species, growing environments, and extraction methods, which can lead to inconsistent mechanical properties in the final composites. Additionally, their strong affinity for moisture due to their hydrophilic nature can compromise dimensional stability and long-term durability. The relatively low thermal resistance of palm fibers further limits their suitability for high-temperature applications. Furthermore, the weak interfacial adhesion between untreated palm fibers and polymer matrices often requires chemical treatments or compatibilizers, adding complexity and cost to the manufacturing process. To align with the stringent safety, performance, and environmental requirements of the automotive sector, continued research and standardization efforts are essential to improve fiber treatment, processing techniques, and composite design.

## References

Ahmad, H., et al.: A review of carbon fiber materials in automotive industry. In: IOP Conference Series: Materials Science and Engineering. IOP Publishing. (2020)

Akovali, G.: Advances in polymer coated textiles. Smithers Rapra. (2012)

Allafi, F., et al.: Advancements in applications of natural wool fiber. J. Nat. Fibers. p. 1–16. (2020)

Anuar, H., Zuraida, A.: Improvement in mechanical properties of reinforced thermoplastic elastomer composite with kenaf bast fibre. Compos. B Eng. **42**(3), 462–465 (2011)

Anuar, N.I.S., et al.: Comparison of the morphological and mechanical properties of oil Palm EFB fibres and kenaf fibres in nonwoven reinforced composites. Ind. Crops Prod. **127**, 55–65 (2019)

Arenas, J.P.: Applications of acoustic textiles in automotive/transportation, In: Acoustic Textiles. Springer. p. 143–163. (2016)

Arumugaprabu, V., et al.: Polymer-Based Composites: Design, Manufacturing, and Applications. CRC Press (2021)

Asyraf, M.R., et al.: Filament-wound glass-fibre reinforced polymer composites: Potential applications for cross arm structure in transmission towers. Polym. Bull. **80**(2), 1059–1084 (2023)

Atakan, R., Sezer, S., Karakas, H.: Development of nonwoven automotive carpets made of recycled PET fibers with improved abrasion resistance. J. Ind. Text. **49**(7), 835–857 (2020)

Atiqah, A., et al.: Effect of alkali and silane treatments on mechanical and interfacial bonding strength of sugar palm fibers with thermoplastic polyurethane. J. Nat. Fibers. **15**(2), 251–261 (2018)

Azmami, O., et al.: Mechanical and aging performances of Palm/Wool and Palm/Polyester nonwovens coated by waterborne polyurethane for automotive interiors. Ind. Crops Prod. **170**, 113681 (2021a)

Azmami, O., et al.: Sol-gel and polyurethane based flame retardant and water repellent coating for Palm/PES nonwovens composite. J. Sol-Gel Sci. Technol. **97**, 92–105 (2021b)

Azmami, O., et al.: Development and application of nonwovens based on palm fiber as reinforcements of unsaturated polyester. J. Compos. Mater. **57**(5), 1035–1054 (2023)

Aznaw, G.M.: Advances in composite structures: a systematic review of design, performance, and sustainability trends. Compos. Mater. **9**(1), 1–17 (2025)

Benzannache, N., et al.: Improving the mechanical performance of biocomposite Plaster/Washingtonian filifira fibres using the RSM method. J. Build. Eng. **33**, 101840 (2021)

Bezazi, A., et al.: Investigation of the date palm fiber for green composites reinforcement: Quasi-static and fatigue characterization of the fiber. Ind. Crops Prod. **146**, 112135 (2020)

Bhong, M., et al.: Review of composite materials and applications. Mater. Today: Proceed. (2023)

Bouasria, A., et al.: Advances in polymer coating for functional finishing of textiles. Front. Text. Mater. Polym. Nanomater. Enzym. Adv. Modif. Tech. 61–86. (2020)

Chaudhary, M., Garg, A.P.: Agricultural residue to wealth: Strategies and Planning. (2023)

Çinçik, E., Günaydin, E.: The influence of calendering parameters on performance properties of needle-punched nonwoven cleaning materials including r-PET fiber. J. Text. Inst. **108**(2), 216–225 (2017)

Danihelová, A., et al.: Usage of recycled technical textiles as thermal insulation and an acoustic absorber. Sustainability. **11**(10), 2968 (2019)

Deng, J., et al.: The surface modification effect on the interfacial properties of glass fiber-reinforced epoxy: a molecular dynamics study. Nanotechnol. Rev. **11**(1), 1143–1157 (2022)

El Messiry, M., Mito, A.B., Issa, S.: Multi-criteria decision analysis for supporting the selection of car upholstery fabric under degradation due to UV exposure. J. Eng. Fibers Fabr. **17**, 15589250221109934 (2022)

Elmarakbi, A.: Advanced composite materials for automotive applications: structural integrity and crashworthiness. John Wiley & Sons. (2013)

Endlerová, D.: Surface modification of polymeric fibres and nonwovens textiles. (2021)

Farahmand, B.: Polymer matrix composites (PMCs) and nanocomposites (methods of manufacturing PMC parts), In: Fundamentals of Composites and Their Methods of Fabrications: PMCs, MMCs, and CMCs. Springer. p. 71–105. (2025)

Fatma, N., et al.: The effect of doum palm fibers on the mechanical and thermal properties of gypsum mortar. J. Compos. Mater. **53**(19), 2641–2659 (2019)

Ferrández-García, C.E., et al.: Physical and mechanical properties of particleboard made from palm tree prunings. Forests **9**(12), 755 (2018)

Friedrich, K., Almajid, A.A.: Manufacturing aspects of advanced polymer composites for automotive applications. Appl. Compos. Mater. **20**, 107–128 (2013)

Fung, W.: Coated and laminated textiles. Vol. 23. Woodhead Publishing. (2002)

Fung, W., Parsons, K.: Some investigations into the relationship between car seat cover materials and thermal comfort using human subjects. J. Coated Fabr. **26**(2), 147–176 (1996)

Ghosh, G., Bhattacharyya, R., Penumadu, D.: Advances in multifunctional composite materials: applications and opportunities in automotive industry. Available at SSRN 5081906. (2024)

Global Textile Market Size & Share Report, 2022–2030, https://www.grandviewresearch.com/industry-analysis/textile-market/methodology. (2025)

Gonçalves, F.A., et al.: Influence of fillers on epoxy resins properties: a review. J. Mater. Sci. **57**(32), 15183–15212 (2022)

Gu, H.R., Kim, S.J., Kim, H.A.: Physical properties of eco-friendly kenaf fiber imbedded nonwoven for automotive pillar trim. Proced. Eng. **200**, 45–52 (2017)

Guna, V., et al.: Engineering sustainable waste wool biocomposites with high flame resistance and noise insulation for green building and automotive applications. J. Nat. Fibers. **18**(11), 1871–1881 (2021)

Guo, Y., et al .: Comparing properties of the warp-knitted spacer fabric instead of sponge for automobile seat fabric. In: Journal of Physics: Conference Series. IOP Publishing. (2021)

GuoH, Z.X., Liu, Z.: Advanced lightweight structural materials for automobiles: properties, manipulation, and perspective. Sci. Adv. Mater. **16**(5), 563–580 (2024)

Gurudatt, K., et al.: Dope-dyed polyester fibers from recycled PET wastes for use in molded automotive carpets. J. Ind. Text. **34**(3), 167–179 (2005)

Hamidon, M.H., et al.: Effects of fibre treatment on mechanical properties of kenaf fibre reinforced composites: a review. J. Market. Res. **8**(3), 3327–3337 (2019)

Hamouda, T., Aly, N.M.: Date palm fiber composites for automotive applications. Date Palm Fiber Composites: Processing, Properties and Applications. p. 387–405. (2020)

Hardcastle, J.: Woven textiles for automotive interiors and other transportation applications, In: Woven Textiles. Elsevier. p. 395–422. (2020)

Hassabo, A.G., et al.: An overview of carpet manufacture: design, dyeing, printing and finishing. J. Text. Color. Polym. Sci. **19**(2), 269–290 (2022)

Isham, A.B.: SMC and BMC press molding technology, In: Delaware Composites Design Encyclopedia. Routledge. p. 131–145. (2017)

Islam, S., Mominul Alam, S.M.: Investigation of the acoustic properties of needle punched nonwoven produced of blend with sustainable fibers. Int. J. Clothing Sci. Technol. **30**(3), 444–458 (2018)

Ismail, W.N.W.: Sol–gel technology for innovative fabric finishing—a review. J. Sol-Gel Sci. Technol. **78**(3), 698–707 (2016)

Kačarević, ŽP., et al.: Biodegradable magnesium fixation screw for barrier membranes used in guided bone regeneration. Bioact. Mater. **14**, 15–30 (2022)

Kalia, S., Avérous, L.: Biopolymers: biomedical and environmental applications. In: John Wiley & Sons. (2011)

Kamath, M., et al.: Cotton fiber nonwovens for automotive composites. Int. Nonwovens J. **1**, 1558925005os-os1561400105 (2005)

Kaniapan, S., et al.: The utilisation of palm oil and oil palm residues and the related challenges as a sustainable alternative in biofuel, bioenergy, and transportation sector: a review. Sustainability. **13**(6), 3110 (2021)

Khan, M.N. et al.: Impact of agricultural wastes on environment and possible management strategies, In: Frontier Studies in Soil Science. Springer. p. 79–108 (2024)

Khan, F., et al.: Advances of composite materials in automobile applications–a review. J. Eng. Res. (2024)

Khatri, H., et al.: Potential of natural fiber based polymeric composites for cleaner automotive component production-a comprehensive review. J. Mater. Res. Technol. **25**, 1086–1104 (2023)

Kiehbadroudinezhad, M., Merabet, A., Hosseinzadeh-Bandbafha, H.: Landfill source of greenhouse gas emission, In: Advances and Technology Development in Greenhouse Gases: Emission, Capture and Conversion. Elsevier. p. 123–145. (2024)

Kim, S.J., Kim, H.A.: Characteristics of eco-friendly Kenaf fiber-imbedded nonwoven for automotive application, in Generation, Development and Modifications of Natural Fibers. IntechOpen. (2019)

KJ, A.: Application of PVC–a superior material in the fields of science and technology. Polym. Plast. Technol. Mater. **63**(15), 2089–2107 (2024)

Kondo, M.Y., et al.: Recent advances in the use of Polyamide-based materials for the automotive industry. Polímeros. **32**, e2022023 (2022)

Lazzari, L.K., et al.: Surface modification of mineral fibers and their composites, In: Surface Modification and Coating of Fibers, Polymers, and Composites. Elsevier. p. 99–124. (2025)

Liu, S., Mao, S., Zhang, P.: Toward on integrally-formed knitted fabrics used for automotive seat cover. J. Eng. Fibers Fabr. **16**, 15589250211006544 (2021)

Madushika, J.W.A., Lanarolle, W.D.G.: A review on novel approaches to enhance sound absorbing performance using textile fibers. J. Text. Inst. **113**(2), 341–348 (2022)
Maity, S., Singha, K., Pandit, P.: Speciality coatings and laminations on textiles. Smart Funct. Text. 151. (2032)
Mamdouh, F., et al.: Overview of different fabric structures. J. Text. Color. Polym. Sci. **19**(2), 291–306 (2022)
Manu, T., et al.: Biocomposites: a review of materials and perception. Mater. Today Commun. **31**, 103308 (2022)
Mastura, M.T., et al.: Environmentally conscious hybrid bio-composite material selection for automotive anti-roll bar. Int. J. Adv. Manuf. Technol. **89**, 2203–2219 (2017)
Mishra, R.K.: Advances in textile structural composites. MDPI. p. 808. (2023)
Mukhopadhyay, S.K., Partridge, J.: Automotive Textiles. Text. Prog. **29**(1–2), 1–125 (1999)
Muthalagu, R., et al.: Tensile attributes and material analysis of kevlar and date palm fibers reinforced epoxy composites for automotive bumper applications. Mater. Today Proceed. **46**, 433–438 (2021)
Muthukumar, N., et al.: Sound and thermal insulation properties of flax/low melt PET needle punched nonwovens. J. Nat. Fibers. **16**(2), 245–252 (2019)
Naidu, L., Moorthy, R.: A review of key sustainability issues in Malaysian palm oil industry. Sustainability. **13**(19), 10839 (2021)
Naik, V., Kumar, M.: A review on natural fiber composite material in automotive applications. Eng. Sci. **18**(18), 1–10 (2021)
Nick, A., Becker, U., Thoma, W.: Improved acoustic behavior of interior parts of renewable resources in the automotive industry. J. Polym. Environ. **10**(3). (2002)
Nonwoven Fabrics Market, https://www.marketsandmarkets.com/Market-Reports/non-woven-fabrics-market-101727296.html. Accessed 20 Apr 2025
Noreikaitė, G., Benaitytė, R., Stanelytė, D.: Analysis of carbon fiber usage in the automotive industry. Taikomieji Tyrimai Studijose Ir Praktikoje. **20**(1), 91–94 (2024)
Ooi, Z.X., et al.: Oil palm frond as a sustainable and promising biomass source in Malaysia: a review. Environ. Prog. Sustain. Energy **36**(6), 1864–1874 (2017)
Osorio, L.L.D.R., Flórez-López, E., Grande-Tovar, C.D.: The potential of selected agri-food loss and waste to contribute to a circular economy: applications in the food, cosmetic and pharmaceutical industries. Molecules **26**(2), 515 (2021)
Palak, H., Karagüzel Kayaoğlu, B.: Analysis of the effect of production parameters on sound absorption and abrasion resistance performance of needlepunched nonwovens for automotive carpet applications using Taguchi method. J. Ind. Text. 1528083719889691. (2019)
Pamuk, G., Çeken, F.: Research on the breaking and tearing strengths and elongation of automobile seat cover fabrics. Text. Res. J. **79**(1), 47–58 (2009)
Patil, A., Patel, A., Purohit, R.: An overview of polymeric materials for automotive applications. Mater. Today Proceed. **4**(2), 3807–3815 (2017)
Poorzeinolabedin, M.: A new production methodology for vacuum infusion process of advanced composite structures. Middle East Technical University (Turkey). (2017)
Pourdeyhimi, B., et al.: Nonwovens—structure-process-property relationships, In: Structure and Mechanics of Textile Fibre Assemblies. Elsevier. p. 109–143. (2019)
Precedence Research (2024) High performance composites market size, share and trends 2024 to 2034. https://www.precedenceresearch.com/high-performance-composites-market. Accessed 30 Mar 2025
Rahman, M.Z., et al.: Advanced biopolymers for automobile and aviation engineering applications. J. Polym. Res. **30**(3), 106 (2023)
Rasheed, A.: Classification of technical textiles, In: Fibers for Technical Textiles. Springer. p. 49–64. (2020)
Ratna, D., Chakraborty, B.C.: Polymer matrix composite materials: structural and functional applications. Walter de Gruyter GmbH & Co KG. (2023)
Richaud, M., Vermeersch, O., Dolez, P.: Specific testing of textiles for transportation, In: Advanced Characterization and Testing of Textiles. Elsevier. p. 399–432. (2018)
Romero, P.E., et al.: Manufacture of polyurethane foam parts for automotive industry using FDM 3D printed molds. CIRP J. Manuf. Sci. Technol. **32**, 396–404 (2021)
Sadashiva, M., et al.: Experimental investigation of bending characteristics of hybrid composites fabricated by hand layup method. In: Journal of Physics: Conference Series. IOP Publishing. (2021)
Sajid, L., et al.: Extraction and application of cellulose microfibers from Washingtonia palm as a reinforcement of starch film. J. Plast. Film Sheeting **37**(4), 529–558 (2021a)
Sajid, L., et al.: Introduction of raw palm fibers in the textile industry by development of nonwoven composite materials based on Washingtonia palm fibers. J. Text. Inst. **112**(11), 1717–1729 (2021b)
Sajid, L., et al.: Extraction and characterization of palm fibers and their use to produce wool-and polyester-blended nonwovens. J. Ind. Text. **51**(2), 177–205 (2021c)
Sakib, M.N., Iqba, A.A.: Epoxy based nanocomposite material for automotive application-a short review. Int. J. Automot. Mech. Eng. **18**(3), 9127–9140 (2021)
Santos, R.D., et al.: Influence of alkaline hornification treatment cycles on the mechanical behavior in curaua fibers. In: Macromolecular Symposia. Wiley Online Library. (2018)
Sarfraz, M.S., Hong, H., Kim, S.S.: Recent developments in the manufacturing technologies of composite components and their cost-effectiveness in the automotive industry: a review study. Compos. Struct. **266**, 113864 (2021)
Sari, P., et al.: Recycling of polyurethanes, in polyurethane chemistry: renewable polyols and isocyanates. ACS Publications. p. 413–427. (2021)
Saricam, C., Okur, N.: Polyester usage for automotive applications. Polyester-Production, Characterization and Innovative Applications, 1st (ed.) Camlibel, No, Ed: p. 69–85. (2018)
Schwartz, M.: Resin transfer molding and associated closed molding and infusion processes. Innovations in Materials Manufacturing, Fabrication, and Environmental Safety. p. 503–529. (2010)
Selvaraj, P.S., et al.: Novel resources recovery from anaerobic digestates: current trends and future perspectives. Crit. Rev. Environ. Sci. Technol. **52**(11), 1915–1999 (2022)
Shanmugam, D., Thiruchitrambalam, M.: Influence of alkali treatment and layering pattern on the tensile and flexural properties of Palmyra palm leaf stalk fiber (PPLSF)/jute fiber polyester hybrid composites. Compos. Interfaces **21**(1), 3–12 (2014)
Shanmugam, V., et al.: Applications of various types of polymer-based composites, In: Polymer-Based Composites. CRC Press. p. 139–149. (2021)
Shekar, H.S., Ramachandra, M.: Green composites: a review. Mater. Today Proceed. **5**(1), 2518–2526 (2018)
Shen, R., et al.: An enhanced vacuum-assisted resin transfer molding process and its pressure effect on resin infusion behavior and composite material performance. Polymers **16**(10), 1386 (2024)
Shinoj, S., et al.: Oil palm fiber (OPF) and its composites: a review. Ind. Crops Prod. **33**(1), 7–22 (2011)
Sikdar, P., et al.: Polyurethane (PU) based multifunctional materials: Emerging paradigm for functional textiles, smart, and biomedical applications. J. Appl. Polym. Sci. **139**(38), e52832 (2022)
Automotive Textiles Market Size, Share & Trends Report, https://www.grandviewresearch.com/industry-analysis/automotive-textiles-market. Accessed 15 Apr 2025. (2030)
Skosana, S.J., Khoathane, C., Malwela, T.: Driving towards sustainability: a review of natural fiber reinforced polymer composites for

eco-friendly automotive light-weighting. J. Thermoplast. Compos. Mater. **38**(2), 754–780 (2025)

Sreenivas, H., Krishnamurthy, N., Arpitha, G.: A comprehensive review on light weight kenaf fiber for automobiles. Int. J. Lightweight Mater. Manuf. **3**(4), 328–337 (2020)

Srivastava, S., et al.: Effects of ageing on aesthetics performance of automotive non-woven fabrics. SAE Technical Paper. (2016)

Thilagavathi, G., et al.: Development of natural fiber nonwovens for application as car interiors for noise control. J. Ind. Text. **39**(3), 267–278 (2010)

Thirumurugan, V., Kumar, M.R., Das, S.: Thermal insulation properties of jute, polypropylene and recycled polyester nonwoven fabrics for automotive textiles. Indian J. Fibre Text. Res. (IJFTR). **46**(2), 186–190 (2021)

Verma, D., Senal, I.: Natural fiber-reinforced polymer composites: feasibiliy study for sustainable automotive industries, In: Biomass, biopolymer-based materials, and bioenergy. Elsevier. p. 103–122. (2019)

Voncina, B.: Recycling of textile materials. MDT Recycl. **1**, 1–37 (2016)

Wang, L.K., et al.: Composting processes for disposal of municipal and agricultural solid wastes. Solid Waste Eng. Manag. **1**, 399–523 (2021)

Wang, B., Gao, H.: Fibre reinforced polymer composites, In: Advances in Machining of Composite Materials: Conventional and Non-Conventional Processes, pp. 15–43. Springer (2021)

Yin, Y., Wang, C.: Water-repellent functional coatings through hybrid SiO2/HTEOS/CPTS sol on the surfaces of cellulose fibers. Colloids Surf. A **417**, 120–125 (2013)

Zhai, S., et al.: Tensile strength of windmill palm (Trachycarpus fortunei) fiber bundles and its structural implications. J. Mater. Sci. **47**(2), 949–959 (2012)

Zhang, Y., et al.: Processing and characterizations of Short fluoroalkyl chain/polyurethane-polylactic acid/low melt polylactic acid Janus nonwoven Medical covers using spray coating. Prog. Org. Coat. **147**, 105736 (2020)

Zhang, Y., et al.: Advances in colored carbon-based fiber materials and their emerging applications. SusMat. **4**(6), e243 (2024)

Zhu, X., et al.: Current advances of Polyurethane/Graphene composites and its prospects in synthetic leather: a review. Eur. Polym. J. **161**, 110837 (2021)

Zielińska, M., Bułkowska, K.: Sustainable management and advanced nutrient recovery from biogas energy sector effluents. Energies **17**(15), 3705 (2024)

# Technological Valorization of Gum Arabic in Arab Countries: Patent Status and Prospects for Sustainable Innovation

Houria Hemdani, Soumia Belouafa, Nada El Arrach, Ayoub Grouli, and Mohammed Berrada

**Abstract**

Gum Arabic (GA) is a natural exudate mainly derived from the African Acacia plant, recognized for its exceptional properties: solubility, biocompatibility, stability, and film-forming ability. Long used in the agrifood and pharmaceutical sectors, GA is now attracting growing interest in developing multifunctional materials, such as hydrogels, nanostructures, and bioadhesive devices. This chapter proposes a cross-analysis between scientific advances in GA and regional technology, as seen through the analysis of patents filed by researchers or inventors from the Arab world. The results reveal a predominance of classic applications (binder, texturizing agent, carrier) in the food, cosmetics, and phytotherapeutic fields. Few patents explore the advanced functionalization of GA or its integration into complex devices. Only a few recent examples, notably in Morocco and Qatar, testify to an emerging orientation towards innovative applications. This panorama highlights the region's still largely under-exploited potential for commercialization, calling for greater support for applied research, a stronger culture of patentability, and more effective structuring of technology transfers, integrating GA into new-generation bioactive platforms.

**Graphic Abstract**

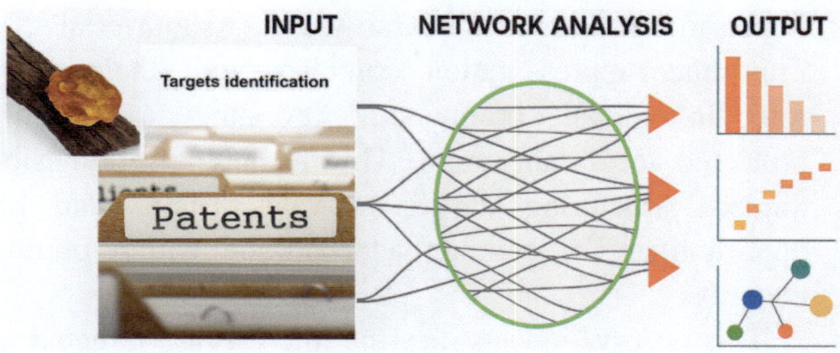


**Keywords**

Gum Arabic · Innovation · Patents · Technological valorization · The Arab world

## 1 Introduction

Gum Arabic (GA), a natural exudate derived mainly from the Acacia Senegal and Acacia Seyal species (Ali et al. 2012; FAO 1997), has been a strategic resource in trade between Africa and the rest of the world for several centuries (Djetoyom et al. 2024; Mujawamariya et al. 2013). Appreciated since ancient times for its functional properties (Andrade et al. 2025; Abdelwahab et al. 2025; Shalini et al. 2025; Choudhary et al. 2024; Sharkawy and Rodrigues 2024; Sharma et al. 2024; Al-Hamayda et al. 2023; Froelich et al. 2023; Prasad et al. 2022), today, it continues to occupy an essential place in various industries, including agrifood (Abdelwahab et al. 2025; Sharkawy and Rodrigues 2024; Sharma et al. 2024; Al-Hamayda et al. 2023; Oluba et al. 2025; Tayebi-Khorrami et al. 2025; Kadirvel and Narayana 2023; Shi et al. 2023; Jiang et al. 2023), pharmaceuticals (Abdelwahab et al. 2025; Shalini et al. 2025; Wang et al. 2025; Choudhary et al. 2024; Ozsoysal and Bilgili 2024; Sharkawy and Rodrigues

H. Hemdani · S. Belouafa (✉) · N. El Arrach · A. Grouli · M. Berrada
Laboratory of Biology and Health, Department of Chemistry, Faculty of Sciences Ben M'Sick, University Hassan II of Casablanca, Casablanca, Morocco
e-mail: soumia.belouafa@univh2c.ma



O. Jamal Eddine et al. (eds.), *Plant Pathology, Fungal Diversity, and Biotechnological Advances in Agriculture*, Sustainable Landscape Planning and Natural Resources Management, https://doi.org/10.1007/978-3-031-97666-7_12

2024; Al-Hamayda et al. 2023; Froelich et al. 2023; Ding et al. 2025), cosmetics (Abdelwahab et al. 2025; Sharkawy and Rodrigues 2024), and textiles (Abdelwahab et al. 2025; Wimmler 2019; Wijesirigunawardana and Perera 2018). Its use extends to more specialized fields such as biomedicine (Sharma et al. 2024; Al-Hamayda et al. 2023; Safaei et al. 2025; Venkatesan et al. 2023).

Despite its versatility, GA, although mass-produced in Arab and African countries, remains under-exploited mainly at the local level, particularly in terms of technological valorization and transformation into high-value-added products (Prasad et al. 2022; Eltahir et al. 2025; Tadesse et al. 2020; Mohammed 2018; Mujawamariya et al. 2013). Furthermore, despite growing scientific interest, the dynamics of GA-related intellectual protection remain highly uneven globally. A review published in 2022 reports, for example, the absence of GA-related patent filings in countries of the Arab world (Ashour et al. 2022). This under-representation contrasts with their strategic role in global production, revealing a deficit of protected local innovation. This technological invisibility hinders structuring innovative value chains and limits opportunities for creating added value within producing countries.

This observation justifies the interest in a targeted analysis of patent dynamics in Arab producer countries to understand the levers and obstacles to innovation better. Against this backdrop, this chapter explores a hitherto unexplored area of the literature: local technological intelligence through the analysis of patents filed in the Arab world or by researchers and/or applicants from Arab countries. It aims to identify the gaps between the abundance of natural resources and their effective regional valorization while identifying perspectives for stimulating protected innovation within an eco-responsible framework.

## 2 Methodology

This research seeks to analyze the field of patents concerning gum Arabic in the Arab world, scrutinizing innovation trends, technological uses, and opportunities for progress. We have opted for a hybrid methodological approach (Sarkis-Onofre et al. 2021), combining a quantitative patent data study with a qualitative technological content appraisal.

– Search strategy and data sources

Our search strategy was deployed in four sequential phases:

**Phase 1:** Identification of databases and search parameters.

International databases: ESPACENET, PATENTSCOPE, WIPO, USPTO, JUSTIN.

Regional databases: EGYPO (Egypt), SAIP (Saudi Arabia), OMPIC (Morocco), INNORPI (Tunisia), COSQC (Iraq).

Keywords used: “Arabic gum,” “gomme arabique,” “Talha gum,” “Acacia Senegal,” and “Acacia Seyal” in combination with relevant technical terms in the title, abstract, and claims fields.

**Phase 2:** Filtering and selection of patents.

Inclusion criteria: patents filed by Arab inventors or institutions, or patents filed in Arab countries involving local participation.

Exclusion criteria: patents concerning other plant gums, patents with no identifiable local contribution.

The initial search was reduced to 23 after applying the inclusion/exclusion criteria.

**Phase 3:** Data extraction and categorization.

Data extracted for each patent: number, country of origin, year, source database, applicants, inventors, functional role of gum Arabic, field of application, level of originality.

Classification of patents according to a double-entry taxonomy: by field of application (agrifood, pharmaceutical, textile, etc.) and by the functional role of GA (binder, stabilizer, film-forming agent, etc.).

**Phase 4:** Analysis and interpretation.

Bibliometric analysis: chronological trends, geographical distribution, collaboration networks.

Lexical analysis: identification of recurring and emerging technical terms.

Innovation mapping: positioning of patents according to their degree of innovation (incremental vs. disruptive) and technological complexity.

Benchmarking: identification of gaps between patented applications and the scientific state of the art, revealing opportunities for commercialization (Fig. 1).

## 3 Overview of Arab Patents

An analysis of GA-related patent filings in Arab countries provides a regional overview of the technological valorization of this biomolecule. Based on a targeted technology watch, this section highlights the dominant trends in protected inventions’ innovation. It also identifies unexploited development opportunities, revealing under-exploited potential for GA in high-value-added sectors, notably biotechnologies, advanced materials, and healthcare.

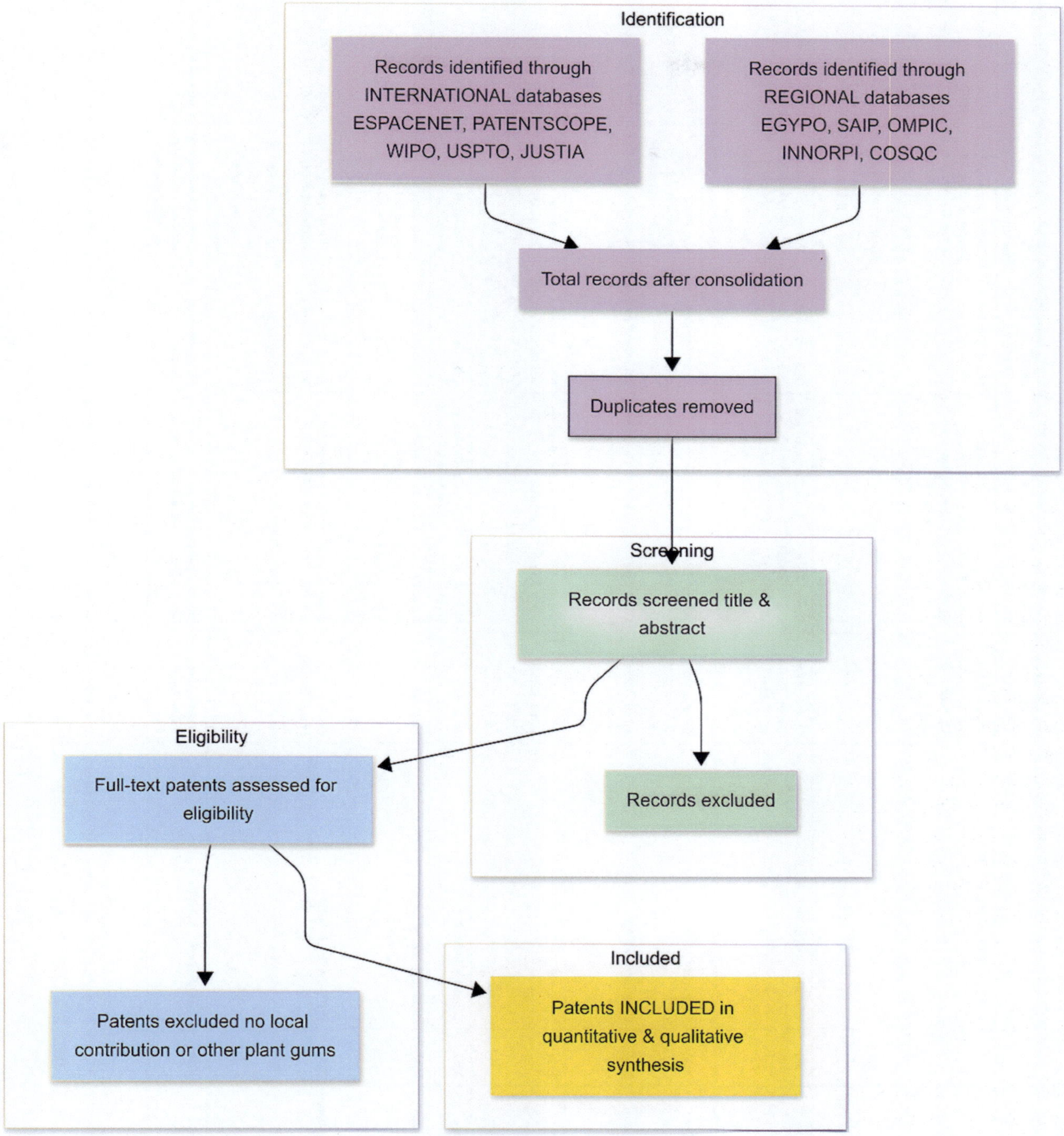


**Fig. 1** Methodological Framework of the chapter

## 3.1 General Results of Patent Analysis

The results of the technology watch are presented in Table 1, which lists the main characteristics of the patents collected. A total of 27 patents were identified on the various platforms analyzed. Of these, three patents were excluded from the detailed analysis (Fuyama 1984; Barrett et al. 2002; Kubo and Matsuda 2007): although they were filed in Arab countries, they came from exclusively foreign inventors and applicants with no identifiable local contribution. In addition, one further patent was specifically excluded (Lakkab et al. 2019), as it did not concern GA but another plant gum: gum extracted from locust bean kernels.

**Table 1** List of patents related to the analysis

| Title | Patent number | Country | Year | Database | Applicants | Inventors | Role of Gum Arabic | Application Domain | Originality/observations | References |
|---|---|---|---|---|---|---|---|---|---|---|
| A process for preparing a formaldehyde ammonium hydroxide complex to be used as a fixative for printing inks and tanning is proposed | EG20420B1 | Egypt | 1999 | WIPO-PATENTSCOPE-EGYPO | Ahmed Bakry Shahat, Khalid Ali Nabhan | Ahmed Bakry Shahat, Khalid Ali Nabhan | GA as a natural binder in synergy with an innovative fixative | Screen printing | High-performance screen-printing ink using GA in synergy with formaldehyde and ammonium-based fixative, offering exceptional adhesion and high resistance on various materials | (Shahat and Nabhan 1999) |
| Using formaldehyde-ammonium hydroxide complex to color sands used in decorative paints | EG21011B1 | Egypt | 2000 | WIPO-PATENTSCOPE-EGYPO | Ahmed Bakry Shahat, Khalid Ali Nabhan | Ahmed Bakry Shahat, Khalid Ali Nabhan | GA as a binder and pigment fixative for colored sands in decorative paints | Decorative materials industry | Ecological process for pigment fixation on sand, with superior performance compared to epoxy-based methods | (Shahat and Nabhan 2000) |
| Valorization of thermosetting waste | MA28318A1 | Morocco | 2006 | Espacenet-OMPIC | Zakarya Driss; El-Morabit M Hamed | Zakarya Driss; El-Morabit M Hamed | Synergistic formulation: GA + plant antioxidants | Food/functional supplement | Innovative application but simple formulation | (Zakarya and El-Morabit 2006) |
| Medicinal preparation for the natural therapy of renal lithiasis and urinary infections | MA28319A1 | Morocco | 2006 | Espacenet-OMPIC | Alami Ouahabi Fouziya | Alami Ouahabi Fouziya | GA as a binder for natural extracts | Traditional medicine | Ethnopharmacological use | (Alami Ouahabi 2006) |
| Flame retardant and termite fighting composition and articles coated therewith | WO2011091806A1 | Egypt | 2011 | Espacenet | Elbaialy El Sayed Kamel Morsy | Elbaialy El Sayed Kamel Morsy | GA as film-forming support | Flame retardant and anti-termite coating for various supports (wood, textiles, cables, etc.) | Protection against fire and anti-termite action, with no toxic or health-hazardous components | (Elbaialy 2011) |
| Biodegradable composite material | TN2016000018A1 | Tunisia | 2017 | Espacenet-INNORPI | Abdesslem bochra | Abdesslem Bochra | GA as a film-forming agent in a hair mask | Cosmetics | Traditional formulation | (Abdesslem 2017) |
| Pharmaceutical or dietary composition based on plants for dry mouth treatment | MA37045A1 | Morocco | 2016 | Espacenet-OMPIC | Univ Mohammed V Souissi | Zaoui Fatima; Faouzi Moulay Abbas; Cherrah Yahia; Etto Bruno | GA as an oral film-forming agent | Pharmaceutical | Innovative mucoadhesive application | (Zaoui et al. 2016) |
| Emulsified food product composition | EG27864B1 | Egypt/Japan | 2016 | WIPO-PATENT SCOPE | Otsuka Pharmaceutical Factory, INC | Endo, Naoyuki; Mizukai, Kazuya | GA as a natural emulsifier | Functional food/ Medical nutrition (prevention of gastroesophageal reflux) | Heat-resistant emulsified food composition allows targeted gastric gelation to prevent gastroesophageal reflux | (Endo and Mizukai 2016) |
| Alcohol free dampening | EG27631B1 | Egypt | 2016 | WIPO-PATENT SCOPE | Ahmed Mmtaz Bakry Ajjur | Ahmed Mmtaz Bakry Ajjur | GA as a desensitizing agent | Offset printing industry | Improved dampening solution formulation for offset printing | (Ajjur 2016) |

(continued)

**Table 1** (continued)

| Title | Patent number | Country | Year | Database | Applicants | Inventors | Role of Gum Arabic | Application Domain | Originality/ observations | References |
|---|---|---|---|---|---|---|---|---|---|---|
| Production of low-calorie value potato chips by coating them with almond gum | TN2017000395A1 | Tunisia | 2019 | Espacenet-INNORPI | Ecole Nat Dingenieurs De Sfax | Fatma Bouaziz; Raoudha Ellouz Ghorbel; Semia Ellouz Chaabouni; Soumaya Zouari Ellouzi | GA as stabilizer | Food/Functional | Natural sweet preparation | (Bouaziz et al. 2019) |
| New experimental drug product for cutaneous leishmaniasis treatment in Iraq, consisting of violet dye solution with gum Arabic in vivo | IQ5167 | Iraq | 2017 | COSQC | Prof. Dr. Nada Mohammed Taha Al-Bashir | Prof. Dr. Nada Mohammed Taha Al-Bashir | GA as binder/ stabilizer | Pharmaceutical; Infectious dermatology | Topical formulation for cutaneous leishmaniasis treatment with superior efficacy to conventional therapy and rapid healing | (Al-Bashir 2017) |
| Polymer membrane with Acacia gum (gum Arabic) additive | WO2018027014A1 | Qatar/USA | 2018 | Espacenet | Qatar Found Education Science & Community Dev | Hussein Mutaz Ali Atieh; Manawi Yehia; Kochkodan Viktor | GA as a functional additive in polymer membrane formulation | Polymer separation membranes, with multiple applications | Polymer membrane enriched with gum Arabic for water and gas purification, with improved performance in terms of porosity, hydrophilicity, and antifouling through incorporation of this natural biopolymer | (Hussein et al. 2018) |
| Aroma-loaded microcapsules with antibacterial activity for eco-friendly applications | WO2018183150A1 | Egypt/USA/ Portugal | 2018 | Espacenet | The American Univ In Cairo; Univ Do Porto; Inst Politecnico De Braganca | Sharkawy Asma; Shoeib Tamer; Rodrigues Alirio; Barreiro Maria; Fernandes Isabel | GA as a microencapsulation vector for prolonged delivery and stability of bioactive compounds | Functional textile | Innovative green grafting process (esterification and thermalization) and controlled release of active ingredients | (Sharkawy et al. 2018) |
| Transformation of camel urine into powder | WO2018182466A1 | Sudan | 2018 | WIPO-PATENT SCOPE | Amjad Taj Elsir, Elhadi | AMJAD TAJ ELSIR, Elhadi | GA as a multifunctional excipient (binder, stabilizer, encapsulation agent) | Alternative/traditional medicine | Gentle process of drying and compaction into tablets or capsules | (Amjad 2018) |
| Resistance Pastes applied on different surfaces of fabrics | EG29790B1 | Egypt | 2020 | WIPO-PATENT SCOPE | Reham Ahmed El Sebaie Shams | Reham Ahmed El Sebaie Shams | GA as a natural binder and thickener | Textile industry/ Fabric printing | Gum Arabic is integrated into modern formulations for artistic or industrial textile printing. It is compatible with different types of dyes (pigments, acrylics, dispersed, etc.) | (Shams 2020) |

(continued)

**Table 1** (continued)

| Title | Patent number | Country | Year | Database | Applicants | Inventors | Role of Gum Arabic | Application Domain | Originality/ observations | References |
|---|---|---|---|---|---|---|---|---|---|---|
| Polymer membrane with Arabic gum for water treatment | US20200108355A1 | Qatar | 2020 | Espacenet-USPTO | Qatar Found For Education, Science, And Community Development | Hussien Muataz; Sabri Souhir; Kochkodan Viktor; Manawi Yehia; Alsuliti HudA | Antifouling agent in polymer membranes | Water treatment | Integration of 0.1 to 7% GA in polysulfone membranes to reduce bacterial biofouling (E. coli, S. aureus, etc.); improvement of porosity and hydrophilicity | (Hussein et al. 2020) |
| Aroma-Loaded Microcapsules with Antibacterial Activity for Eco-friendly Applications | US20200315168A1 | Egypt/ Portugal/ USA | 2020 | Espacenet-USPTO | The American Univ In Cairo; Univ Do Porto; Inst Politecnino De Braganca | Sharkawy Asma Mohamad Abdallah; Shoeib Tamer; Rodrigues Alirio Egidio; Barreiro Maria Filomena Filipe; Fernandes Isabel Patrícia Martins | GA as shell material in a complex coacervation system | Functional and smart textile | Functional textile formulation based on microencapsulation to confer durable antimicrobial and fragrant properties, thanks to a green fixation technology by esterification | (Sharkawy et al. 2020) |
| Composite reverse osmosis membrane, a water filtration system, and a method of manufacturing the same | US10946338B2 | Qatar | 2021 | Espacenet-Justia-USPTO | Qatar Foundation | Hussien Muataz; Manawi Yehia; Kochkodan Viktor | GA as a functional component in the second layer of a composite reverse osmosis membrane | Water treatment technologies | Innovative manufacturing process by interfacial polymerization | (Hussein et al. 2021) |
| Food supplements containing chia seeds and other organic compounds for optimizing digestion, absorption, and metabolism | WO2022060256A1 | Saudi Arabia | 2022 | Espacenet-SAIP | Alaffari Saleh Omar Ahmed | Alaffari Saleh Omar Ahmed | GA as a prebiotic fiber | Food supplement/ Digestive Health | Stimulation of beneficial microbiota | (Alaffari 2022) |
| Method for inhibiting metal corrosion during acid cleaning or pickling | US20220316073A1 | Saudi Arabia | 2022 | Espacenet-Justia-SAIP-USPTO | Univ King Fahd Pet & Minerals | Solomon Moses M Umoren Saviour, A | GA is part of a synergistic polysaccharide mixture | Metal treatment: Corrosion inhibition | Corrosion inhibition composition forms an ecological and effective synergistic system to protect metals during acid treatments | (Solomon and Umoren 2022) |
| Edible composition to prolong the shelf life of food products | US20220312782A1 | Saudi Arabia | 2022 | Espacenet-Justia-SAIP-USPTO | Univ King Faisal | Salem Sahar Mostafa Kamel; Al-Taebah Haya Sami Salem | GA is a natural film-forming agent that forms a protective edible coating | Food industry: Food preservation | Edible composition forming a natural protective coating designed to extend the shelf life of fresh food products (meat, fruits, vegetables) | (Salem and Al-Taebah 2022) |

(continued)

**Table 1** (continued)

| Title | Patent number | Country | Year | Database | Applicants | Inventors | Role of Gum Arabic | Application Domain | Originality/ observations | References |
|---|---|---|---|---|---|---|---|---|---|---|
| Compositions comprising natural extracts for stimulating the immune response | US20230057065A1 | Lebanon/ Quebec | 2023 | Espacenet-Justia-USPTO | Lebanese Arabian Company for Alternative Therapeutics; Shiban Canada Inc | Al-Shaibani Abdulhameed Abdulridha K; Maallah Abderrahim | GA as a natural excipient in proposed dosage forms | Pharmaceutical; Immunostimulation | Natural immunostimulant composition designed to strengthen the immune response against infections, particularly viral, via a gentle, natural, and versatile formulation | (Al-Shaibani and Maallah 2023) |
| Cementitious mortar composition and methods of preparation thereof | US20250051226A1 | Saudi Arabia | 2025 | Espacenet-WIPO-Patentscope-Justia-SAIP-USPTO | Univ Imam Abdulrahman Bin Faisal | Nasir Muhammad; Al-Kutti Walid; Islam A B M Saiful; Alotaibi Khalid Sager | GA as a structure-modifying additive | Construction and building materials | Modified cementitious composition, generated after curing a unique porous microstructure with fibrillar granules and controlled microcracks, paving the way for innovative applications in technical and sustainable building materials | (Nasir et al. 2025) |

This initial selection highlights regional dynamism around GA and structural shortcomings, particularly local participation in intellectual property and technological diversification. The following subsections will explore these aspects in greater depth.

## 3.2 Patent Mapping

A chronological and geographical analysis of GA-related patent filings in Arab countries reveals contrasting dynamics between local potential and actual technological valorization. As shown in Fig. 2, the trend in filings over the period 1999–2025 remains irregular overall, with peaks in activity observed between 2016 and 2022. Linear modeling suggests a slight increase over time, while polynomial fitting reveals occasional cycles of interest rather than sustained growth. These observations reflect an intermittent interest in exploiting GA in technological innovation, often dependent on isolated projects or cyclical policies.

In spatial terms, Fig. 3 illustrates the collaboration networks between Arab countries and their international partners. Egypt is a central player, maintaining partnerships with technologically advanced countries such as the USA, Japan, and Portugal. Other countries such as Morocco, Saudi Arabia, Qatar, and Tunisia have an active but less connected presence. On the other hand, some major GA producers, such as Sudan, are poorly represented, underlining the gap between raw production and intellectual valorization.

These data suggest that the use of GA in national innovation systems in the Arab world is still fragmented. They also reveal a strategic need for regional structuring around collaborative platforms to strengthen the filing of high-value-added patents.

## 3.3 Target Application Areas

An analysis of patents filed around GA reveals a wide range of industrial applications, although some fields are more exploited. Figure 4 shows a typology of the application areas identified.

The health and medicine sector dominates (5 patents), with patents on pharmaceutical formulations, controlled-release systems, and bioactive coatings for therapeutic purposes.

Next come industrial applications (4 patents), including printing materials, decorative formulations, and multifunctional composites. These uses reflect a gradual shift from GA to higher-value-added technical matrices.

Functional textiles also occupy an important place (4 patents), with the integration of GA for intelligent or antibacterial textiles.

The agrifood sector (nutrition, coating, preservation) remains present (3 patents), although less innovative in patented formulations, reflecting a still classic orientation.

An emerging field, water treatment, and separation membranes (3 patents), is beginning to appear on the technological landscape, opening up new interdisciplinary perspectives.

Patent analysis reveals emerging application sectors for gum Arabic beyond conventional uses. Traditional medicine applications (2 patents) and cosmetic formulations (1 patent)

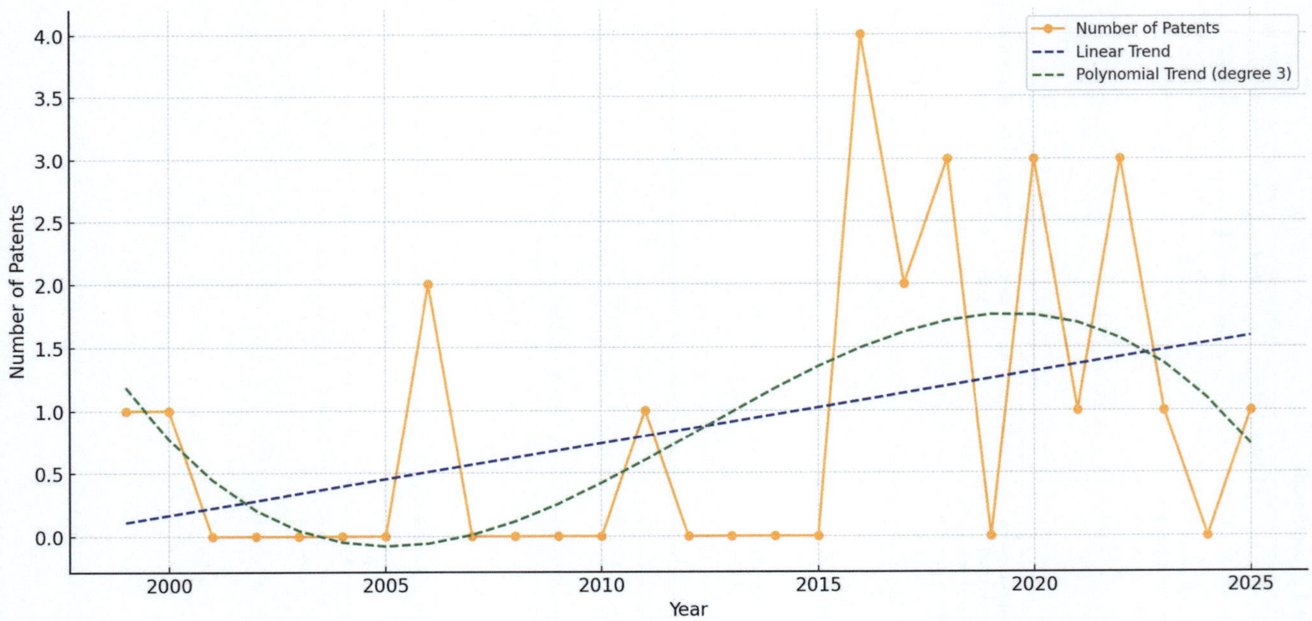


**Fig. 2** Patent filing trend (1999–2025) with linear and polynomial fits

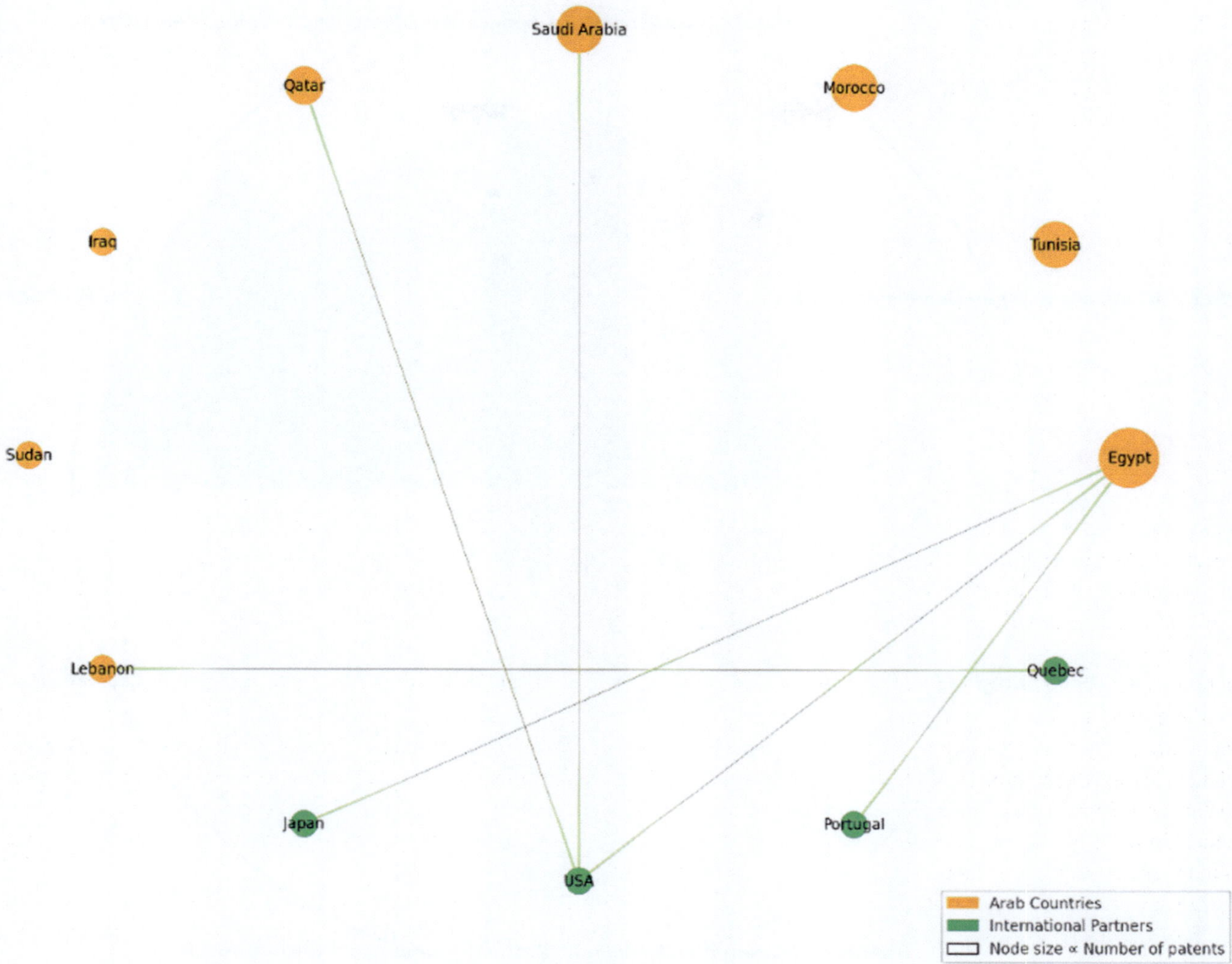


**Fig. 3** Enhanced collaboration network—Arab countries and international partners

represent specialized niches where ancestral knowledge is transformed into protected intellectual property, bridging artisanal techniques with industrial processes.

Similarly, the development of gum Arabic-based protective and anti-corrosion coatings (1 patent) showcases an innovative technical direction with potential applications in materials science.

## 3.4 Functional Roles of GA in Patents

Beyond application areas, analyzing the functional roles attributed to GA in patents reveals a high degree of technological versatility. Figure 5 highlights the diversity of functions performed by GA in the innovative formulation field in Arab countries.

The role of binder/fixer is the most frequently mentioned (28%), confirming GA's ability to ensure the cohesion of materials, fix pigments, or stabilize natural extracts in solid or semi-solid matrices.

Next comes the stabilizer/functional additive function (24%), which exploits GA's colloidal properties. This function often improves formulations' physical, chemical, or microbiological stability.

The role of film-former/coating agent appears in 16% of the patents, mainly in the context of edible coating, surface protection, or surface encapsulation.

Encapsulation/controlled release and synergistic galenic agent functions represent 12% of the total patents analyzed, reflecting an interest in advanced pharmaceutical and nutraceutical applications. These functions aim to improve active ingredients' bioavailability, release kinetics, or stability.

Finally, the role of texturizer/functional nutrition agent is the least exploited (8%), despite GA's polysaccharide structure offering interesting rheological properties and a potential prebiotic effect.

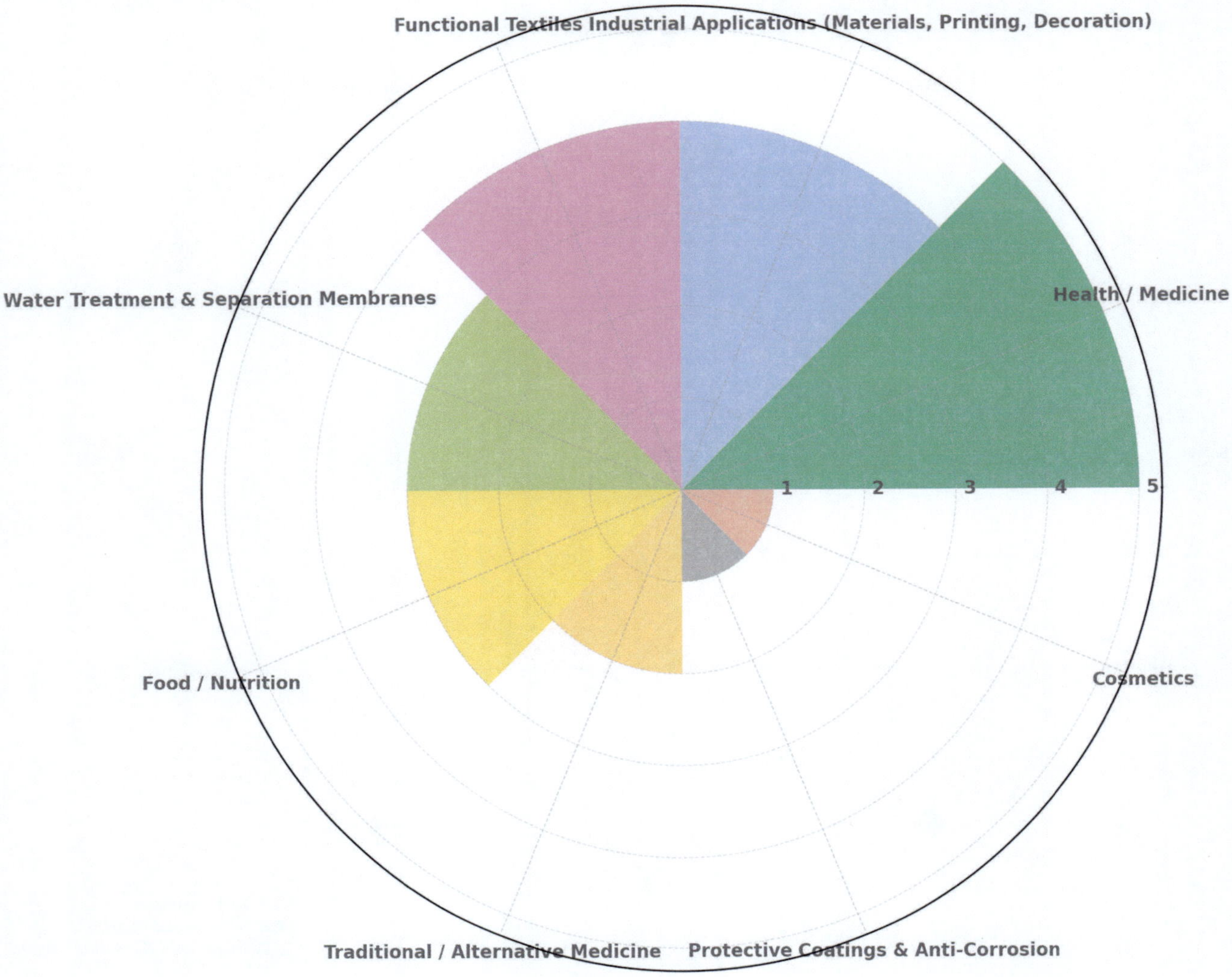


**Fig. 4** Patent count by application domain (Rosace diagram—improved readability)

This functional breakdown highlights the multifunctional potential of gum Arabic but also reveals a lack of synergistic exploitation between several functions within a single patent, which could represent an avenue for future innovation.

## 3.5 Innovation Levels and Emerging Trends

An examination of GA-related patents in the Arab world reveals the dominant technological directions and emerging trends in innovation (Fig. 6).

The term "Green process," highlighted by its frequency and size, reflects a growing desire to anchor inventions in a sustainability and green chemistry approach. This positioning is reinforced by the recurrence of terms such as "Grafting," "Esterification," or "Interfacial polymerization," testifying to the increasing chemical complexity of GA-based formulations.

Words like "Controlled release," "Mucoadhesive," or "Smart textiles" illustrate the emergence of high-value-added applications in the biomedical, pharmaceutical, or innovative materials fields. These trends reflect a gradual shift towards a more advanced use of GA, away from its traditional use as a simple additive or binder.

The appearance of words like "Antiviral protection," "Immunostimulant," or "Leishmaniasis treatment" suggests a refocusing on public health applications, particularly about regional challenges. Conversely, certain words reflect more traditional or artisanal uses, such as "Edible coating," "Sweet preparation," and "Offset printing," showing the persistence of conservative approaches in particular patents.

This lexical mapping highlights a fragile balance between classic technological improvements and more innovative approaches. It also highlights significant potential for developing multidisciplinary innovations around GA at the interface of materials engineering, biotechnology, and sustainable chemistry.

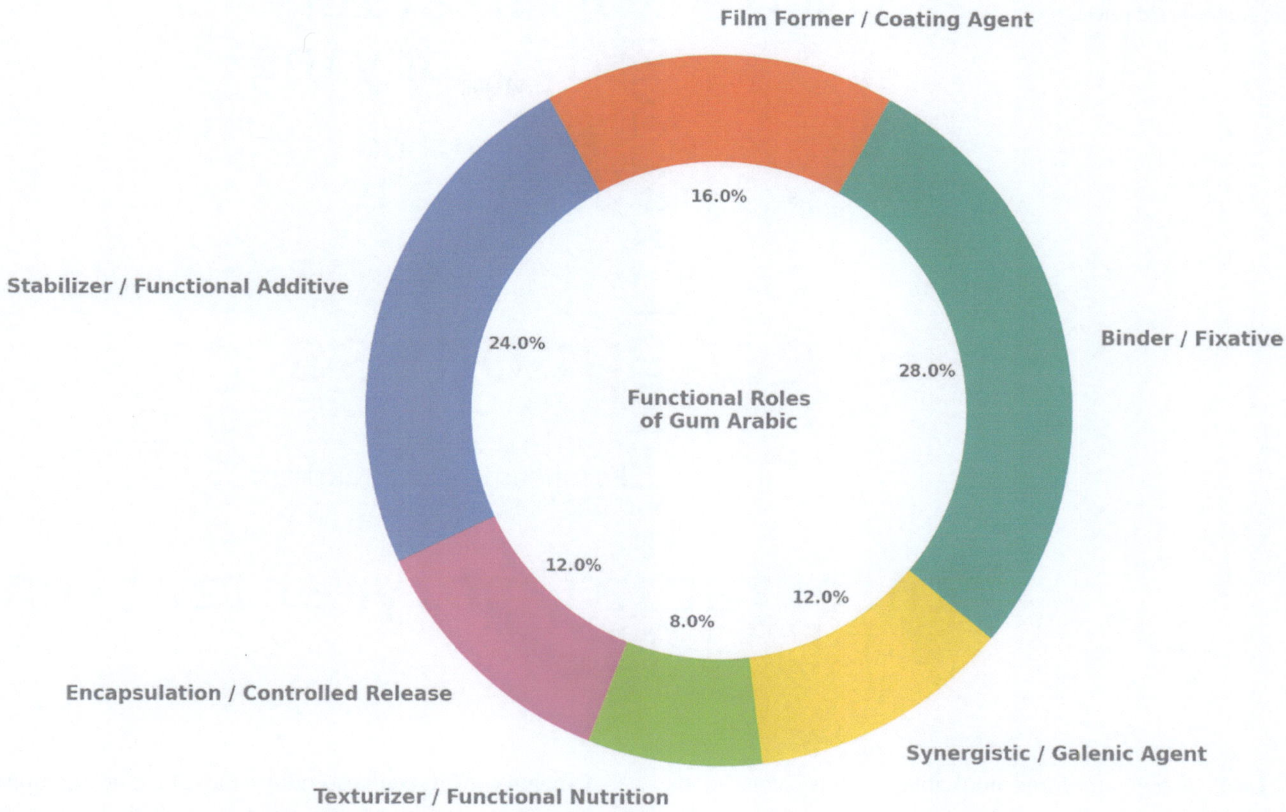


**Fig. 5** Distribution of functional roles of gum Arabic in Arab patents

## 4 Limits and Prospects for Development

An analysis of Arab patents for GA reveals several limitations that hamper the optimal exploitation of this strategic resource.

### 4.1 Observed Limits

Despite the abundant production of GA in several Arab countries, an analysis of patents reveals several significant obstacles to its technological valorization.

- The predominance of traditional uses and simple formulations: Many patents examined remain focused on classic applications such as binders, texturizing agents, or stabilizers without radical innovation or significant technological transformation. This is confirmed by the absence of advanced techniques in many patents, reflecting a still artisanal or unsophisticated approach, which limits GA's access to high-value-added markets.
- Limited integration into complex technological systems: Few patent applications integrate GA into next-generation devices, such as nanomaterials, advanced biomedical devices, or innovative coatings. This observation is corroborated by the scarcity of smart textiles, mucoadhesives, or interfacial polymerization in patents, suggesting an underexploitation of GA's advanced physicochemical properties in emerging technologies.
- Lack of multidisciplinary synergies: Patents often appear compartmentalized by sectors such as food, pharmaceuticals, and textiles, with no real cross-disciplinary synergies. The low co-occurrence of cross-disciplinary fields illustrates this segmentation, even though GA has excellent potential as a hybrid biomaterial that can be developed through integrative approaches combining several fields of expertise.
- Lack of a structured value chain strategy: Although several countries are major producers, the value chain lacks an integrated vision of local valorization based on applied research, technology transfer, and industrialization. This deficit is indirectly highlighted by the absence of scale-ups, industrial implementations, or local value-adding, which would signal an ambition for large-scale development.

**Fig. 6** Innovation keywords in gum Arabic-based patents

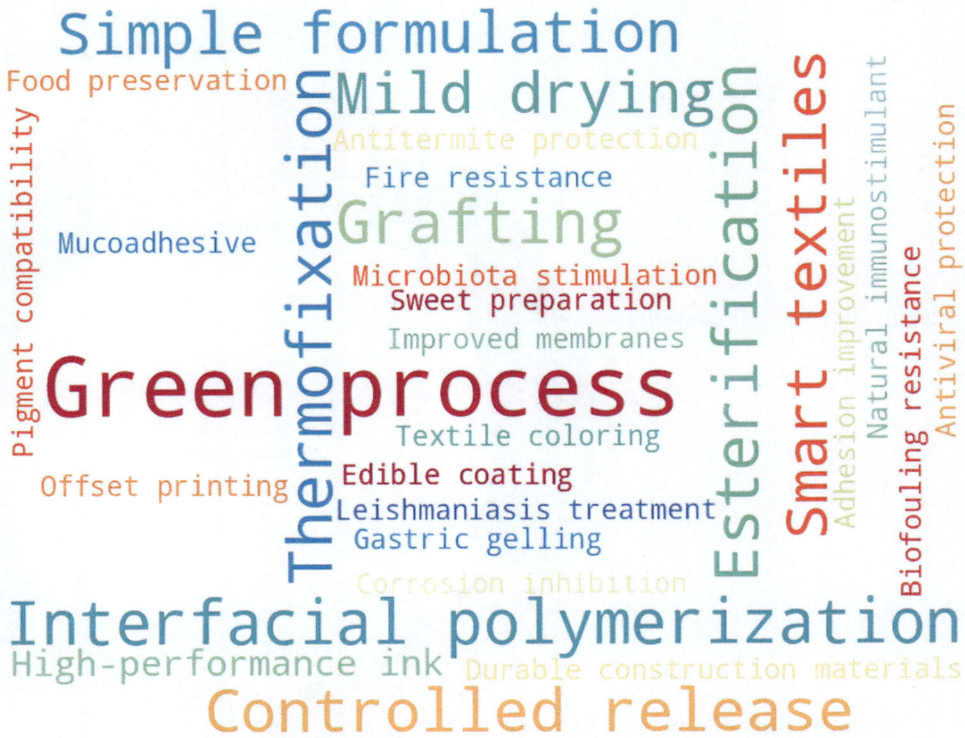


- Lack of regulatory and normative aspects: None of the patents examined explicitly referenced regulatory requirements, safety standards, or quality certifications such as ISO, pharmacopeias, or international food standards. This absence reflects a neglect of normative dimensions crucial in highly regulated sectors such as agrifood and biomedical. This can act as a brake on the marketing and export of innovative GA-based products.

## 4.2 Valuation Prospects

An analysis of the limitations identified in Arab patents relating to GA highlights concrete areas for improvement that can be used to enhance its technological value. These prospects are structured around two levels: targeted actions directly linked to the shortcomings identified and a broader vision integrating regional and international dynamics.

### 4.2.1 Targeted Prospects: Responding to Current Obstacles

Faced with the challenges identified in the patent analysis, several avenues of action appear to be priorities for strengthening the technological valorization of GA in Arab countries. These targeted prospects aim to fill the gaps observed while mobilizing existing levers at the crossroads of applied research, industrial innovation, and regional development.

To begin with, it seems essential to move beyond traditional uses and focus on high-tech applications. This implies the development of GA structural modification processes, such as grafting, esterification, or encapsulation, to optimize their performance in complex systems such as biomaterials, active films, or controlled-release devices.

Integrating GA into advanced technological systems is also a priority. Research initiatives should explore their association with smart polymers, nanomaterials, or bioactive components to design innovative functional objects, such as injectable hydrogels, medical textiles, or selective membranes. Pilot platforms or demonstrator projects could accelerate the transition from proof of concept to functional prototype.

The low level of cross-disciplinary cooperation observed calls for creating cross-sector networks, bringing together expertise in chemistry, biotechnology, materials engineering, nutrition, and health. An integrated approach would encourage co-designing multifunctional GA-based solutions, breaking with silo-based innovation logic.

One of the main challenges is establishing a thorough regional value chain for gum Arabic that extends from industrial processing to agricultural cultivation. This calls for implementing national innovation support policies, including mechanisms to assist technological transfer and intellectual property protection, specialized incubators, and financial programs. These frameworks would support the growth of regional industry leaders who can turn patents into commercially viable goods.

Integrating quality, safety, traceability, and certification requirements from a project's beginning is essential to guarantee market acceptability and commercialization of gum Arabic developments. Gum Arabic usage in delicate industries like food processing and healthcare would be secured by creating regional norms that comply with worldwide standards (ISO, EFSA, FDA).

### 4.2.2 General Outlook: Towards a Sustainable Bio-economic Strategy

In the medium and long term, GA could become a strategic biomaterial for sustainable innovation in Arab countries. Its valorization should align with the principles of the circular bioeconomy and sustainable development objectives by contributing to environmentally-friendly solutions in health, functional food, water treatment, etc.

The internationalization of innovations is also a key focus. Arab institutions would do well to raise their profile by co-producing patents with foreign partners, notably through collaborative projects (Horizon Europe, PRIMA, AUF, etc.), to amplify their work's scientific and commercial impact.

Last but not least, human capacity building is essential—targeted training programs in biopolymer valorizing, patenting, industrialization, and prototyping need to be developed. Creating regional centers of excellence dedicated to natural gums and bio-sourced materials would represent a structuring lever to support local innovation and position Gum Arabic as a flagship product of the knowledge economy in producing countries.

## 5 Conclusion

This chapter highlights the discrepancy between Arab countries' position as major producers of GA and the weakness of their technological valorization in patents filed. The analysis reveals a predominance of traditional uses, little integration into innovative systems, a lack of disciplinary transversality, and limited consideration of regulatory standards.

In light of these findings, several avenues are being explored: promoting GA's functional transformation, encouraging interdisciplinary approaches, structuring regional value chains, strengthening research and intellectual property capacities, opening up to international partnerships, and integrating standards requirements right from the design stage.

By adopting a progressive but ambitious approach, GA can become much more than a natural exudate; it can become a vector for sustainable innovation and a strategic pillar of the regional bioeconomy.

**Acknowledgements** The Authors sincerely thank everyone who participated in this chapter.

**Competing Interests** The authors have no conflicts of interest relevant to the content of this chapter to declare.

## References

Abdelwahab, S.I., Taha, M.M.E., Mariod, A.A.: Performance analysis, conceptual mapping, and emerging trends for Gum Arabic research: a comprehensive bibliometric analysis from 1916 to 2023. Food Prod. Process. Nutr. **7**(1), 4 (2025)

Abdesslem, B.: Biodegradable composite material. Tunisian Patent No. TN2016000018A1. National Institute for Standardization and Industrial Property. (2017)

Ajjur, A.M.B.: Alcohol free dampening. Egyptian Patent No. EG27631B1. Egyptian Patent Office. (2016)

Alaffari, S.O.A.: Food supplement containing chia seeds and other organic compounds for optimizing digestion, absorption, and metabolism. International Patent No. WO2022060256A1. World Intellectual Property Organization. (2022)

Alami Ouahabi, F.: Medicinal preparation for natural therapy of renal lithiasis and urinary infections. Moroccan Patent No. MA28319A1. Moroccan Industrial and Commercial Property Office. (2006)

Al-Bashir, N.M.T.: A new experimental drug product for cutaneous leishmaniasis treatment in Iraq consists of a violet dye solution with gum Arabic in vivo. Iraqi Patent No. IQ5167. Central Organization for Standardization and Quality Control. (2017)

Al-Hamayda, A., Abu-Jdayil, B., Ayyash, M., Tannous, J.: Advances in microencapsulation techniques using Arabic gum: a comprehensive review. Ind. Crops Prod. **205**, 117556 (2023). https://doi.org/10.1016/j.indcrop.2023.117556

Ali, N.E.S., Elkarim, A.M.A., Fageer, A.S.M., Nour, A.A.M.: Physico-chemical characteristics of some acacia gums. Int. J. Agric. Res. **7**, 406–413 (2012). https://doi.org/10.3923/ijar.2012.406.413

Al-Shaibani, A.A.K., Maallah, A.: Compositions comprising natural extracts for stimulating the immune response. U.S. Patent Application No. US20230057065A1. U.S. Patent and Trademark Office. (2023)

Amjad, T.E.E.: Transformation of camel urine into powder. International Patent No. WO2018182466A1. World Intellectual Property Organization. (2018)

Andrade, L.A., Nunes, C.A., Pereira, J.: Gums from different botanical sources: physicochemical characterization is seldom found in the literature. Food Chem. **465**, 142134 (2025). https://doi.org/10.1016/j.foodchem.2024.142134

Ashour, M.A., Fatima, W., Imran, M., Ghoneim, M.M., Alshehri, S., Shakeel, F.: A review of the main phytoconstituents, traditional uses, inventions, and patent literature of gum Arabic, emphasizing Acacia seyal. Molecules **27**(4), 1171 (2022)

Barrett, L., Geddes, J.E., Mangano, S.F., Schmick, F., Whitehouse, A.S.: Soft confectionery. Tunisian Patent No. TNSN00019A1. National Institute for Standardization and Industrial Property. (2002)

Bouaziz, F., Ellouz Ghorbel, R., Ellouz Chaabouni, S., Zouari Ellouzi, S.: Production of low-calorie value potato chips by coating with almond gum. Tunisian Patent No. TN2017000395A1. National Institute for Standardization and Industrial Property. (2019)

Choudhary, A., Bains, A., Sridhar, K., Dhull, S.B., Goksen, G., Sharma, M., Chawla, P.: Recent advances in modifications of exudate gums: functional properties and applications. Int. J. Biol. Macromol. (2024). https://doi.org/10.1016/j.ijbiomac.2024.132688

Ding, Y., Zong, Q., Zhang, Q., Wang, Y., Wang, J., Huang, W., Sun, W., Zhai, Y.: Gum Arabic-based multifunctional antibacterial adhesion hydrogel dressings loaded with doxycycline hydrochloride for wound healing. Int. J. Biol. Macromol. **306**, 141284 (2025). https://doi.org/10.1016/j.ijbiomac.2025.141284

Djetoyom, D., Gang, T., Ngueilbaye, A., Lubari, J.Y.R.: Assessing potential and impact factors driving Chadian gum Arabic international trade based on the augmented gravity model. Sustainability **16**(23), 10215 (2024). https://doi.org/10.3390/su162310215

Elbaialy, E.S.K.M.: Flame retardant and termite fighting composition and articles coated therewith. International Patent No. WO2011091806A1. World Intellectual Property Organization. (2011)

Eltahir, M.E.S., Musa, F.I., Mohammed, E.M.I., Mohamed, A.A.A.: Situation analysis of gum Arabic producers associations (GAPAs) under IAMDP Sudan. Soc. Sci. Human. Open **11**, 101350 (2025). https://doi.org/10.1016/j.ssaho.2025.101350

Endo, N., Mizukai, K.: Emulsified food product composition. Egyptian Patent No. EG27864B1. Egyptian Patent Office. (2016)

FAO: Compendium of food additive specifications. FAO Food and Nutrition Paper 52/5. Food and Agriculture Organization of the United Nations, Rome, Italy. ISBN: 92-5-104023-0. (1997)

Froelich, A., Jakubowska, E., Jadach, B., Gadziński, P., Osmałek, T.: Natural gums in drug-loaded micro- and nanogels. Pharmaceutics **15**(3), 759 (2023). https://doi.org/10.3390/pharmaceutics15030759

Fuyama, H.: Oil-in-water insecticidal compositions. Egyptian Patent No. EG14220B1. Egyptian Patent Office. (1984)

Hussein, M.A., Manawi, Y., Kochkodan, V.: Polymer membrane with Acacia gum (gum Arabic) additive. International Patent No. WO2018027014A1. World Intellectual Property Organization. (2018)

Hussein, M., Sabri, S., Kochkodan, V., Manawi, Y., Alsuliti, H.: Polymer membrane with Arabic gum for water treatment. U.S. Patent Application No. US20200108355A1. U.S. Patent and Trademark Office. (2020)

Hussein, M., Manawi, Y., Kochkodan, V.: Composite reverse osmosis membrane, a water filtration system, and a manufacturing method are the same. U.S. Patent No. US10946338B2. U.S. Patent and Trademark Office. (2021)

Jiang, H., Wang, F., Ma, R., Tian, Y.: Advances in gum Arabic utilization for sustainable applications as food packaging: reinforcement strategies and applications in food preservation. Trends Food Sci. Technol. **142**, 104215 (2023). https://doi.org/10.1016/j.tifs.2023.104215

Kadirvel, V., Narayana, G.P.: Edible gums—an extensive review of their diverse applications in various food sectors. Food Bioeng. **2**(4), 384–405 (2023). https://doi.org/10.1002/fbe2.12067

Kubo, M., Matsuda, H.: Amplifier of the effect of an adrenocorticoid containing gum Arabic. Moroccan Patent No. MA29105. Moroccan Industrial and Commercial Property Office. (2007)

Lakkab, I., Lachkar, M., El Hajaji, H., El Bali, B.: New manufacturing processes for carob gum. Moroccan Patent No. MA41080A1. Moroccan Industrial and Commercial Property Office. (2019)

Mohammed, H.M.E.: Gum Arabic: certification and assessment of marketing opportunities. In: Gum Arabic (pp. 45–64). Academic Press. (2018)

Mujawamariya, G., Madi, O.P., Zoubeirou, A.M., Sene, A., Maisharou, A., D'Haese, M.: Common challenges in gum Arabic production and commercialization in West Africa: a comparative study of Cameroon, Niger, and Senegal. Int. for. Rev. **15**(2), 182–199 (2013). https://doi.org/10.1505/146554813806948477

Nasir, M., Al-Kutti, W., Islam, A.B.M.S., Alotaibi, K.S.: Cementitious mortar composition and methods of preparation thereof. U.S. Patent Application No. US20250051226A1. U.S. Patent and Trademark Office. (2025)

Oluba, O.M., Owoso, T.O., Bayo-Olorunmeke, A.O., Erifeta, G.O., Josiah, S.J., Ojeaburu, S.I., Subbiah, N., Palanisamy, T.: Probing the role of ginger starch on physicochemical and thermal properties of gum Arabic hybrid biocomposite for food packaging applications. Carbohydr. Polym. Technol. Appl. **9**, 100650 (2025). https://doi.org/10.1016/j.carpta.2024.100650

Ozsoysal, S., Bilgili, E.: Non-traditional natural stabilizers in drug nanosuspensions. J. Pharm. Biotech Ind. **1**(1), 38–71 (2024). https://doi.org/10.3390/jpbi1010005

Prasad, N., Thombare, N., Sharma, S.C., Kumar, S.: Gum Arabic–a versatile natural gum: a review on production, processing, properties and applications. Ind. Crops Prod. **187**, 115304 (2022). https://doi.org/10.1016/j.indcrop.2022.115304

Safaei, M., Mobarakeh, M.S., Azizi, B., Shoohanizad, E., Wong, L.S., Nikkerdar, N.: Optimization of cellulose/gum Arabic/Ag nanocomposite synthesis for antibacterial applications. AIMS Mater. Sci. **12**(2), 278–300 (2025)

Salem, S.M.K., Al-Taebah, H.S.S.: Edible composition to prolong the shelf life of food products. U.S. Patent Application No. US20220312782A1. U.S. Patent and Trademark Office. (2022)

Sarkis-Onofre, R., Catalá-López, F., Aromataris, E., Lockwood, C.: How to properly use the PRISMA statement. Syst. Rev. **10**(1), 1–3 (2021)

Shahat, A.B., Nabhan, K.A.: A process for preparing a formaldehyde ammonium hydroxide complex to be used as a fixative for printing inks and tanning is proposed. Egyptian Patent No. EG20420B1. Egyptian Patent Office. (1999)

Shahat, A.B., Nabhan, K.A.: Using formaldehyde-ammonium hydroxide complex to color sands in decorative paints. Egyptian Patent No. EG21011B1. Egyptian Patent Office. (2000)

Shalini, G., Revad, T., Pandya, H., Solanki, H.: Phytochemical and pharmacological perspectives on natural edible gums: a review bridging traditional medicine and modern therapeutics. Int. J. Sci. R. Tech, 2(4). (2025)

Shams, R.A.E.S.: Resistance pastes are applied to different surfaces of fabrics. Egyptian Patent No. EG29790B1. Egyptian Patent Office. (2020)

Sharkawy, A., Shoeib, T., Rodrigues, A., Barreiro, M., Fernandes, I.: Aroma-loaded microcapsules with antibacterial activity for eco-friendly applications. International Patent No. WO2018183150A1. World Intellectual Property Organization. (2018)

Sharkawy, A.M.A., Shoeib, T., Rodrigues, A.E., Barreiro, M.F.F., Fernandes, I.P.M.: Aroma-loaded microcapsules with antibacterial activity for eco-friendly applications. U.S. Patent Application No. US20200315168A1. U.S. Patent and Trademark Office. (2020)

Sharkawy, A., Rodrigues, A.E.: Plant gums in Pickering emulsions: a review of sources, properties, applications, and future perspectives. Carbohyd. Polym. **332**, 121900 (2024). https://doi.org/10.1016/j.carbpol.2024.121900

Sharma, S.C., Prasad, N., Pandey, S.K., Singh, R.: Mechanization in processing commercially significant natural resins and gums: a review. Ind. Crops Prod. **222**, 119630 (2024). https://doi.org/10.1016/j.indcrop.2024.119630

Shi, Y., Li, J., Gu, L., Su, Y., Chen, W., Zhang, M., Chang, C., Yang, Y.: Synergistic effect of gum Arabic and xanthan gum on improving rheological properties of low-fat mayonnaise with egg white protein microparticle as a fat mimetic. Int. J. Food Sci. Technol. **58**(3), 1037–1048 (2023). https://doi.org/10.1111/ijfs.16237

Solomon, M.M., Umoren, S.A.: Method for inhibiting metal corrosion during acid cleaning or pickling. U.S. Patent Application No. US20220316073A1. U.S. Patent and Trademark Office. (2022)

Tadesse, W., Dejene, T., Zeleke, G., Desalegn, G.: Underutilized natural gum and resin resources in Ethiopia for future directions and commercial utilization. World J. Agric. Res. **8**(2), 32–38 (2020). https://doi.org/10.12691/wjar-8-2-2

Tayebi-Khorrami, V., Shahgordi, S., Dabbaghi, M.M., Fadaei, M.S., Shahrbabak, S.M., Fallahianshafiei, S., Fadaei, M.R., Hasnain, M.S., Nayak, A.K., Askari, V.R.: From nature to nanotech: harnessing the power of electrospun polysaccharide-based nanofibers as sustainable packaging. Int. J. Biol. Macromol. (2025). https://doi.org/10.1016/j.ijbiomac.2025.140127

Venkatesan, J., Hur, W., Gupta, P.K., Son, S.E., Lee, H.B., Lee, S.J., Hyeon, C., Cheon, S.H., Kim, D.H., Seong, G.H.: Gum Arabic-mediated liquid exfoliation of transition metal dichalcogenides as photothermic anti-breast cancer candidates. Int. J. Biol. Macromol. **244**, 124982 (2023). https://doi.org/10.1016/j.ijbiomac.2023.124982

Wang, W., Dai, Y., Cheng, C., Wang, R., Ma, J., Jing, Q.: Development and functional evaluation of curcumin-loaded zein-gum Arabic-flaxseed gum complex nanoparticles for anti-fatigue applications. Int. J. Biol. Macromol. (2025). https://doi.org/10.1016/j.ijbiomac.2025.142998

Wijesirigunawardana, P.B., Perera, B.G.K.: Development of a cotton smart textile with medicinal properties using lime oil microcapsules. Acta Chim. Slov. **65**(1), 150–159 (2018). https://doi.org/10.17344/acsi.2017.3727

Wimmler, J.: From Senegal to Augsburg: gum Arabic and the central European textile industry in the eighteenth century. Text. Hist. **50**(1), 4–22 (2019). https://doi.org/10.1080/00404969.2019.1582580

Zakarya, D., El-Morabit, M.H.: Valorization of thermosetting waste. Moroccan Patent No. MA28318A1. Moroccan Industrial and Commercial Property Office. (2006)

Zaoui, F., Faouzi, M.A., Cherrah, Y., Etto, B.: Pharmaceutical or dietary composition based on plants for dry mouth treatment. Moroccan Patent No. MA37045A1. Moroccan Industrial and Commercial Property Office. (2016)

# Analysis of the Seasonal Succession of Phytoplankton Assemblages with Environmental Parameters: The Case of the Oualidia Lagoon

Lamia Daghor, Touria Hssaida, Oumaima Jamal Eddine, Sanaa Berrada, Mohamed Fraikech, and Keltoum Elbouhmadi

**Abstract**

According to Margalef's model of phytoplankton community succession, algal communities follow a specific seasonal dynamic, influenced by ecological factors such as variations in temperature, salinity, and nutrients. In this study, we carried out a quantitative and qualitative analysis of phytoplankton species during 2014–2015, using seawater samples taken weekly from the Oualidia lagoon. The aim was to study the variation in phytoplankton assemblages as a function of physico-chemical parameters, according to the Margalef'smodel. Phytoplankton communities are dominated by diatoms in winter/spring and by dinoflagellates in summer/autumn. Diatoms are mainly represented by the genera: *Nitzshia*, *Pseudonitzshia*, *Navicula*, *Chaetoceros* and *Thalssiosira*, and dinoflagellates by the genera: *Prorocentrum, Protoperidinium quinquecorne*, *scripsiella*, *Gonyaulax and Ceratium.* Overall, this dynamic is conditioned by nutrient availability (nitrite, nitrate and orthophosphate), light conditions and seasonal variations in temperature and turbulence.

**Keywords**

Phytoplankton · Algal communities · Diatoms · Dinoflagellats · Physico-chemical parameters

L. Daghor (✉)
Faculty of Health, University of Mundiapolis, Casablanca, Morocco
e-mail: l.daghor@gmail.com

T. Hssaida
Geosciences and Applications Laboratory (LGA), Faculty of Sciences Ben M'sick, University Hassan II of Casablanca, Casablanca, Morocco

O. Jamal Eddine
Chemistry Department, Laboratory of Engineering and Materials (LIMAT), Faculty of Sciences Ben M'sick, University Hassan II of Casablanca, Casablanca, Morocco

S. Berrada
Biotechnology, Environment, Agro-Food and Health Laboratory, Faculty of Science Dhar El Mahraz, Sidi Mohamed Ben Abdellah University, Fez, Morocco

M. Fraikech
National Institute for Fisheries Research, Casablanca, Morocco

L. Daghor · K. Elbouhmadi
Laboratory of Ecology and Environment, Faculty of Sciences Ben M'sick, University Hassan II of Casablanca, Casablanca, Morocco

## 1 Introduction

The phytoplankton assemblages succession, their responses to the physicochemical parameters of the environment, and their development strategies have been discussed by several authors (Margalef 1961; Frontier 1969; Travers, 1971; Margalef 1997; Pierre and Lassus 1983; Reynolds 1988; Del Amo et al. 1997b; Cullen et al. 2002; Winder and Sommer 2012; Hernández Fariñas et al. 2020).

The most widely applied model is Margalef's Mandala (1978), it represents the distribution of different phytoplankton groups under gradients of turbulence and nutrient concentration.

The combination of these two factors affects the temporal and spatial regulation of the various phytoplankton morphotypes. Different phytoplankton morphotypes have similar shapes and physiological characteristics. Margalef defines these groups as having r versus K strategies. r: are rapidly reproducing species favored by unstable, turbulent environments and relatively high nutrient concentrations (e.g. diatoms), whereas, organisms with K strategies reproduce more slowly and dominate more stable environments (e.g., dinoflagellates).

Several regions worldwide have applied this model to understand the succession of phytoplankton assemblages: in France, in the Bay of Biscay (Herbland 2012); the Bay of Brest (Del Amo et al. 1997b); the Odet estuary (Laguerre



O. Jamal Eddine et al. (eds.), *Plant Pathology, Fungal Diversity, and Biotechnological Advances in Agriculture*, Sustainable Landscape Planning and Natural Resources Management, https://doi.org/10.1007/978-3-031-97666-7_13

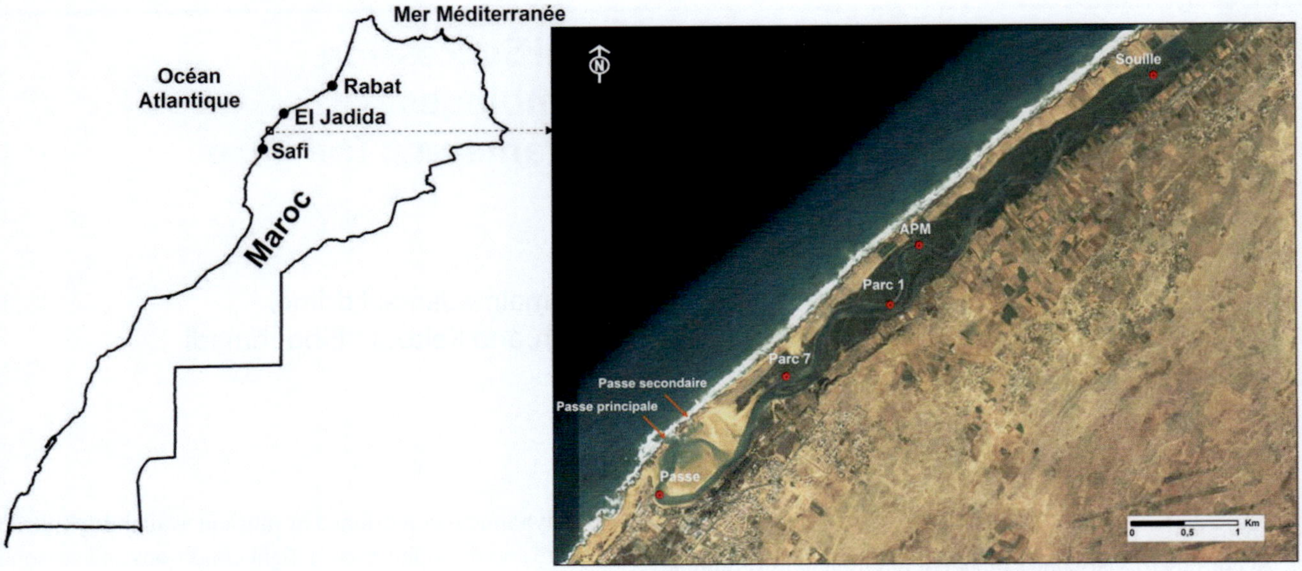


**Fig. 1** Geographical location of the Oualidia lagoon and sampling sites (Images Geo Eye from Google Earth 2014)

et al. 2013), and in Morocco in the Oualidia lagoon (the present study). In this lagoon system, monitoring covers the period from April 2014 to March 2015. Thus, this period can be divided into 4 intervals "below" according to the classic diagram of the succession of phytoplankton assemblages (Figs. 2, 3 and 4); only species showing seasonal variation and significant abundances are presented.

## 2 Materials and Methods

### 2.1 Study Area

The Oualidia lagoon, located on the Atlantic coast, lies along the shoreline of the Doukkala province, approximately 79 km south of El Jadida and 64 km north of Safi (Fig. 1).

Its geographical coordinates are 09°02′50″ West and 32°44′42″ North.

The lagoon covers a total area of approximately 4 km$^2$, with a length of 7 km and an average width of 0.5 km. Maximum and minimum depths are around 5 m and 1.5 m, respectively.

Separated from the Atlantic Ocean by a consolidated dune barrier, the lagoon maintains communication with the ocean through two natural channels.

It does not receive input from any major river.

Oualidia Lagoon was selected as a study site due to its significant aquaculture and shellfish farming activities. However, these activities are increasingly threatened by colored water events and harmful algal blooms. The study of phytoplankton communities, particularly toxic species, represents an essential tool to ensure the sanitary quality of shellfish products from the lagoon.

The site is under regular surveillance by the RSSL (Marine Environment Surveillance Network) of the National Institute of Fisheries Research (INRH), with the monitoring station initially based in Casablanca and later relocated to El Jadida to ensure better oversight and efficiency. Groups all employ similar enzymatic tools like tryptamine and indole-3-acetonitrile for these conversions (Romanenko et al. 2015) (Figs. 1 and 2). This shared biochemical machinery suggests auxin production evolved early and has been conserved throughout algal diversification.

### 2.2 Collecting Samples

Water samples for phytoplankton analysis were collected at four stations ($P^1$, $P^2$, $P^3$) within the lagoon. The geographical coordinates of the sampling stations are as follows:

$P^1$: Passe (32°44′04.2"N 009°02′44.4"W),
$P^2$: Parc 7 (32°44′,950N 09°01′,520W),
$P^3$: Ancien Parc Marost (32°45′33.13"N 9°00′43.27"W).

At each station, water samples were collected for both the identification and enumeration of the phytoplankton population using an inverted optical microscope, and for the physico-chemical analysis of the Oualidia lagoon water. In-situ measurements included temperature, pH, and salinity using specific probes, while nitrate and suspended matter were analyzed ex-situ.

#### 2.2.1 Phytoplankton Biomass Characteristics

The Oualidia lagoon is characterized by a fairly high algal biomass, with significant spatiotemporal variability.

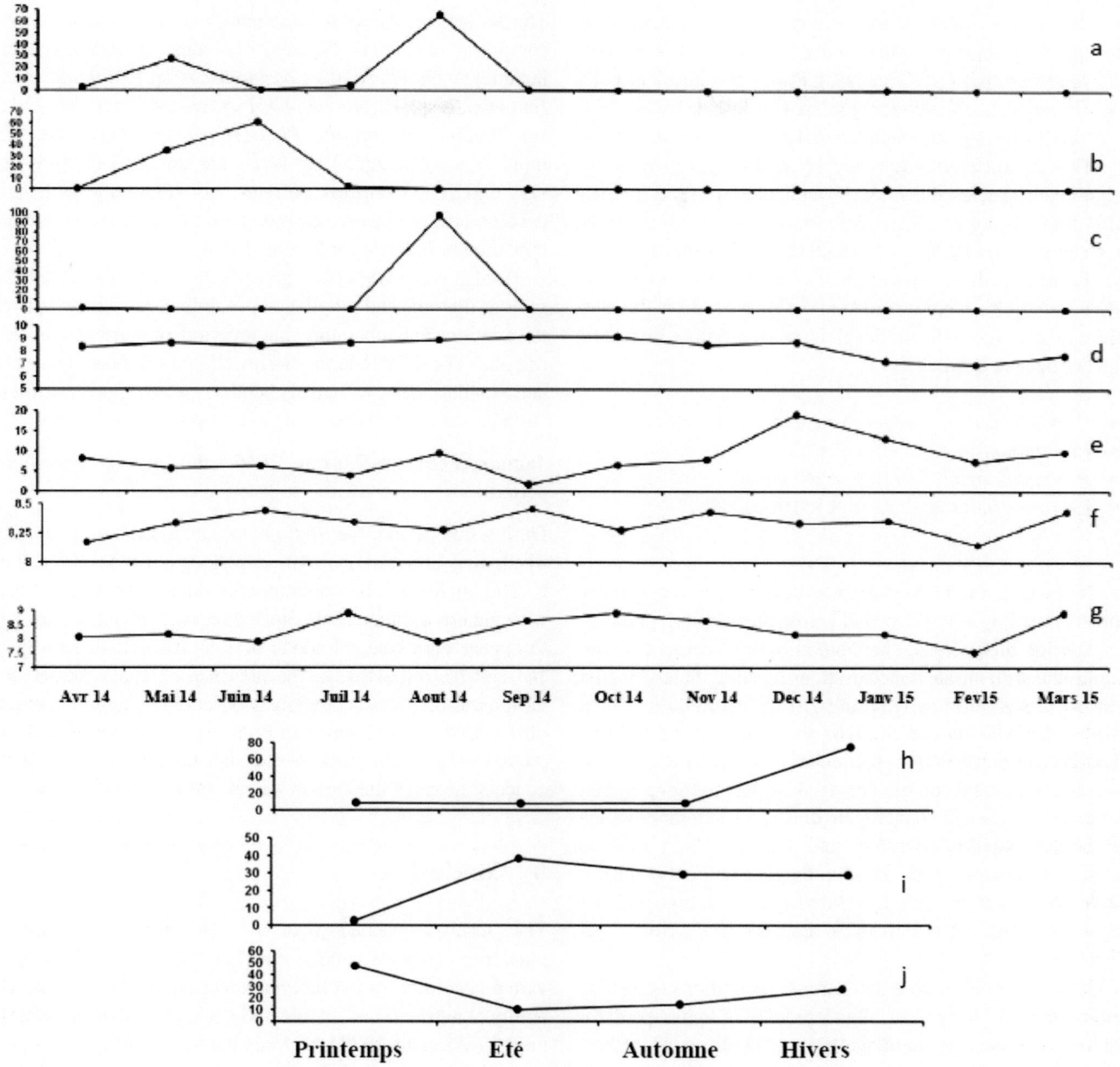


**Fig. 2** Diagram of evolution of phytoplankton assemblages with the various chemical parameters (expressed as relative percentage %) at the station PARC 7: **a** Total flora, **b** Diatoms, **c** Dinoflagellates, **d** Specific richness, **e** Temperature, **f** Turbidity, **g** Salinity, **h** Orthophosphates, **i** Nitrates, **j** Nitrite

Indeed, the total flora ranges from $1.10^2$ and $9.10^5$ cells per liter, and blooms exceeding a threshold of $10^7$cells per liter, were recorded in summer 2014.

**Mid-March/April/May 2014:** Spring bloom.

Two phytoplankton blooms marked this period. The first bloom was caused by a species of dinoflagellate (unidentified), observed at all three sampling stations: Passe, Parc 7, and APM, with concentrations of $1.8 \times 10^5$, $2 \times 10^5$, and $1.1 \times 10^5$ cells/L, respectively. The second bloom, caused by *Thalassiosira* sp., occurred at the beginning of May, with a peak concentration exceeding $5.9 \times 10^6$ cells/L at the APM station. However, this species was absent at the Parc 7 station. Meanwhile, a significant increase in orthophosphate concentrations was observed only at the Parc 7 station. According to previous results by Herbland et al. (1998), spring blooms and/or spring algal growth in general can be limited by phosphorus. A similar study in France confirmed this hypothesis, demonstrating that winter blooms can be limited by phosphorus (Labry 2001).

During the same period, dinoflagellate species began to appear, primarily represented by the following species: *Alexandrium* sp. ($3 \times 10^2$ cells/L), *Prorocentrum micans* ($1.6 \times 10^3$ cells/L), and *Heterocapsa* sp. ($7 \times 10^2$ cells/L).

June/July/August/September 2014: Secondary Bloom.

This period is characterized by rapid algal growth and significant species diversity. Several bloom events marked this phase. In June, *Chaetoceros* sp. reached average cellular concentrations of $1.8 \times 10^5$ cells/l across all stations.

Then, in July, more than $1.3 \times 10^5$ cells/l of Pseudonitzschia were recorded at the Parc 7 station. Towards the end of this period, dinoflagellate concentrations increased, including species such as:

- *Protoperidinium quinquecorne* ($2.5 \times 10^4$ cells/l),
- *Katodinium* sp. ($3 \times 10^4$ cells/l),
- *Prorocentrum micans* ($8.9 \times 10^4$ cells/l),
- *Prorocentrum scutellum* ($6.8 \times 10^4$ cells/l).

In August, a red tide event was detected, with concentrations exceeding $1 \times 10^4$ cells/l across all sampling points.

During this period, the lagoon waters became highly stratified, with nitrate concentrations reaching 662.47 μg/l at the APM station in early spring 2014. According to Shafee (1989) and Hassou et al. (2016), the waters of the Oualidia Lagoon are nutrient-rich, particularly in inorganic nitrogen, due to agricultural runoff. This leads to intense development of macrophytes and phanerogam during the summer months.

Several studies (Berdalet and Estrada 1993; Sullivan et al. 2003) demonstrate that dinoflagellate blooms typically develop in waters with low turbulence and high nutrient concentrations (Margalef 1979; Estrada 1997; Solé et al. 2016).

It is important to note that summer conditions (sunshine, calm winds) favor the development of phytoplankton blooms, especially dinoflagellates (Cogels et al. 1993). According to the same authors, episodic phytoplankton blooms are regularly reported in the central area of Lake Guiers (Senegal) under calm and very sunny weather conditions.

**October/November/December 2014**: precipitation/winter.

In winter and early spring, the total phytoplankton biomass is at its lowest, with concentrations not exceeding 5280 cells/L. The diatoms are the most dominant group. They are primarily composed of central diatoms. The most abundant genera include: *Achnanthes* sp., *Coscinodiscus* sp., *Diploneis* sp., *Entomoneis* sp., *Grammatophora* sp., *Nitzschia* sp., *Navicula* sp., *Paralia sulcata*, and *Pleurosigma* sp. While nutrient inputs are significant, light levels are low, and turbidity is very high (due to storms, currents, etc.), diatom growth and development are favored by high nutrient concentrations and moderate turbulence (Solé et al. 2016).

Finally, some taxonomic groups are observed year-round, such as diatoms and dinoflagellates, which are less sensitive to variations in temperature, photoperiod, or nutrient concentrations. These include the Scripsiella group, Naviculaceae, and Fragilariaceae, with many benthic species appearing after strong winds or tides (known as tychoplanktonic species).

**January/February/March 2015**: the spring transition period:

During this period, the total phytoplankton biomass is low, fluctuating between $7 \times 10^2$ cells/L and $1.4 \times 10^4$ cells/L. The phytoplankton community is dominated by diatoms, with genera such as Thalassiosira sp. and Navicula starting to appear with concentrations of $1 \times 10^3$ cells/L and $7 \times 10^2$ cells/L, respectively. The presence of Thalassiosira sp. during this period was previously observed by Laguerre et al. (2013) in the Odet Estuary in France, and in Brest Bay (Del Amo 1997). At this time, we do not have data on nutrient concentrations in the lagoon waters (Figs. 3 and 4).

## 3 Conclusion

The results revealed that the distribution of phytoplankton classes—namely dinoflagellates and diatoms—varied according to hydroclimatic conditions and the availability of nutrient salts. In this study, we adopted the Margalef model (Margalef 1978a, b; Margalef et al. 1979), a widely recognized conceptual framework that effectively explains the relationship between the physico-chemical environment and phytoplankton dynamics. According to this model, phytoplankton can be grouped based on their adaptive strategies: the "r-strategy," which includes species like diatoms that thrive in turbulent, nutrient-rich waters, and the "K-strategy," associated with dinoflagellates that are better suited to stable, oligotrophic, and stratified environments.

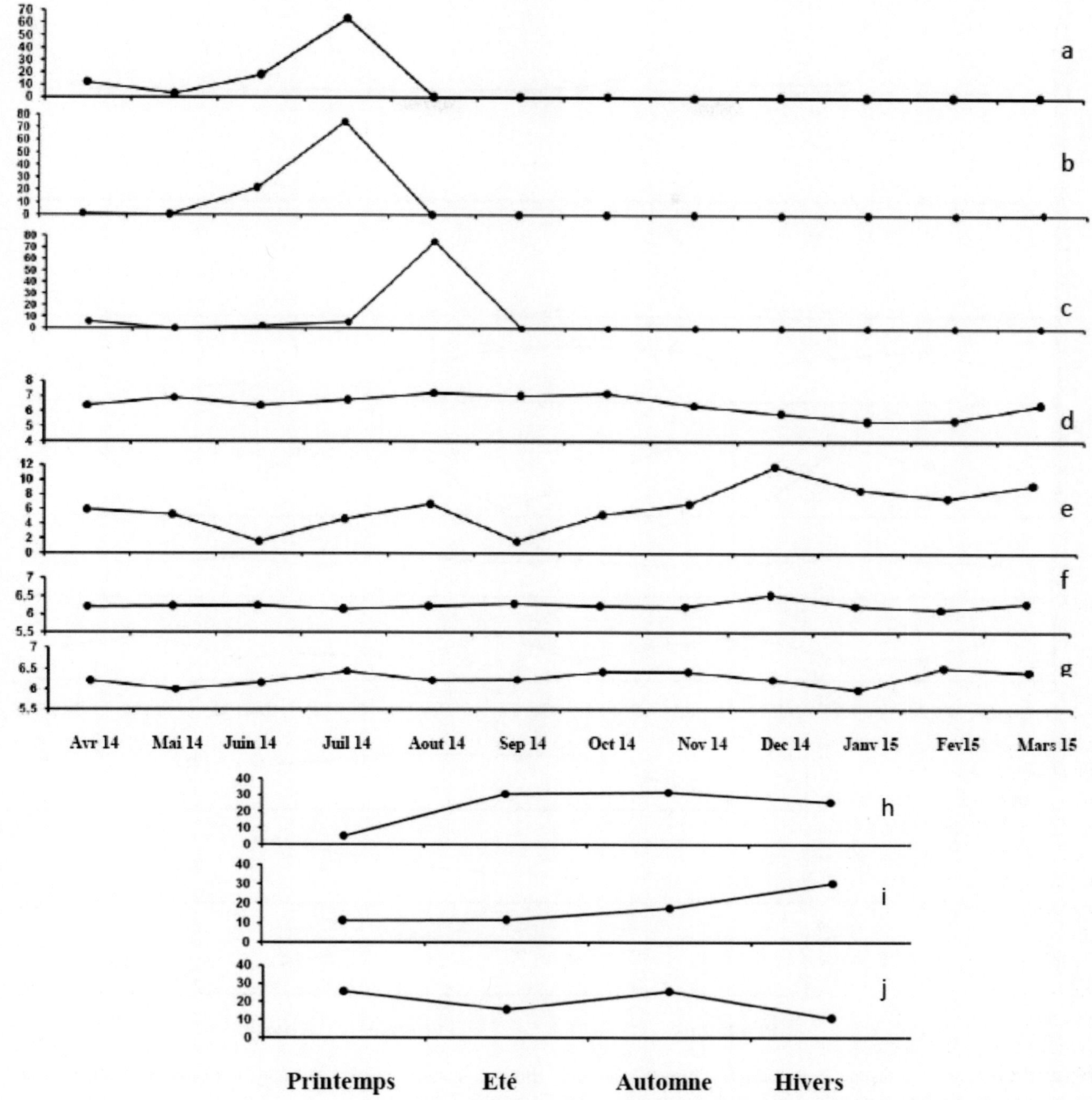


**Fig. 3** Diagram of evolution of phytoplankton assemblages with the various chemical parameters (expressed as relative percentage %) at the station PARC 7: **a** Total flora, **b** Diatoms, **c** Dinoflagellates, **d** Specific richness, **e** Temperature, **f** Turbidity, **g** Salinity, **h** Orthophosphates, **i** Nitrates, **j** Nitrite

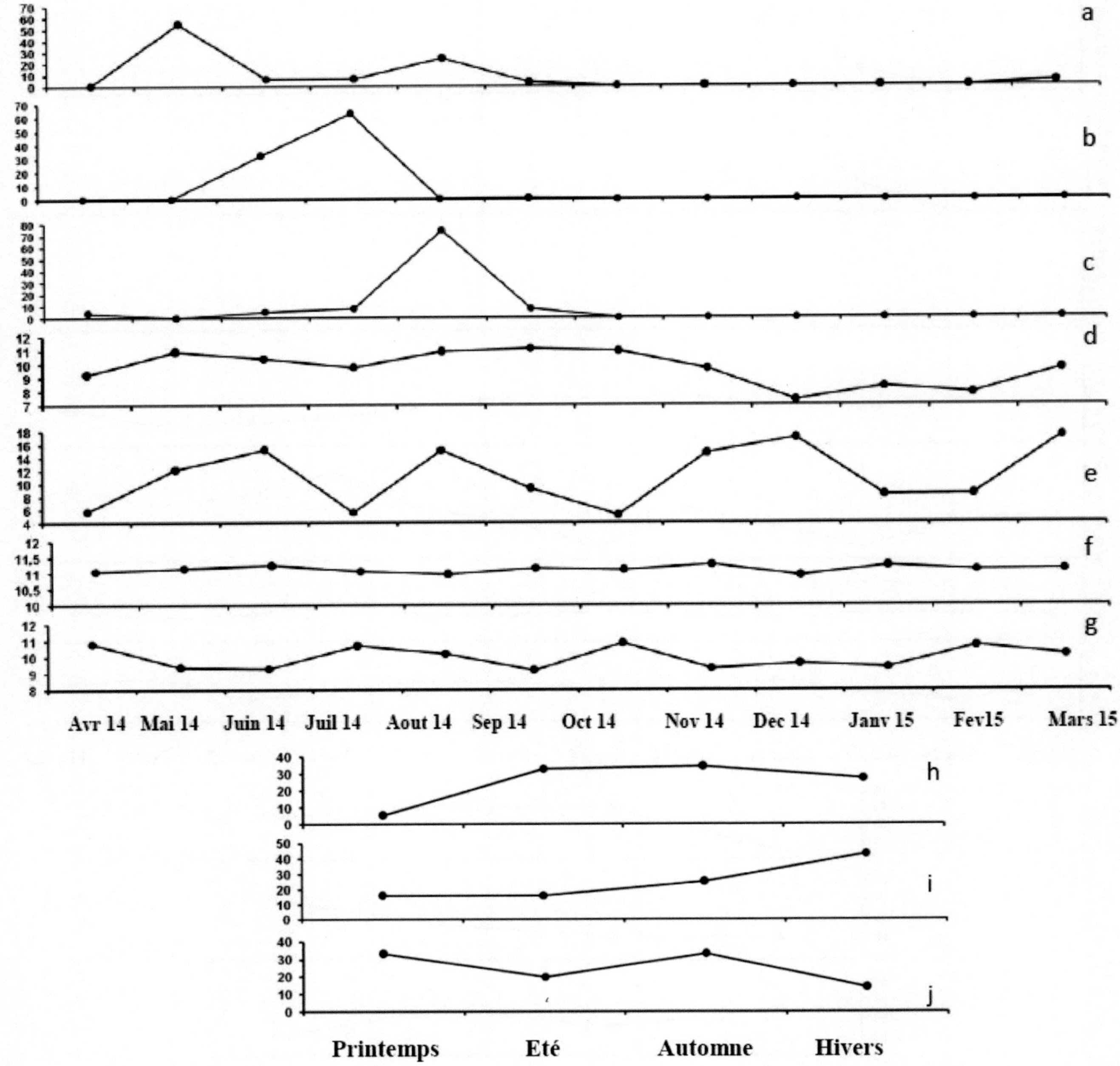


**Fig. 4** Diagram of the evolution of phytoplankton assemblages with the various chemical parameters (expressed as relative percentage %)) at station level APM: **a** Total flora, **b** Diatoms, **c** Dinoflagellates, **d** Specific richness, **e** Temperature, **f** Turbidity, **g** Salinity, **h** Orthophosphates, **i** Nitrates, **j** Nitrite

## References


Del Amol, Y., Le pape, O., Treguer, P., Queguiner, B., Menesguen, A., Rnino, A.: (1997b)

Del Amol, Y., Le Pape, O., Treguer, P., Queguiner, B., Menesguen, A., Rnino, A.: Impacts of high nitrate freshwater inputs on macrotidal ecosystems. I. Seasonal evolution of nutrient limitation for the diatomdominated phytoplankton of the Bay of Brest (France). Mar Ecol Prog Ser, 161: 213 2241 (1997b)

Cogels, F.X., Thiam, A., Gac, J.Y.: Premiers effets des barrages du fleuve Sénégal sur le lac de Guiers. Rev. Hydrobiol. Trop. **26**(105), 117 (1993)

Cullen, J.J., Franks, P.J.S., Karl, D.M., Longhurst, A.: Physical influences on marine ecosystem dynamics. In: Robinson, A.R., McCarthy, J.J., Rothchild, B.J. (eds.), The Sea Wiley, New York. 12. 297 336. (2002)

Frontier, S.: La structure de la communauté et l'analyse de la diversité. In J. Jouenne (Ed.), Colloque sur les relations entre le peuplement et l'environnement (pp. 83-95) (1969). Paris: CNRS.

Hassou, N., Boujaber, N., Oumaskour, K., Lakhdar, F., Benba, J., Zourarah, B., Assobhei, O., Etahiri, S.: Nutrients variation in the coastal waters of oualidia lagoon morocco in relation to the contribution of watershed and hydro chemical characteristics. Int. J. Adv. Res. **4**(12), 1697–1706 (2016)

Hassou, N., Boujaber, N., Oumaskour, K., Lakhdar, F., Benba, J., Zourarah, B., Assobhei, O., Etahiri, S.: Nutrients variation in the coastal waters of oualidia lagoon, morocco, in relation to the contribution of watershed and hydro chemical characteristics. Int. J. Adv. Res. **4**(12), 1697–1706 (2016)

Herbland, A. Phytoplankton communities and their ecological significance. In: J. Smith (Ed.), Marine Ecology Advances (pp. 45–67) (2012). Elsevier.

Hernández Fariñas, T., Menet, N., M'Zari, L., Hernández Fariñas, T., Menet, N., M'Zari, L., Courtay, G., Lampert, L.: Etude de la dynami Courtay, G., Lampert, L. (2020). Que et de la composition du phytoplancton via l'approche des pigments appliquée au littoral Normand. Rapport Ifremer, RST ODE/UL/LERN/20-11, p. 62 (2020)

Laguerre, H., Czamanski, M., et Arnaud, F.: Mieux connaître et faire connaître la. (2013)

Margalef, R: Ideas for a synthetic approach to the ecology of water bodies. Internationale Revue der gesamten Hydrobiologie Hydrographie **46**(1), 139–164 (1961). https://doi.org/10.1002/iroh.19610460112

Margalef, R.: Distribución horizontal del fitoplancton marino a escala media (1 a 10 Km) ilustrada con un ejemplo del área de afloramiento del NW de África. Mem. Real Acad. Cienc. Artes Barcelona **43**(131), 148 (1976)

Margalef, R.: Lifeforms of phytoplankton as survival alternatives in an unstable environment. Oceanol. Acta **1**(493), 509 (1978a)

Margalef, R.: Phytoplankton communities in upwelling areas. The example of NW Africa. Oecol. Aquat. **3**(97), 132 (1978b)

Margalef, R.: Our biosphere. New York: Scientific American Library. ISBN: 978-0716738059 (1997)

Margalef, R.: Life-forms of phytoplankton as survival alternatives in an unstable environment. Oceanol. Acta **1**(4), 493–509 (1978)

Pierre, M.J., Lassus, P.: Dinoflagelles toxiques sur les cotes francaises pendant l'été 1983. Perturbations des ecosystemes en baie de Vilaine. Analyse des successions phytoplanctoniques precedant l'apparition d'un dinoflagelle toxique. Revue des Travaux de l'Inst. des Pêches Maritimes (0035 2276) (ISTPM), 12, 47. 134–148. (1983)

Romanenko, V. A., et al.: Phylogenetic relationships of marine bacteria based on 16S rRNA gene sequences. J Microbiol. Methods **110**, 1–7 (2015). https://doi.org/10.1016/j.mimet.2015.02.007

Qualité de l'eau, de la rivière à l'estuaire, à partir du suivi des sels nutritifs et du plancton, pour répondre aux enjeux de reconquête de qualité de l'eau et de protection d'espèces endogènes Etude sur la rivière de l'Odet (Finistère). Rapport d'études «Projet ECOESTUA), p. 40

Reynolds, C.S.: Functional morphology and the adaptive st rategies of freshwater phytoplankton, p. 388–433. (1988)

Shafee, M.S.: Carpet shell clam culture in Morocco European. Aquac. Soc. Spec. Publ. **10**(233), 234 (1989)

Solé, J., Berdalet, E., Arin, L., Cros, L., Delgado, M., Kuwata, A., Llebot, C., Marrasé, C.: The ecology of plankton in biological oceanography: a tribute to Marta Estrada's task (2016)

The ecology of plankton in biological oceanography: a tribute to Marta Estrada's task SCI. MAR., 80S1, pp. 33–38

Travers, M.: Diversité du phytoplancton du Golfe de Marseille en 1964. Mar. Biol. **8**(4), 308–343 (1971)

Winder, M., et Sommer, U.: Phytoplankton response to a changing climate. (2012)

MIX
Papier aus verantwortungsvollen Quellen
Paper from responsible sources
FSC® C105338

If you have any concerns about our products,
you can contact us on
**ProductSafety@springernature.com**

In case Publisher is established outside the EU,
the EU authorized representative is:
**Springer Nature Customer Service Center GmbH**
**Europaplatz 3, 69115 Heidelberg, Germany**

Printed by Libri Plureos GmbH
in Hamburg, Germany